IFS

Conditionals, Belief, Decision, Chance, and Time

Edited by

WILLIAM L. HARPER
The University of Western Ontario

ROBERT STALNAKER
Cornell University

and

GLENN PEARCE
The University of Western Ontario

D. REIDEL PUBLISHING COMPANY

DORDRECHT : HOLLAND / BOSTON : U.S.A.
LONDON : ENGLAND

Library of Congress Cataloging in Publication Data

CIP

Main entry under title:

Ifs : conditionals, belief, decision, chance, and time.

(The University of Western Ontario series in philosophy of science ;
v. 15)
Includes bibliographies and index.
1. Conditionals (Logic) – Addresses, essays, lectures.
2. Probabilities – Addresses, essays, lectures.   3.   Decision-making –
Addresses, essays, lectures.   4.   Chance – Addresses, essays, lectures.
5.   Time – Addresses, essays, lectures.   I.   Harper, William Leonard,
1943–   II.   Stalnaker, Robert.   III.   Pearce, Glenn.
IV.   Series: University of Western Ontario. University of Western Ontario
series in philosophy of science ; v. 15.
BC199.C56I38   160   80-21638
ISBN 90-277-1184-4
ISBN 90-277-1220-4 (pbk.)

Published by D. Reidel Publishing Company,
P.O. Box 17, 3300 AA Dordrecht, Holland

Sold and distributed in the U.S.A. and Canada
by Kluwer Boston Inc.,
190 Old Derby Street, Hingham, MA 02043, U.S.A.

In all other countries, sold and distributed
by Kluwer Academic Publishers Group,
P.O. Box 322, 3300 AH Dordrecht, Holland

D. Reidel Publishing Company is a member of the Kluwer Group

Printed in The Netherlands

TABLE OF CONTENTS

PART 5:INDICATIVE VS. SUBJUNCTIVE CONDITIONALS

PART 6:CHANCE, TIME, AND THE SUBJUNCTIVE CONDITIONAL

SERIES PREFACE

With publication of the present volume, The University of Western Ontario Series in Philosophy of Science enters its second phase. The first fourteen volumes in the Series were produced under the managing editorship of Professor James J. Leach, with the cooperation of a local editorial board. Many of these volumes resulted from colloguia and workshops held in connection with the University of Western Ontario Graduate Programme in Philosophy of Science. Throughout its seven year history, the Series has been devoted to publication of high quality work in philosophy of science considered in its widest extent, including work in philosophy of the special sciences and history of the conceptual development of science. In future, this general editorial emphasis will be maintained, and hopefully, broadened to include important works by scholars working outside the local context.

Appointment of a new managing editor, together with an expanded editorial board, brings with it the hope of an enlarged international presence for the Series. Serving the publication needs of those working in the various subfields within philosophy of science is a many-faceted operation. Thus in future the Series will continue to produce edited proceedings of worthwhile scholarly meetings and edited collections of seminal background papers. However, the publication priorities will shift emphasis to favour production of monographs in the various fields covered by the scope of the Series.

THE MANAGING EDITOR

W. L. Harper, R. Stalnaker, and G. Pearce (eds.), Ifs, vii.
Copyright © 1980 by D. Reidel Publishing Company.

PREFACE

This volume is intended for students and professionals in philosophy of language, linguistics, decision theory and logic, and for any others who are interested in the exciting recent developments relating conditionals to subjective probability, chance and time. A number of the papers make use of the probability calculus, but they are accessible to the general philosophical reader. Indeed, these papers constitute a good introduction for students of philosophy to those aspects of probability theory that have been used to illuminate conditionals. The volume also contains a good introduction to the possible worlds approach to conditionals for philosophy students and for decision theorists and probability theorists who want to exploit the newly articulated relationships between conditionals and probability.

Included are the classic papers by Robert Stalnaker and David Lewis on the possible worlds approach to conditionals, and their classic papers on the relationship between conditionals and conditional probability. In addition there are a number of more recent papers which extend this work and develop new and exciting ways in which research on conditionals and research on probability can illuminate each other. Most of these more recent papers were delivered at the May 1978 *University of Western Ontario* workshop on pragmatics and conditionals. It became clear at this workshop that the interactions between probability and conditionals have resulted in interrelated new work on the roles of English subjunctive and indicative conditionals and on the proper relationships among subjective probability, chance and time. The papers are arranged so as to display these themes along the lines on which they naturally developed. An introduction traces the development of these themes through the various papers in the volume.

We are grateful to the following for permission to include the articles contained in this volume:

Robert Stalnaker, " A Theory of Conditionals" which first appeared in *Studies in Logical Theory, American Philosophical Quarterly* Monograph Series, No. 2, Oxford: Blackwell, 1968. It was reprinted in Ernest Sosa (ed.), *Causation and Conditionals,* Oxford: University Press, 1975.

David Lewis, "Counterfactuals and Comparative Possibility" which first appeared in *Journal of Philosophical Logic*, 2, 1973. It was reprinted in

W. L. Harper, R. Stalnaker, and G. Pearce (eds.), Ifs, ix.
Copyright © 1980 by D. Reidel Publishing Company.

D. Hockney *et al.* (eds.), *Contemporary Research in Philosophical Logic and Linguistic Semantics*, Western Ontario Series in the Philosophy of Science, Vol. 4, Dordrecht: Reidel, 1975.

Robert Stalnaker, "A Defense of Conditional Excluded Middle" which was presented at the 1978 U.W.O. Conference and will also appear in *Journal of Philosophical Logic.*

Robert Stalnaker, "Probability and Conditionals" which first appeared in *Philosophy of Science*, 37, 64–80, 1970.

David Lewis, "Probabilities of Conditionals and Conditional Probabilities" which first appeared in *Philosophical Review*, 85, 297–315, 1976.

Robert Stalnaker, "Letter to David Lewis" which is previously unpublished.

Gibbard and Harper, "Counterfactuals and Two Kinds of Expected Utility", which first appeared in C. Hooker *et al.* (eds.), *Foundations and Applications of Decision Theory*, Western Ontario Series in the Philosophy of Science, Vol. 13, Dordrecht: Reidel, 1978.

Robert Stalnaker, "Indicative Conditionals", which first appeared in *Philosophia*, 5, 1975. It also appeared in A. Kasher (ed.), *Language in Focus*, Boston Studies in the Philosophy of Science, Vol. 43, Dordrecht: Reidel, 1976.

Allan Gibbard, "Two Recent Theories of Conditionals", which was presented at the 1978 U.W.O. workshop and is not published elsewhere.

John Pollock, "Indicative Conditionals and Conditional Probability", which is a write up of an objection to Gibbard that Pollock raised at the 1978 workshop. Neither this comment nor Gibbard's reply are published elsewhere.

Brian Skyrms, "The Prior Propensity Account of Subjunctive Conditionals", which grew out of Skyrms' talk at the 1978 workshop and is not published elsewhere.

David Lewis, "A Subjectivist's Guide to Objective Chance", which was presented at the 1978 U.W.O. workshop and also appears in R. C. Jeffrey (ed.), *Studies in Inductive Logic and Probability*, Vol. II (pp. 263–93), Berkeley and Los Angeles: University of California Press, 1980.

Thomason and Gupta, "A Theory of Conditionals in the Context of Branching Time", which grew out of a paper by Thomason and comments by Gupta at the 1978 workshop. It also appears in *Philosophical Review.*

Bas van Fraassen, "A Temporal Framework for Conditionals and Chance", which was presented at the 1978 workshop and also appears in *Philosophical Review.*

PART 1

INTRODUCTION

WILLIAM L. HARPER

A SKETCH OF SOME RECENT DEVELOPMENTS IN THE THEORY OF CONDITIONALS

INTRODUCTION

The papers discussed in this sketch represent what I take to be a very exciting stream in recent work on conditionals. The first section includes two classics, Stalnaker's 'A Theory of Conditionals' and Lewis' 'Counterfactuals and Comparative Possibility', together with Stalnaker's new paper 'A Defence of Conditional Excluded Middle'. These papers contrast sharply with the earlier work of Goodman, Chisholm, and others, which attested to the problematic character of talk about alternative possibilities by drawing attention to the ambiguity and extreme context dependence of our linguistic intuitions about counterfactuals.[1] Stalnaker and Lewis proceed by constructing abstract models that take as primitive the very sort of alternative possibilities that these earlier writers found problematic. They use these models to formulate new and interesting questions which can then be used to suggest examples on which to test linguistic intuitions. Whatever one thinks about the ultimate suitability of the possible worlds account, as an analysis of English conditionals, he must agree that the dispute between Stalnaker and Lewis in these papers has considerably sharpened and clarified our linguistic intuitions.

In addition to the direct testing of models against linguistic intuition, Stalnaker attempted, in 'Probability and Conditionals', to defend his account by appeal to intuitions about subjective probability. The connection between these intuitions and conditionals is suggested by Frank Ramsey's test for evaluating conditionals. If this test is correct then one's evaluation of a conditional should be given by his corresponding subjective conditional probability. Stalnaker showed that this Ramsey test idea forced his conditional logic. Lewis responded in 'Probabilities of Conditionals and Conditional Probabilities' with a surprising trivialization result that has proved to be one of the main constraints on all recent attempts to relate probability and conditionals. It looks as though the price of Ramsey's test, as a paradigm, is to either make conditionals radically context dependent or to give up conditional propositions altogether and endorse a conditional assertion account of conditional sentences.

In 'Counterfactuals and Two Kinds of Expected Utility' Allan Gibbard and

3

W. L. Harper, R. Stalnaker, and G. Pearce (eds.), Ifs, 3–38
Copyright © 1980 by D. Reidel Publishing Company.

I developed a suggestion of Stalnaker's to show that Ramsey test reasoning is not appropriate to guide decision making in certain interesting situations like Newcombe's problem and the Prisoner's Dilemma. In these situations an act can count as evidence for something even though the agent knows it can't causally influence it. Hypothetical reasoning appropriate to guide choices must be sensitive to beliefs about causal influence in ways that Ramsey test reasoning is not. This suggests that decision making can provide a rival to the Ramsey test paradigm. It turns out that the Ramsey test paradigm seems appropriate for indicative conditionals while the distinct decision making paradigm seems appropriate for subjunctive conditionals.

In 'Indicative Conditionals' Stalnaker defended his theory as an account of indicative as well as subjunctive conditionals. On this account indicative conditionals are context dependent. Allan Gibbard in 'Two recent Theories of Conditionals' extends the Ramsey test account of indicative conditionals along conditional assertion lines developed by Ernest Adams. He argues against both the classical material truth functional account and against Stalnaker's account. In 'Indicative Conditionals and Conditional Probability' John Pollock attacks Ramsey's test as a paradigm for indicative conditionals and Gibbard responds in 'Reply to Pollock'.

The decision making paradigm for subjunctive conditionals is developed by Brian Skyrms in 'The Prior Propensity Account of Subjunctive Conditionals'. He proposes that the appropriate evaluation of such conditionals ought to be given by the subjective expectation of the relevent objective conditional chance. In 'Subjectivists Guide to Objective Chance' Lewis illuminates the controversial idea of objective chance by calling attention to intuitions relating subjective probability and beliefs about objective chance. These suggest important connections between objective chance and branching time. Thomason and Gupta, 'A Theory of Conditionals in the Context of Branching time', investigate conditionals in the context of worlds with branching time and van Fraassen builds on their investigation to model both conditionals and chance in 'A Temporal Framework for Conditionals and Chance'.

I. THE CLASSIC STALNAKER–LEWIS DEBATE

1. In his 1968 paper 'Theory of Conditionals' Robert Stalnaker motivated his account by calling attention to Frank Ramsey's test for evaluating the acceptability of hypothetical statements (Stalnaker, 1968, this volume, pp. 41–55). With some modifications introduced to handle antecedents

believed to be counterfactual, Ramsey's test can be summed up in the following slogan:

First, hypothetically make the minimal revision of your stock of beliefs required to assume the antecedent. Then, evaluate the acceptability of the consequent on the basis of this revised body of beliefs.

Stalnaker suggested that an appropriate account of truth conditions for conditionals ought to correspond to this Ramsey test account of acceptability conditions (Stalnaker, 1968, this volume, p. 44). The obvious counterpart to the *minimal revision of a body of belief that would be required to make an antecedent acceptable* is *the minimal revision of a world that would be required to make an antecedent true.* So, he makes his conditional true just in case the consequent would be true in the minimal revision of the actual world required to make the antecedent true.

Stalnaker's theory is an application of this idea to the kind of possible worlds semantics used in modal logic. He uses a selection function to represent minimal revisions of worlds. Such a function f assigns to each world w and sentence A a world $f(A, w)$, that is to count as the minimal revision of w required to make A true. In possible world semantics the content of a statement is given by specifying which of the possible worlds would make it true. Where $A > B$ is a conditional with antecedent A and consequent B Stalnaker's account specifies.

$A > B$ is true in w iff B is true in $f(A, w)$.

This makes the semantical content of a conditional sentence relative to selection of a nearest world.

Stalnaker proposes a number of general conditions of any reasonable way of selecting what is to count as the nearest antecedent world (1968, this volume, pp. 46–47). These conditions generate an interesting logic for conditional sentences.[2] A number of features of this logic fit very well with linguistic intuitions about specific arguments involving conditionals. For example, the antecedent strengthening law,

$$\frac{A > B}{\therefore (A \land C) > B}$$

is not valid in Stalnaker's logic, though it is valid for material conditionals. Consider the following English argument:

If I put sugar in this cup of tea it will taste fine
_____
∴ If I put sugar and diesel oil in this cup of tea it will taste fine

This is clearly of the same form as the antecedent strengthening law, and it is equally clearly not valid.

Of special interest for our purposes is the following principle which Stalnaker takes as an axiom for his conditional logic:

(SA) $\Diamond A \supset [\sim(A > B) \equiv (A > \sim B)]$.

Validity of this schema, which we shall call Stalnaker's axiom, corresponds to the idea that the negation of a conditional with a possible antecedent is equivalent to a conditional with the same antecedent and negated consequent. In the presence of the rest of his logic, Stalnaker's axiom is equivalent to conditional excluded middle,

(CEM) $(A > B) \vee (A > \sim B)$.

Validity of both of these principles corresponds directly to the assumption that a single world is selected as the closest antecedent world. If $f(A, w)$ is the world selected, then either B is true there or $\sim B$ is true there and the corresponding conditionals are true or false in w accordingly. Thus, either $(A > B)$ or $(A > \sim B)$ is true, and $\sim(A > B)$ is true just in case $(A > \sim B)$ is true.

2. In his 1972 paper 'Counterfactuals and Comparative Possibility' (Lewis, 1972, this volume, p. 60), Lewis points out that it is not plausible to assume that comparative similarity of possible worlds is such that there will always be a unique nearest antecedent world. For an antecedent such as 'Bizet and Verdi are compatriots' a world where they are both French and a world where they are both Italian might be equally close to the actual world, and both might be closer than any others. If this were so then selecting one over the other would be arbitrary. (Lewis, 1972, this volume, p. 60). He proposes an analysis that does not make the semantical content of a counterfactual depend on any arbitrary tie breaking selection in such cases. (Lewis, 1972, this volume, pp. 63–64).

I shall use Lewis' notation to formulate his truth conditions. Lewis restricts his theory to conditionals of the sort we usually express in English with subjunctive constructions, while Stalnaker intends his theory to apply to indicative as well as to subjunctive conditionals.[3] It will be useful to have a notation for explicitly subjunctive conditionals, so let

$$A \,\square\!\!\rightarrow B$$

abbreviate "If it were that A, then it would be that B". This notation is suggestive of the relationship Lewis proposes between *would* and *might* conditionals, which parallel the relation between necessity and possibility.

Lewis' truth conditions are formulated directly in terms of comparative similarity of possible worlds. Where A is possible, he specifies:

$A \,\square\!\!\rightarrow B$ is true at w iff, some $(A \wedge B)$-world is closer to w than any $(A \wedge \sim B)$ world.

Where the antecedent A is impossible Lewis makes the conditional true vacuously, as does Stalnaker.[4] If there is a unique closest A-world, Lewis' analysis agrees with Stalnaker's. If there are ties, however, Lewis' analysis will make $A \,\square\!\!\rightarrow B$ true just in case every nearest A-world is a B world.

Given the plausible assumption about comparative similarity of worlds where Bizet and Verdi would be compatriots, Lewis' analysis makes both of

(1)   "It is not the case that if Bizet and Verdi were compatriots Bizet would be Italian."

and

(2)   "It is not the case that if Bizet and Verdi were compatriots Bizet would not be Italian."

come out true. This would be a counter-example to Stalnaker's axiom, and conditional excluded middle; but, as Lewis himself suggests (Lewis, 1973, p. 80), asserting both these sentences together sounds like a contradiction.[5]

According to both Stalnaker and Lewis (Stalnaker, 1968, this volume, p. 49; Lewis, 1972, this volume, p. 61) one reason this sounds like a contradiction is that, in English, a common way of negating a conditional is to assert the corresponding conditional with negated consequent. Stalnaker's axiom corresponds directly to this practice.

Lewis counters this appeal to common usage by introducing the might conditional. Let,

$$A \,\diamond\!\!\rightarrow B$$

abbreviate "If it were that A then it might be that B". According to Lewis, it is very natural to have

$$\sim (A \,\square\!\!\rightarrow B) \text{ iff } (A \,\diamond\!\!\rightarrow \sim B)$$

so that the might conditional provides an idiomatic way of negating a would conditional. It does not sound at all contradictory to assert both

(1')   "If Bizet and Verdi were compatriots Bizet might be Italian"

and

(2')   "If Bizet and Verdi were compatriots Bizet might not be Italian".

If (1') is an idiomatic way of asserting what (1) asserts in a contrived way, and similarly for (2') and (2), then asserting (1) and (2) together is, perhaps, not really contradictory after all. On his side, Lewis challenges Stalnaker to provide an analysis of English might conditionals. (Lewis, 1972, this volume, p. 63).

3. In 'Defense of Conditional Excluded Middle', (Stalnaker, 1979, this volume, pp. 87–103), a new paper delivered at the *1978 U.W.O. Spring Workshop on Pragmatics and Conditionals,* Stalnaker responds to Lewis' criticisms. He grants that it is not reasonable to assume there will always be a unique nearest antecedent world, but likens his relativization of conditionals to selection functions to the relativization of quantifiers to a universe of discourse and the relativization of predicates to propositional functions. In each case the abstract semantic theory assumes well defined sets with sharp boundaries, but realistic applications present indeterminacies. Stalnaker suggests that an account of vagueness adequate to handle this problem for the usual semantics for quantifiers and predicates will also handle it for his semantics for conditionals. (Stalnaker, 1979, this volume, pp. 89–90).

He gives substance to this suggestion by endorsing van Fraasen's general account of vagueness (Stalnaker, 1979, this volume, p. 90).[6] On this account a context with indeterminacies is represented by a set of sharp valuations – all the various possible ways of arbitrarily settling the indeterminacies. A sentence is true (false) in such a vague context just in case it would come out true (false) in every one of these corresponding sharp valuations; otherwise, it is indeterminate. When this account is applied to the Bizet–Verdi example the sentences

(1a)   "If Bizet and Verdi were compatriots then Bizet would be Italian"

and

(2a)   "If Bizet and Verdi were compatriots then Bizet would not be Italian" are both indeterminate; but, their disjunction is true, because it comes out true on both ways of arbitrarily breaking the tie by selecting a world as nearest.

Lewis discussed this way of reconciling Stalnaker's selection function with ties. In addition to arguing that it cannot meet his challenge to account for the might conditional, Lewis faults it because it required for its application another assumption about nearness which he calls the limit assumption (Lewis, 1972, this volume, p. 63; Lewis, 1973, pp. 30, 81). This is the assumption that there always be some antecedent worlds that are at least as close as any others. Lewis suggests that there may be cases where there are no closest A-worlds at all, but only an infinite sequence of ever closer A-worlds (Lewis, 1972, this volume, pp. 63, 69, 70; Lewis, 1973, pp. 19, 20). Perhaps, worlds where I am over seven feet tall, are ordered according to how much my height in them differs from my actual height. If so, then for any world where I am $7 + \epsilon$ feet tall there will be a closer world where I am $7 + (\epsilon/2)$ feet tall. Lewis' analysis is designed to work even in this kind of case, but without some closest worlds to select from, a selection function would be worse than arbitrary. It would have to strongly violate the nearness relation by choosing some world over closer worlds.

One counterintuitive feature of the example was pointed out by Nollaig MacKenzie when Lewis gave a talk at Toronto in 1971. For every positive ϵ, the counterfactual "If I were over 7 feet tall then I would be under $7 + \epsilon$ feet tall" comes out true. In recent years John Pollock (Pollock, 1976, pp. 18–21), and Hans Herzberger (Herzberger, 1979), as well as Stalnaker (Stalnaker, 1979, this volume, pp. 97–98) have argued that this leads to serious difficulty for Lewis' rejection of the limit assumption. Let

$$\Gamma(A) = \{B : A \Box\!\!\rightarrow B \text{ is true}\}.$$

If A is "I am over 7 feet tall" and worlds are ordered as we have supposed, then, according to Lewis' account, $\Gamma(A)$ will contain A together with all sentences of the form "I am under $7 + \epsilon$ feet tall", for — say every positive rational. But, this insures that for every real number x, my height would not equal x feet if I were over 7 feet tall. Stalnaker points out that Lewis' relation between would and might conditionals makes this yield the even more damaging result that there is no real number x such that my height *might* be x if I were over seven feet tall (Stalnaker, 1979, this volume, p. 97).[7] Problems of this kind are always available to plague the intelligibility of any situation where an antecedent is purported to lead to violation of the limit assumption.

If the limit assumption is justified, then Lewis' case against Stalnaker depends on his challenge to account for might conditionals. Stalnaker responds by proposing

$$\Diamond(A > B)$$

as a rendering of "If it were that A then it might be that B". This makes the might conditional a special case of more general might constructions and Stalnaker suggests that 'might' has the same range of senses, in his conditional constructions, as it has generally. Normally it expresses epistemic possibility, but sometimes it expresses non-epistemic possibility of one sort or another (Stalnaker, 1979, this volume, p. 99).

Lewis has already rejected this proposal on the basis of the following counter example (Lewis, 1973, pp. 80–81): Suppose there was no penny in my pocket, but I did not know it. Consider,

"If I had looked, I might have found a penny".

According to Lewis, this English might conditional is plainly false, but the corresponding $\Diamond(A > B)$ construction comes out true whether $\Diamond$ is interpreted as epistemic or as alethic possibility. On the epistemic interpretation it comes out true; because, for all I know, I would have found a penny if I had looked. On the alethic interpretation it comes out true; because, having no penny in my pocket was merely a contingent fact, not a necessary truth.

Stalnaker grants that the non-epistemic sense Lewis intends is one possible interpretation of the example, and he proposes a quasi-epistemic reading of might on which his construction captures it (Stalnaker, 1979, this volume, pp. 100–101). Consider what would be compatible with an idealized state of knowledge in which all the relevant facts were known (Stalnaker, 1979, this volume, p. 101). On this reading the example comes out plainly false, as Lewis intends. Moreover, so long as there are indeterminacies in the language this kind of possibility does not collapse into truth (Stalnaker, 1979, this volume, p. 100). Indeed, in cases like the Bizet and Verdi example it makes Stalnaker's might construction behave very like Lewis' (Stalnaker, 1979, this volume, p. 101).

One advantage Stalnaker attributes to his construction is that it includes the epistemic *might* as another special case (Stalnaker, 1979, this volume, p. 101). This lets it explain why denying a *would* conditional while affirming the corresponding *might* conditional has the same air of anomaly while affirming the corresponding *might* conditional has the same air of anomaly as Moore's paradox (Stalnaker, 1979, this volume, pp. 99, 100, 102).[8] This runs counter to the prediction Lewis' construction gives. Moreover, the pervasiveness of the anomaly suggests that in might conditionals, just as for might generally, the epistemic reading is usually dominant.

II. CONDITIONALS AND CONDITIONAL BELIEF

(The Ramsey Test Paradigm)

1. In his 1970 paper 'Probability and Conditionals' (1970, this volume, pp. 107–128), Stalnaker defended his conditional logic by developing connections between the Ramsey test idea and the Bayesian account of rational conditional belief. He suggests that a rational agent's subjective probability assignment to a conditional he evaluates by Ramsey's test ought to be the same as the subjective conditional probability he assigns to its consequent given its antecedent. This suggestion has come to be known as Stalnaker's Hypothesis:

(SH) $\quad P(A > B) = P(B \mid A),$

where $P(A > B)$ is the agent's degree of belief in the conditional $A > B$, and $P(B \mid A)$ is his conditional degree of belief in B given A. In the Bayesian model, an agent's conditional degree of belief in B given A is just the evaluation of B he would deem appropriate on the basis of the minimal revision of what he now believes required to assume A. But, such an evaluation is just Ramsey's test, so that $P(B \mid A)$ measures how well $A > B$ does on this test, and if degree of belief in $A > B$ is guided by this test performance then $P(A > B)$ should be equal $P(B \mid A)$.

Consider Stalnaker's axiom and conditional excluded middle for cases where the agent's degree of belief in an antecedent A is non-zero. For such cases, the basic conditional probability law for negations is

(PN) $\quad P(\sim B \mid A) = 1 - P(B \mid A).$

Given Stalnaker's Hypothesis, this corresponds to Stalnaker's Axiom

(SA) $\quad \Diamond A \supset (\sim(A > B) \equiv (A > \sim B)).$

We have

(1) $\qquad P(\sim(A > B)) = 1 - P(A > B) = 1 - P(B \mid A)$

and

(2) $\qquad P(A > \sim B) = P(\sim B \mid A),$

so that (PN) insures that $\sim(A > B)$ and $(A > \sim B)$ always get the same subjective probability assignment. This in turn insures that

(CEM) $(A > B) \vee (A > \sim B)$

always gets assigned subjective probability one.

In the orthodox Bayesian model conditional belief on A is only defined when the agent's degree of belief in A is non-zero. In order to extend the Stalnaker Hypothesis to cover Ramsey tests for counterfactuals, the Bayesian model must be extended so that $P(B \mid A)$ can be well defined even when $P(A)$ is zero. Stalnaker constructs such an extension by explicating the idea of minimal revisions of bodies of belief. Let K be the content of all the sentences to which an agent assigns subjective probability 1, and let $K(A)$ be the content of all the sentences to which he assigns subjective conditional probability 1, given A. The core of Stalnaker's construction is a set of conditions motivated by the idea that K represents what the agent accepts (we may call it the agent's acceptance context), and $K(A)$ represents the minimal revision of K required to accept A (we may call this the agent's A-assumption context). These conditions allow the A-assumption context, $K(A)$, to be consistent even when A is inconsistent with K.[9]

Stalnaker relativises the orthodox Bayesian coherence requirement to assumption contexts, so that

$P(B \mid A)$

is to be a coherent guide to decisions where the possible outcomes are exactly those consistent with $K(A)$. For cases where A is treated as impossible, so that no outcomes are consistent with $K(A)$, he adopts the convention that $P(B \mid A) = 1$ for every B. The result of these restrictions is a representation theorem that allows as extended conditional belief functions on the sentential calculus (sentences closed under negation and conjunction) exactly those functions that satisfy Karl Popper's axioms for conditional probability.[10] The important feature of Popper's axiomatization is that $P(B \mid A)$ can be non-trivial even when $P(A)$ is zero. Stalnaker's representation theorem allows Popper functions to be interpreted as possible subjective belief states of a Bayesian agent who can entertain non-trivial conditional beliefs relative to counterfactual assumptions.

In order to formulate the Ramsey test restriction on extended conditional belief for languages closed under a primitive conditional connective, Stalnaker introduces subfunctions. For each sentence C and extended conditional belief function P, the subfunction of P with respect to C is defined as follows:

$$P_C(B \mid A) = P(A > B \mid C).$$

The restriction then is that P, all subfunctions of P, subfunctions of subfunctions, and so on, must satisfy the Stalnaker Hypothesis as well as all the other conditions. This adds the Ramsey test evaluation of $P(A > B)$ to the general requirements on rational extended conditional belief.

A sentence A for a language L with a primitive conditional connective is Ramsey Test valid just in case it receives conditional belief 1 on every sentence in L for every Ramsey-Test extended belief function on L. That is to say,

> A is Ramsey Test valid iff for every Ramsey test P on L, $P(A \mid C) = 1$ for all sentences C of L.

Stalnaker was able to prove that the Ramsey Test valid sentences are exactly the theorems of his conditional logic.

2. Stalnaker's conditional belief semantics generated his conditional logic from the Ramsey test paradigm for evaluating the acceptability of conditionals. This provided a strong argument in favor of his side of the debate with Lewis on the validity of the Stalnaker axiom and conditional excluded middle. Lewis' counter argument was a surprising trivialization result. In 'Probability of Conditionals and Conditional Probability' (Lewis, 1975, p. 129 below) he proved that no probability function meeting Stalnaker's conditions can assign positive probability to three or more pairwise incompatible statements. This is a devistating result, for it shows that Stalnaker's conditional belief functions cannot represent non-trivial belief states.

Lewis' trivialization does not depend on Stalnaker's extension of conditional belief to include counterfactual conditional probabilities. Any probability function (a classical one where $P(B \mid A)$ is undefined when $P(A) = 0$ as well as any Popper function) which satisfies the Stalnaker hypothesis and

(CSH) $P(A \rightarrow B \mid C) = P(B \mid A \wedge C)$, if $P(A \wedge C) \neq 0$

for any binary connective $\rightarrow$, is shown to be trivial by Lewis' result. The condition (CSH) is just the result of extending the Stalnaker hypothesis to conditional probabilities and requiring that conditional probabilities satisfy the standard laws.

Lewis first presented his result in June 1972 and began circulating it in manuscript shortly afterwards.[11] It received so much attention that a

number of papers (and part of one book) devoted to working out its consequences for the Ramsey test idea were in print (or forthcoming) before
Lewis' paper was finally published in 1976.[12] van Fraassen (1976a)
developed a Stalnaker Hypothesis conditional that can be incorporated
in non-trivial belief systems, by making the conditional radically context
dependent so that the conditional proposition evaluated for assigning the
unconditional probability

$$P(A > B)$$

need not be the same as the conditional proposition evaluated in assigning
the conditional probability

$$P(A > B \mid C).$$

van Fraassen proved that the Stalnaker hypothesis can hold non-trivially
if the Stalnaker logic is weakened by dropping a conditional antecedent
substitution principle

$$(\text{CAS}) \quad (A > B) \wedge (B > A) \wedge (A > C) \supset (B > C)$$

which both Lewis and Stalnaker share. He also showed that the Stalnaker
hypothesis can hold non-trivially with the full Stalnaker logic if compounds
involving conditionals are limited to

$$A > B$$

$$(A > B) > C$$

$$A > (B > C)$$

where none of A, B, or C contains conditionals. Stalnaker (1974, pp. 302–
306) proved that the trivialisation cannot be avoided for the full Stalnaker
logic if one allows arbitrary truth functional compounds with conditionals
in them, such as

$$\sim A \wedge (A > \sim B).^{13}$$

as consequents.

Ernest Adams had been working on the Ramsey test for some years before
these developments (Adams, 1966). He used Lewis' result to argue that
English indicative conditionals are used to make conditional assertions
rather than to assert conditional propositions (Adams, 1975, pp. 7, 8).
According to Adams, the Ramsey test paradigm is right in that assertability
of indicative conditionals goes by the corresponding conditional probability,
but this is not to be construed as the probability that a conditional

proposition is true. For Adams, it may well be that there is no such thing as an indicative conditional proposition.[14]

3. Lewis' paper (Lewis, 1975, this volume, pp. 129–147) also contains a probability argument supporting H. P. Grice's (*William James Lecture* manuscript) defense of the truth functional material conditional as an interpretation of the English indicative conditional. According to Grice's account of conversation, it would be misleading to assert a material conditional $(A \supset B)$ to the extent that the probability of vacuity, $P(\sim A)$, is high and the probability of falsity, $P(A \wedge \sim B)$, is a large fraction of the probability of non-vacuity, $P(A)$. Lewis calls attention to the following probability result, which holds so long as $P(A) > 0$:

$$P(B \mid A) = P(A \supset B) - [P(\sim A) \cdot (P(A \wedge \sim B)/P(A))]$$

This tells us that the conditional probability of the consequent on the antecedent is just the result of diminishing the probability of $P(A \supset B)$ by the factors which would make its assertion misleading. According to Lewis, the Ramsey test idea is right in that the conditional probability $P(B \mid A)$ is the appropriate evaluation of the assertability of the indicative conditional "If A then B". Moreover, Adams is right that this is not the same as the probability that the proposition asserted is true. But, this need not be because there is no indicative conditional proposition, as Adams suggests. Rather, the conditional probability is just that diminution of probability of truth needed to account for factors which would render assertion misleading.

Lewis suggests that an important advantage of this account over Adam's conditional assertion account is that the material conditional interpretation automatically includes embedded conditionals while the conditional assertion account fails to do so.

III. DECISION MAKING CONDITIONALS

(Another Paradigm)

1. When Lewis first presented his trivialization result in 1972, Stalnaker was on hand to respond. His reaction was to give up the Stalnaker hypothesis and to go on to argue that conditionals appropriate to guide decisions ought to violate the Stalnaker hypothesis in certain interesting kinds of decision contexts. These contexts are ones where an act is epistemically relevent to a desired outcome without being able to causally influence the

outcome. Newcomb's problem and the prisoner's dilemma provide examples of such contexts. The following example from 'Counterfactuals and Two Kinds of Expected Utility' (Gibbard and Harper, 1978, this volume, pp. 153–190), which develops Stalnaker's suggestion, should make it clear that decision making provides a role for conditionals for which Ramsey's test is not the appropriate guide.

Robert Jones is one of several rising young executives competing for a very important promotion. The company brass have found the candidates so evenly matched that they have employed a psychologist to break the tie by testing for personality qualities correlated with long run success in the corporate world. The test was administered to Jones, and the other candidates, on Thursday morning. The promotion was decided Thursday afternoon on the basis of the test scores, but it will not be announced until Monday. On Friday morning Jones learnt, through a reliable company grapevine, that the promotion went to the candidate who scored highest on a factor called ruthlessness; but, he is unable to discover which of them this is.

It is now Friday afternoon and Jones is faced with a decision. Fred Smith has failed to meet his minimum output quota for the third straight assessment period, and a long standing company policy rules that he should be fired on the spot. Jones, however, has discovered that Smith's recent work difficulties have been due to a fatal illness suffered by his now deceased wife. Jones is quite sure that Smith's work will come up to a very high standard after he gets over his loss, provided he is treated leniently. Jones also believes he can convince the brass that leniency to Smith will benefit the company. Unfortunately, he has no way to get in touch with them before they announce the promotion on Monday.

We may represent Jones' preferences by the following matrix, where J is the event of Jones getting his promotion, $\sim J$ is the event of Jones failing to get his promotion, F is the act of firing Smith, and $\sim F$ is the act of Jones not firing Smith.

	J	$\sim J$
F	100	0
$\sim F$	105	5

Jones' utilities reflect the fact that he would rather not fire Smith in the event that he got his promotion, and that he would also rather not fire

Smith in the event that he did not get his promotion. These utilities also show that Jones cares twenty times as much about his promotion as he does about firing Smith.

Jones believes that his behaviour in the decision he now faces will provide evidence about how well he scored on the ruthlessness factor in Thursday's personality test. These beliefs are summed up in the following subjective conditional probability assignments he makes:

$$P(J \mid F) = 0.75 \qquad P(\sim J \mid F) = 0.25$$
$$P(J \mid \sim F) = 0.25 \qquad P(\sim J \mid \sim F) = 0.75.$$

If Jones were to use these conditional probabilities to guide his expected utility reasoning in this decision he would fire Smith. Using these conditional probabilities and Jones' utilities the expectation of firing Smith is

$$(100 \times 0.75) + (0 \times 0.25) = 75$$

and the expectation of not firing smith is

$$(105 \times 0.25) + (5 \times 0.75) = 30.$$

But, Jones believes that his promotion is causally independent of his firing Smith or not firing Smith. He knows that his promotion is already decided on the basis of the test yesterday, and that nothing he does today can change the relevant test scores.

These important causal beliefs of Jones are not reflected in the conditional probabilities that appear in his matrix. Let

$$A \mathbin{\square\!\!\rightarrow} O$$

abbreviate the subjunctive conditional

> If I were to perform A then O would obtain.

According to Stalnaker's suggestion, the appropriate probabilities for computing expected utility in Jones' matrix ought to be

$$P(F \mathbin{\square\!\!\rightarrow} J) \qquad P(F \mathbin{\square\!\!\rightarrow} \sim J)$$
$$P(\sim F \mathbin{\square\!\!\rightarrow} J) \qquad P(\sim F \mathbin{\square\!\!\rightarrow} \sim J).$$

In place of the conditional probability of the event given the act, we propose the unconditional probability of the corresponding subjunctive conditional. This gives a simple way of representing Jones' belief that what he does now cannot causally influence whether he got the promotion. He will accept, assign probability 1 to, the following biconditionals

$$(F \mathbin{\square\!\!\rightarrow} J) \equiv (\sim F \mathbin{\square\!\!\rightarrow} J)$$

and

$$(F \mathbin{\square\!\!\rightarrow} \sim J) \equiv (\sim F \mathbin{\square\!\!\rightarrow} J).$$

But, if he accepts a biconditional he must assign the same subjective probability to each side. So $P(F \mathbin{\square\!\!\rightarrow} J)$ is the same as $P(\sim F \mathbin{\square\!\!\rightarrow} J)$ while $P(F \mathbin{\square\!\!\rightarrow} \sim J)$ is the same as $P(\sim F \mathbin{\square\!\!\rightarrow} J)$. With such agreement in probabilities the expected utility of not firing Smith will be higher than that of firing Smith. Not firing Smith seems, clearly, to be the rational decision for Jones, given the utilities and beliefs we have ascribed to him.

Using conditional probability to compute expected utility in cases like Jones' is to confuse stochastic dependence with the causal dependence that ought to count in decision making. Such cases are not common, and subjective stochastic dependence (and independence) is often a good guide to the agent's relevant causal beliefs. Therefore, our model ought to insure that Stalnaker's hypothesis will hold when conditional probability reasoning would be appropriate. Given certain weak constraints on the logic of a decision making conditional we have $P(A \mathbin{\square\!\!\rightarrow} O) = P(O \mid A)$, provided that $P(A \mathbin{\square\!\!\rightarrow} O \mid A) = P(A \mathbin{\square\!\!\rightarrow} O)$ (Gibbard and Harper, 1979, this volume, pp. 155–156). This restricts violations of the Stalnaker hypothesis to just those cases, like Jones', where whether an act is performed or not counts as evidence about what would happen if it were to be performed.

2. Situations of the sort faced by Jones provide contexts where a conditional appropriate to guide decision making should not be evaluated by Ramsey's test. One can give a rough and ready account of conditions under which such a decision making conditional ought to be accepted. One ought to accept

$$A \mathbin{\square\!\!\rightarrow} O$$

just in case he believed that one of the following two cases obtained:

(a)   Doing A would bring it about that O.

(b) O will obtain, and doing A would not change this.

With a little generalization to handle antecedents other than actions, one can obtain a rival to the Ramsey test paradigm for evaluating the acceptability of conditional sentences. On this rival account

$$P \mathbin{\square\!\!\rightarrow} Q$$

would be acceptable just in case one accepted either of

(a′)   *P* would bring about *Q*.

or

(b′)   *Q,* and *P* would not change this.

These claims are infected with the vague and difficult notion of causal bringing about; but, as the Jones example illustrates, this may be all to the good, for, as such examples show, some way of bringing in an agent's beliefs about what he can causally influence is required by an adequate account of rational decision making.

Let us try out these two acceptability accounts, the Ramsey test and our rough causally sensitive paradigm, on a version of an example used by Ernest Adams (Adams, 1970) to illustrate differences between English indicative and subjunctive conditionals.

Consider:

(I)   If Oswald didn't shoot Kennedy, then someone else did.
(S)   If Oswald hadn't shot Kennedy, then someone else would have.

The Ramsey test seems to accord quite well with the way we evaluate the acceptability of the indicative conditional (I). For most of us, the claim that Kennedy was shot is a salient piece of what we take to be our accepted body of knowledge. When each of us hypothetically revises his body of knowledge to assume the antecedent that Oswald didn't shoot Kennedy, he retains this salient claim that Kennedy was shot. This in turn, forces high creedence for the consequent that someone shot Kennedy. This Ramsey test reasoning seems to be the right account of the high creedence most of us place in the indicative conditional (I).

When we turn to the subjunctive conditional (S) the causal paradigm is much more appropriate than Ramsey's test. Presumably, we would not accept (S) unless we believed something like the following story:

> Other marksmen were in position to shoot Kennedy if Oswald missed or failed to fire, or were there to shoot Kennedy regardless of Oswald.

This is exactly the kind of story that would render (S) acceptable on the causally sensitive paradigm.

These examples suggest that the differences between these two paradigms for evaluating the acceptability of conditional sentences correspond, in some case at least, to a difference one wants to have between acceptability evaluations appropriate for English indicative and subjunctive conditionals. When one realizes that the kind of hypothetical reasoning exemplified by Ramsey's test plays an important role in organizing our knowledge and planning for knowledge change, and that the kind of hypothetical reasoning exemplified in our causally sensitive paradigm plays an important role in guiding action, then he should not be surprized to discover that English provides conditional constructions appropriate for each of these important kinds of job.[15]

IV. INDICATIVE CONDITIONALS

1. In 'Indicative Conditionals' Stalnaker defends his theory as an account of English indicative conditionals by developing pragmatic principles to explain away the apparent validity of the following argument: "Either the butler or the gardener did it. Therefore, if the butler didn't do it, the gardener did it". Stalnaker calls this the direct argument and warns that its validity would establish that the indicative conditional is logically equivalent to the truth functional material conditional (Stalnaker, 1975, this volume, p. 193).[16] In order to have its apparent validity not count against his theory he argues that the inference is reasonable even though it is not valid. (Stalnaker, 1975, this volume, p. 194).

Reasonable inference is a pragmatic relation between speech acts. According to Stalnaker, an inference is reasonable just in case, in every context in which the premises could be appropriately asserted, it is impossible to accept the premises without being committed to the conclusion (Stalnaker, 1975, this volume, p. 195). An argument is valid, on the other hand, just in case the propositions asserted in its premises semantically entail its conclusion. The two notions can differ because what some sentences assert is context dependent. Thus,

$$\therefore \text{ "I am here now"}$$

is a zero premissed reasonable argument, but its conclusion does not assert a semantically valid proposition.

For Stalnaker's purposes, the most important feature of a context is the presumed common knowledge or common assumption of the participants in the discourse. Using possible worlds, these background presuppositions can be summarized as a set of worlds. Stalnaker calls this the context set. An assertion is appropriate in a given context only if the proposition is neither inconsistent with nor entailed by the context set. When an appropriate assertion is made the context set is reduced to just those worlds in it that are consistent with the proposion asserted. Think of the context set as the relevent candidate situations which the presumed background knowledge leaves open. An appropriate assertion of proposition A acts as the claim that the actual situations is one of the A-candidates.

Stalnaker imposes a Grice-like (H. P. Grice *William James Lectures* manuscript) condition on appropriate assertion of disjunctions

(G)   Asserting a disjunction is appropriate only in a context which allows either disjunct to be true without the other.

He also imposes a condition on selection functions that makes indicative conditionals context dependent

(C)   If an antecedent is compatible with a context set then the antecedent world selected in that context should also be compatible with it.

This principle need not apply to subjunctive conditionals, according to Stalnaker, because the subjunctive mood in English is a conventional device for indicating that presuppositions are being suspended (this volume, p. 200).

These conditions are sufficient to render the direct argument reasonable.[17] Suppose that $A \lor B$ is appropriate to a given context. Then, according to condition (G), $B \land \sim A$ is compatible with its context set. Now add $A \lor B$ to the presumed background knowledge, thus reducing the context set to just its $A \lor B$-worlds. Consider the conditional,

If $\sim A$ then B.

The antecedent $\sim A$ is compatible with the new context set, since all the $B \land \sim A$-worlds of the old context set will remain after $A \lor B$ is assumed. By condition (C) therefore, the A-world selected to interpret the conditional must also be in the new context set. But, since all these are $A \lor B$-worlds, any $\sim A$-world in it must be a B-world.

2. In 'Two Recent Theories of Conditionals' Gibbard shows that Adams' version of the Ramsey test account and Stalnaker's theory give the same logic on the limited domain to which Adams' account applies. But, he argues that the interpretations given by the two accounts make them appropriate for quite different jobs. Roughly, Adams' account is appropriate for indicative conditionals, while Stalnaker's is appropriate for subjunctives. Indeed,

Gibbard's paper gives substance to the suggestion that the decision making and Ramsey test paradigms correspond to the different ways in which subjunctive and indicative conditionals are typically used. He explicates the grammatical features that distinguish subjunctive from indicative conditionals and argues that these correspond to the important semantical differences one would expect from our two paradigms. He interprets the subjunctive conditional as a proposition, but defends Adams' conditional assertion interpretation of indicative conditionals.

Adams' version of the Ramsey test paradigm is based on his definition of probability consequence. This definition applies to a propositional calculus L of truth functions, together with the result of adding compound of the form

$$(A \to B)$$

where A and B are propositions in L and $\to$ is a primitive conditional connective. This *simple conditional extension* of L has no conditionals constituents, nor does it have any truth functional compounds of conditionals. The Stalnaker hypothesis,

(SH) $P(A \to B) = P(B \mid A)$,

uniquely determines any extension of any classical probability function P on L to all conditionals in the simple conditional extension of L for which P assigns non-zero probability to the antecedent. Where B is in the simple conditional extension of L and Γ is a finite subset of the simple conditional extension of L,

> B is *a probability consequence* of Γ iff for every $\epsilon > 0$ there is a $\delta > 0$ such that for any probability function P on L with $P(A) > 1 - \delta$ for every A in Γ we have $P(B) > 1 - \epsilon$.

On the limited domain to which this definition applies, *probability consequence* agrees, not only, with Stalnaker's logic, but with Lewis' logic as well. Compounds, such as conditional excluded middle

$$(A \to B) \lor (A \to \sim B)$$

which distinguish Lewis' logic from Stalnaker's are not represented in Adams' simple conditional extension of a propositional system. But, there is a natural constraint on negations of conditionals

$$P(\sim(A \to B)) = 1 - P(B \mid A)$$

which will extend the domain enough to support Stalnaker's logic against Lewis'.

Gibbard proposes examples which show that Stalnaker's contextual propositional account of indicative conditionals will require sometimes making the propositional content of an indicative conditional depend on the utterer's subjective probabilities (pp. 29–33). He then goes on to use a variant of Lewis' argument to show that such radical context dependence must be required if assertability goes by Ramsey's test and also corresponds to probability that the conditional proposition asserted is true (pp. 33–34). He suggests that when context dependence becomes this radical there is little to choose between Stalnaker's contextual conditional proposition account and Adam's non-propositional conditional assertion account.

The advantage one expects a propositional account to have is a natural way of handling embedded conditionals. Gibbard proves that if we want the indicative conditional to be a propositional function (even if only relative to a given context) and want it to conform to the fact that we treat embedded indicative conditionals of the form

$$A \rightarrow (B \rightarrow C)$$

as logically equivalent to

$$A \wedge B \rightarrow C,[18]$$

then the indicative conditional must be logically equivalent to the material conditional. The only way a Stalnaker type account can avoid this result is to make the propositional function depend on place in the sentence so that the two arrows in $A \rightarrow (B \rightarrow C)$ correspond to different propositional functions.[19] According to Gibbard, this suggests that Stalnaker's account of such embedding would have to be just as ad hoc as an attempt to extend Adams' account in an appropriate way.

Against the material conditional propositional interpretation defended by Grice and Lewis, Gibbard suggests the following embedded indicative conditional sentence of the form $(D \rightarrow B) \rightarrow F$:

If the cup broke if dropped, then it was fragile.

This idiomatic English sentence is assertable by someone who knows that the cup was being held at a moderate height over a carpeted floor, even if he assigns very low subjective probability to the cup's being dropped and also assigns very low subjective probability to its being fragile. Interpreted truth functionally, the sentence would be logically equivalent to: $(D \wedge \sim B) \vee F$,

Either the cup was dropped and didn't break, or it was fragile.

But, a speaker might very well assign low subjective probability to this disjunction, while assigning quite high subjective probability to the indicative conditional that is supposed to be equivalent to it. The Grice–Lewis diminution of assertability device does not help here, because this is an example in which the indicative conditional is highly assertable while the corresponding material conditional is not.

In 'Indicative Conditionals and Conditional Probability' John Pollock proposes a counterexample to the thesis that indicative conditionals are to be evaluated by Ramsey's test. In his 'Reply to Pollock' Gibbard argues that Pollock's example does not undercut his Ramsey test thesis.

V. SUBJUNCTIVE CONDITIONALS CHANCE AND TIME

1. In 'The Prior Propensity Account of Subjunctive Conditionals' Brian Skyrms proposes a measure of asssertability for the kind of causally sensitive conditional our decision making paradigm requires (Skyrms, 1979, this volume, pp. 259–265). According to Skyrms the assertability of a subjunctive conditional ought to go by the subjective expectation of the objective conditional chance of the consequent given the antecedent. Let $H_1 \ldots H_n$ be the hypotheses concerning the relevent objective chances that I deem open alternatives, and for each i let C_i be the conditional chance according to H_i. Using Lewis' notation for the subjunctive conditional Skyrms proposal may be rendered

$$P(A \, \Box\!\!\rightarrow B) = \sum_{i=1}^{i=n} P(H_i) \cdot C_i \, (B \text{ given } A),$$

where $P(A \, \Box\!\!\rightarrow B)$ is my present degree of assertability for $A \, \Box\!\!\rightarrow B$, and each $P(H_i)$ is my degree of belief that H_i is true.

Let us try out Skyrms' proposal on Jones' decision problem. In this example the relevent objective conditional chances depend on whether or not he was the candidate with the highest test score, There are only two hypotheses that matter

H_1:   He was the candidate that scored highest on the ruthlessness factor.

H_2:   He wasn't.

where $F(\sim F)$ stand for his firing (not firing) Smith and $J(\sim J)$ stand for his getting (not getting) the new job, the relevent conditional chances are as follows:

$$C_1(J \text{ given } F) \quad = 1 = C_1(J \text{ given } \sim F)$$

$$C_1(\sim J \text{ given } F) = 0 = C_1(\sim J \text{ given } \sim F)$$

$$C_2(J \text{ given } F) \quad = 0 = C_2(J \text{ given } \sim F)$$

$$C_2(\sim J \text{ given } F) = 1 = C_2(\sim J \text{ given } \sim F).$$

On Skyrms' proposal the decision making conditionals are evaluated so that,

$$P(F \,\square\!\!\rightarrow J) = P(H_1) \cdot C_1(J \text{ given } F) + P(H_2) \cdot C_2(J \text{ given } H_2)$$

$$= P(H_1) \cdot 1 + P(H_2) \cdot 0$$

$$= P(H_1).$$

Similarly,

$$P(\sim F \,\square\!\!\rightarrow J) = P(H_1)$$

and

$$P(F \,\square\!\!\rightarrow \sim J) = P(H_2) = P(\sim F \,\square\!\!\rightarrow \sim J).$$

Whatever $P(H_1)$ and $P(H_2)$ are, the agreement of $P(F \,\square\!\!\rightarrow J)$ with $P(\sim F \,\square\!\!\rightarrow J)$ and of $P(F \,\square\!\!\rightarrow \sim J)$ with $P(\sim F \,\square\!\!\rightarrow \sim J)$ insures that the dominate act of not firing Smith will have higher utility than the dominated act of firing Smith. This is exactly the result we want. Skyrms' proposal represents the agent's beliefs about causal influences and independencies by his beliefs about objective conditional chances.[20]

Skyrms does not claim that his measure of assertability for subjunctive conditionals is a probability of truth of a conditional proposition. As far as his account goes, the subjunctive conditional may be just as much a matter of conditional assertion as Adams and Gibbard claim the Ramsey test indicative conditonal is.

2. Propensity or objective chance is about as much a subject for controversy as causality, and it is about as hard to do away with as well.[21] Frequentists, the orthodox theorists who interpret probability as a matter of objective empirical fact (e.g. Von Mises, 1957; Cramer, 1955; Reichenbach, 1949), have not found it easy to make sense of probability assertions about a single case. Yet assertions about chance in a single case, such as my claim that this

newly made die is biased so that its chance of coming up a six when I next toss it is 25%, seem to be paradigms of probability assertions.[22]

Subjectivists have no trouble with the idea of single case probabilities, but orthodox subjectivists such as De Finetti (1975) reject as nonsense interpretations of probability assertions as claims about objective empirical facts. Yet, my assertion about the chance of six on the next throw of my new die seems to be intelligible as just such an empirical claim.

In 'A Subjectivists Guide to Objective Chance' Lewis takes both objective chance and creedence or rational subjective degree of belief as primitive. In this respect he follows Carnap's (Carnap, 1945) policy of making room for both empirical and epistemic probability. Lewis illuminates the difficult concept of objective chance by explicating some clear connections between creedence and beliefs about chance.[23]

Lewis offers a questionnaire, with answers, designed to bring out the way our beliefs about objective chance ought to guide our subjective probability assignments. If you are certain that the chance of my die coming up a six when I throw it later today is 25%, then your degree of belief in the proposition that it comes up a six on that toss ought to be 0.25. Suppose now that you find out that my die is really an ordinary one after all, and that it has been tossed many times in the past with a relative frequency of about one sixth for outcome six, but that you, nevertheless, continue to be certain that the chance of a six on my toss is 25%. Here also you should continue to assign degree of belief 0.25. The degree of belief in the proposition conditional on the chance statement ought to be resilient with respect to this kind of evidence, because this kind of evidence guides creedence by its effect on beliefs about chance. Lewis calls such evidence *admissible*.

Not all evidence is admissible. Suppose after the toss you find out that the actual outcome was a three. Even if you continue to believe that the objective chance just before the toss of its outcome being a six was 25%, you no longer guide your creedence about the outcome by that belief. Now, you assign zero to the proposition that the outcome was six.

Finally, suppose you have no inadmissible evidence and are uncertain about which of several alternative hypotheses correctly describes the objective chance. Perhaps you assign degree of belief .6 to the hypothesis that the chance is 25% and degree of belief 0.4 to the hypothesis that the chance is 1/6. You should now assign

$$(0.6)(0.25) + (0.4)(\tfrac{1}{6}) = 0.217$$

as your degree of belief in the proposition that the die comes up a six. Your subjective creedence in the proposition is to be guided by your subjective expectation of its objective chance.

Lewis sums up the intuitions that guided our answers to his questionnaire in a principle he calls the *Principle Principle* (Lewis, 1978, this volume, pp. 270–271).

Let C be any reasonable initial creedence function. Let t be any time. Let x be any real number in the unit interval. Let X be the proposition that the chance, at time t, of A's holding equals x. Let E be any proposition compatable with X that is admissible at time t. Then,

$$C(A \mid X \wedge E) = x.$$

According to Lewis, this principle seems to capture all we know about chance. He illustrates it by showing how our answers to his questionnair all follow from it and goes on to deduce other interesting consequences. Among these consequences are just the sorts of connection with long run frequencies that have guided the frequency interpretation of objective probability (Lewis, 1978, this volume, pp. 286 ff).

For our purposes the important thing about Lewis' principle is its explicit relativization to time. He uses a Labyrinth story to illustrate the fact that we regard chance as time dependent and contingent (Lewis, 1978, this volume, p. 274).

Suppose you enter a labyrinth at 11:00 a.m., planning to choose your turn whenever you come to a branch point by tossing a coin. When you enter at 11:00, you may have a 42% chance of reaching the center by noon. But in the first half hour you may stray into a region from which it is hard to reach the center, so that by 11:30 your chance of reaching the center by noon has fallen to 26%. But then you turn lucky; by 11:45 you are not far from the center and your chance of reaching it by noon is 78%. At 11:49 you reach the center; then and forevermore your chance of reaching it by noon is 100%.

At each of these times your reaching the center by noon depends on such contingent features of the world as the structure of the Labyrinth, the speed with which you walk through it and where you are in it. (Lewis, 1978, this volume, p. 274).

According to Lewis, propositions entirely about historical facts no later than t are, as a rule, admissible at t. If proposition A is about such matters of fact at some time (or time interval) t_A and t is later than t_A, then A is admissible at t for claims about the objective chance of A. This has the effect that the objective chance of A becomes 100% during t_A and stays that way ever after. This asymmetry of chance with respect to past and future lends itself to tree diagrams.

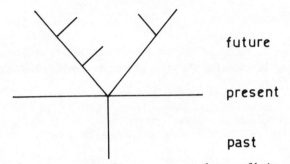

The trunk is the only past that has any present chance of being actual while the branches are the various alternative futures with some present chance of being actual (Lewis, 1978, this volume, p. 277). For Lewis this asymmetry of fixed past and open future is a broad contingent feature of our world, not a matter of logical necessity. (Lewis, 1978, this volume, p. 277)).

3. In 'A Theory of Conditionals in the Context of Branching Time' Thomason and Anil Gupta explore the interactions of conditionals with tense. They begin with a straight forward adaption of Stalnaker's theory. Moments construed as world-states are ordered into a tree structure by a relation earlier than ($<$). The branches (maximal chains relative to $<$) represent alternative possible histories.

An equivalence relation ($\simeq$) picks out moment pairs that are to count as copresent, that is as alternative possible presents to one another. The Stalnaker selection function is relativised to both a moment: and to a possible history h. Where A is a sentence, $s(A\ i\ h)$ must pick out a pair $\langle i'h' \rangle$ that is to count as the A-pair closest to $\langle i\ h \rangle$. The moment i must be co-present with i and the history h' must pass through it. Selection must also satisfy a principle of past predominance so that past similarity outweighs future similarity.

Thomason and Gupta argue that the following inferences relating conditionals and what is settled (i.e. what is true on all branches leading from moment i) ought to be valid.[24]

Let SA symbolize that A is settled as true now.

(Edelberg) $S \sim A$, $S(A > B)$　$\therefore A > S(A \supset B)$

(Weak Edelberg) $S \sim A$, $A > SA$, $S(A > B)$　$\therefore A > SB$.

They show that these inferences are not valid on the simple extension of Stalnaker's theory they began with. In order to insure validity of these principles without giving up conditional excluded middle they add more structure to their model. In place of the history h they propose a future choice function $\mathscr{F}$ which assigns a history $\mathscr{F}(i)$ to every moment i, $\mathscr{F}(i)$ may be regarded as the course of events that would obtain if moment i were actual. With a number of appropriate restrictions to make $\mathscr{F}$ respect the tree structure and to make the settled operator behave properly they are able to produce a model that does what they want.

They suggest that natural extension of such models is to restrict the set of future choice functions to those that satisfy some conditions of causal coherence. This kind of extension seems promising as a way to investigate the causal dependencies that the decision making conditional must respect.

4. In 'A Temporal Framework for Conditionals and Chance', van Fraassen introduces conditionals and chance into a general state space model for tensed language. In this model each possible world x comes with a state function h_x giving its state $h_x(t)$ at each moment t. The function h_x represents the historical development of world x through time. Of considerable interest is the set $H(x, t)$ of worlds that agree historically with x up through time t. This *t-cone* of x plays the same role as Thomason's branching time. A proposition A is settled as true in x at t just in case every world in $H(x, t)$ is an A-world. van Fraassen, also, introduces *backward looking* propositions. A proposition A is backward looking at t just in case for any world x, A is true in x only if every world in $H(x, t)$ is an A-world. If SA symbolizes A is settled, then it turns out that A is backward looking at t just in case $A \supset SA$ holds at t for every world. Backward looking propositions are important both for conditionals and for chance.

van Fraassen works with a weak Stalnaker logic where the conditional antecedent substitution axiom

(CAS) $(A \to B) \wedge (B \to A) \wedge (A \to C) \supset (B \to C)$,

which is valid for both Lewis and Stalnaker, does not hold. Where $s^x A$ is the world selected as closest A-world to x, this corresponds to giving up Stalnaker's condition.

$S^x A$ is in B and $S^x B$ is in A only if $S^x A = S^x B$,

which is to give up the idea that selection respects some notion of closeness that has ordering properties.

When conditionals are evaluated from times as well as worlds some temporal conditions are imposed on selection. Thomason's idea of past predominance is embodied in

(I) $S_t^x A$ is in $H(x, t)$, if A is compatable with $H(x, t)$.

The world selected as closest A-world to x at moment t agrees with x's history through t, if possible. In order to validate the Edelberg inference

(E) $S \sim A$, $S(A \to B)$ $\therefore A \to S(A \supset B)$,

van Fraassen proposes

(II)   If $z = S_t^x A$ and $u \in H(z, t) \cap A$, then $u = S_t^{x'} A$ for some x' in $H(x, t)$.

This has the effect that z is selected as closest A-world to x at t only if every A-world in the t-cone of z is selected as closest A-world for some world in the t-cone of x.

Anil Gupta proposed an additional principle which should hold for any proposition E that is backward looking at all times.

$$A \to \sim S \sim B, \qquad B \to \sim S \sim A, \qquad B \to E \qquad \therefore A \to E.$$

This principle does the job for the conditional substitution axiom van Fraassen rejects, when the consequent E is universally backward looking. van Fraassen validates it by imposing:

(III)   If $x = S_t^w A$ and B is compatable with $H(x, t)$, $z = S_t^w B$ and A is compatable with $H(z, t)$, then $H(z, t) = H(x, t)$.

The world selected as closest A-world to w at t and the world selected as closest B-world to w at t should have the same history through t if possible.

van Fraassen introduced chance in world x at time t as a measure on $H(x, t)$. Where $C_t^x A$ is the chance at time t that A is true in x, he imposes

(1) $C_t^x(A) = C_t^x(H(x, t) \cap A)$.

This assigns to each proposition A the C_t^x measure of its intersection with $H(x, t)$. He also requires

(2)   If $y \in H(x, t)$ then $C_t^y = C_t^x$

so that past history determines chance. This has the effect that any proposition A which is backward looking at t has $C_t^x(A)$ either one or zero.

van Fraassen proposes the following relation between creedence and chance. Where P_t is an agent' subjective probability assignement t:

(Basic Principle)

$$P_t(X/Y) = \frac{1}{P_t(Y)} \int_Y C_t(X)\, dP_t$$

provided Y is backward looking at t and $P_t(Y) \neq 0$.

Conditional creedence is to go by the subjective conditional expectation of the objective chance. When Lewis' admissible evidence at t is interpreted as propositions which are backward looking at t, this principle is equivalent to the principle principle.

The subjective conditional expectation of the objective chance used in van Fraassen's principle is different from the subjective expectation of the objective conditional chance that Skyrms proposes as the appropriate evaluation of a decision making conditional.

One way of formulating the Skyrms' idea in van Fraassen's system would be

$$\int_K [C_t^x(X \wedge Y)/C_t^x(Y)]\, dP_t,$$

where the integral is taken over the whole set of worlds K. Another would be

$$\int_K [C_t^x(X \text{ given } Y)]\, dP_t,$$

where objective conditional chance is taken as a primitive that may or may not always be defined by the classical ratio. When Y is an act that the agent can choose the later formulation seems preferable, since the agent, as decision maker, ought to control, rather than wonder about, the objective chance that he will choose one way rather than another.[25]

When the distinct t-cones compatable with what I accept are a finite set $H_1 \ldots H_n$ then

$$C_t^x(B \text{ given } A)\, dP_t = \sum_{i=1}^{i=n} P_t(H_i) \cdot C_t^i(B \text{ given } A),$$

where each C_t^i is the common C_t^x (B given A) for worlds x in H_i. Here the distinct t-cones play the role of Skyrm's finite set of alternative hypotheses about the relevent conditional chance. Whether C_t^x (B given A) is taken as primitive or defined by usual ratio, Skyrms' subjective expectation of objective conditional chance is representable in van Fraassen's Framework.

VI. CONCLUDING REMARKS

One of the main difficulties in making abstract models illuminate real problems is getting interesting constraints on what features the models ought to have. Linguistic intuitions have long been the main source of constraints on models of conditionals. These papers show a unified and exciting development in which such linguistic intuitions are supplemented by intuitions generated by Ramsey's test and decision making. Each of these paradigms corresponds to an important kind of hypothetical reasoning with its own job to do. Ramsey test reasoning plays an important role in organizing knowledge and accounting for rational belief change. It draws on intuitions generated by this role. The distinct kind of causally sensitive hypothetical reasoning used to guide decisions draws on intuitions about the relations between time and objective chance and especially about the relation between subjective probability and beliefs about objective chance.

Among the fruitful problems suggested by the developments sketched here are the proper account of objective conditional chances and the formulation of a conditional with probability of truth equal to the subjective expectation of the relevant objective conditional chance. Both jobs for hypothetical reasoning are intimately involved in these problems. This interaction suggests that investigating these problems can provide illumination for the fields and problem areas involved in each kind of hypothetical reasoning as well as further advances in our understanding of conditionals.

I want to call attention to three important papers that were not included in our volume. One is David Lewis' paper 'Counterfactual Dependence and Times Arrow' which generates the temporal tree structures so important for both chance and conditionals from some interesting criteria for judging overall similarity of worlds. Another paper is Howard Sobel's long working manuscript 'Probability, Chance and Choice'. This monograph, which unfortunately was too long to be included in our volume, is the most developed account of conditional chance yet available. Nancy Cartwright's paper 'Causal Laws and Effective Strategies' relates the issues raised by our decision making paradigm to causal laws.

University of Western Ontario

NOTES

[1] This comparison with Chisholm and Goodman was suggested to me by Stalnaker's survey lecture at the 1979 American Philosophical Association meetings in New York.

[2] Stalnaker's system is extended to include quantification in a joint paper with Richmond Thomason (Stalnaker and Thomason, 1970). Thomason provided a Fitch Style formulation of Stalnaker's conditional logic (Thomason, 1970).

[3] Lewis endorses Grice's defense of the material conditional as an account of the English indicative conditional (Lewis, 1975, this volume, pp. 137–139). Stalnaker defends his theory as an account of indicative, as well as subjective, conditionals in Stalnaker, 1975, this volume, p. 198).

[4] Where A is impossible in w, Stalnaker makes $f(A\ w)$ an absurd world where every sentence is true. (Stalnaker, 1968, this volume, p. 46).

[5] Lewis' passage only claims that asserting both of these together with "If Bizet and Verdi were compatriots, Bizet either would or would not be Italian" sounds like a contradiction.

[6] The suggestion that van Fraassen's account of vagueness be used to help Stalnaker's theory deal with ties was first proposed in print by Thomason (Thomason, 1970).

[7] Pollock (Pollock, 1976, pp. 18–21; Herzberger, 1979; and Stalnaker, 1979) all argue that giving up the limit assumption requires giving up a desirable feature of counterfactual inferences. Stalnaker and Pollock formulate this as a principle of counterfactual closure:

If $\Gamma(A)$ semantically entails C, then $C \in \Gamma(A)$.

Herzberger argues that Lewis' violations of the limit assumption require also giving up a weaker principle of counterfactual consistency

$\Gamma(A)$ is consistent, if A is possible.

Violating counterfactual consistency would not make Lewis' system formally inconsistent; nevertheless, counterfactual consistency and counterfactual closure, as well, are intuitively desirable features for a conditonal logic.

Neither of these principles corresponds to the validity of any single sentence; therefore, Lewis (Lewis, 1972, this volume, p. 83) was able to point out that giving up the limit assumption does not require giving up any theorem not already lost by giving up uniqueness.

[8] John Mackie has challenged this intuition. In a letter to me he suggests that if my pocket contains a concealed coin it would be appropriate to deny the would conditional

"If I were to look I would find the coin"

while affirming the corresponding might conditional.

[9] A somewhat simplified formulation of these conditions and some additional arguments defending them are to be found in Harper (1975), Harper (1976), and Harper (1978).

[10] In addition to the references in Stalnaker (1970), axiomatizations and various properties of Popper functions are explored in Van Fraassen (1976b), and in Harper et al. (1979).

[11] Lewis dropped his bombshell at the June 1972 meeting of the Canadian Philosophical Association in Montreal. He, Stalnaker and William Rozeboom were commenting on a paper I gave extending Stalnaker's probability semantics to quantified conditional logic. This work was a major part of my Ph.D. dissertation (Harper, 1974) and Lewis' result led to many improvements.

[12] I used the Ramsey test idea restricted to certainty,

$$P(A \to B) = 1 \text{ iff } P(B \mid A) = 1,$$

without the full Stalnaker hypothesis, to construct non-trivial models of iterated rational belief change (Harper, 1976). Stalnaker (Stalnaker, 1976) modified Lewis' result to trivialize an earlier version of this model (Harper, 1974; Harper, 1975). The successful model, which was developed in response, requires making the conditional even more radically context dependent than van Fraassen's. A Ramsey test conditional is interpreted relative to the acceptance context from which it is evaluated. Conditionals nested to the right, e.g. $A \to (B \to C)$, play an important role in modeling iterated rational belief changes. The first conditional is evaluated from the basic acceptance context, the next $(B \to C)$ is evaluated from the point of view of the result of minimally revising the basic context to assume the antecedent of the first conditional. This makes the propositional function corresponding to the first $\to$ different from the propositional function corresponding to the second $\to$. Though this is a very radical kind of context dependence, it results from a quite natural account of right hand nesting as representing hypothetical sequences of assumption. This account has been employed to extend the Bayesian learning model to handle revisions of previously accented evidence (Harper, 1978).

   Peter Gärdenfors (1978) has recently applied this kind of certainty Ramsey test model to explore the effect on conditional logic of making various conditions on the idea of minimal revision of a body of knowledge.

[13] See Gibbard (1979) for an exposition of van Fraassen's construction and Stalnaker's proof that this construction cannot be extended to the Stalnaker logic. See this essay (p. 29) for the difference between van Fraassen's Logic and Stalnaker's.

[14] Brian Ellis (Ellis, 1973), John Mackie (Mackie, 1973), and Issac Levi (Levi, 1977) have also developed conditional assertion accounts of conditionals.

[15] John Mackie (manuscript and letter to me) and Isaac Levi (Levi, 1977) each suggest that a Ramsey Test approach can be made to work for both examples by putting different restriction on subjunctive and indicative suppositions.

[16] Validity of the direct argument would show that $\sim A \supset B$ (which is equivalent to $A \vee B$) entails that If $\sim A$ then B. This would show their logical equivalence, because the converse entailment is obviously valid.

[17] It is interesting to consider the corresponding argument from $A \supset B$ to If A then B:

   "Either the butler didn't do it or the gardener did. Therefore, if the butler did it the gardener didn't."

This argument does not seem reasonable; but, apparently, Stalnaker's account would call it reasonable.

   Stalnaker pointed out (private communication) that it is hard to know just what to make of

   "Either the butler didn't do it or the gardener did."

Perhaps, he suggests, this is because we presuppose that "The butler did it" and "The gardener did it" are exclusive. This presupposition violates (G), because $\sim A \vee B$ is only appropriate in contexts compatable with both $(A \wedge B)$ and $(\sim A \wedge \sim B)$. When the example is changed to make it clear that we are in a context where (G) is obeyed

the inference seems fine. Consider: "Either the butler didn't do it or the gardener was his accomplice. Therefore, if the butler did it the gardener was his accomplice."
[18] This equivalence is challenged by Adams (1975, p. 33). Gibbard answers in a footnote (Gibbard, 1979, p. 246). It is not at all clear that this equivalence ought to hold for examples where $P(B \mid A) = 0$. Consider a variant on Gibbard's poker example. (Gibbard, 1979, p. 241). I know that Pete is a crafty player who has been informed what his opponent's hand is. Consider:

(1)   Given that Pete is going to fold, he will lose if he calls.

and

(2)   If Pete folds and calls he will lose.

The first seems moderately sensible and even assertable while the second seems to be nonsense. Perhaps, (1) only seems sensible because the reading of the second conditional is implicity subjunctive. Moreover, it may be that nested indicatives of the form $A \to (B \to C)$ are not assertable unless $P(A \wedge B) > O$. Even if this does turn out to be a limitation on English indicative conditionals it is not a desirable limitation on the Ramsey test idea of epistemic minimal revision. One ought to be able to distinguish between epistemic and causally sensitive conditional reasoning even when the assumptions are regarded as counterfactual. This is so because, in addition to its role in planning for knowledge aquisition, Ramsey test reasoning is also important in organizing the knowledge we already have. Comparisons of relative immunity to revision are an important part of such organization and such comparisons are elicited by counterfactual assumptions.
[19] This kind of relativisation to nesting is just what happens quite naturally on my account of Ramsey test conditionals were nesting to the right corresponds to sequential assumption making. (See note 10 and references therein.)
[20] Several other proposals for using subjective expectation of objective conditional chances as the relevent probabilities for decision making have been put forward (Jeffrey, 1979; Sobel, 1979). Sobel's proposal is the most developed yet available. He presents it in a very long and very interesting manuscript. (Sobel, 1979), which is the latest version of a series of revisions going back to 1977.
[21] Since Popper proposed a propensity interpretation of probability (Popper, 1959) controversy about single case chances has been sharp and extensive. Kyburg (1974) provides an overview of some of this literature. Perhaps the most extensive development of an approach taking propensity as primitive is that of Mellor (Mellor, 1971).
[22] Kyburg and van Fraassen propose modal frequency accounts of single case chance (Kyburg, 1976; van Fraassen, 1977). These attempts to make empirical sense out of the idea that hypothetical frequencies can explicate propensities.
[23] Skyrms proposes an analysis of chance as resilient subjective conditional belief (Skyrms, 1977). Such resilience will play a large role in Lewis' account, but, unlike Skyrms, Lewis doe not attempt to reduce chance to resilient belief.
[24] Thomason's student Edelberg first proposed that these inferences ought to be valid.
[25] If we use the latter version and set $C_t^{\vec{x}}$ (B given A) to be the measure according to C_t^x of the set of worlds z in $H(x, t)$ where $S_t^z A$ is a B-world, while keeping van Fraassen's constraints on chance and selection, then we have

$$P_t(A \ \Box\!\!\rightarrow B) = \int_K C_t^x \, (B \text{ given } A) \, dP_t$$

as is required for a conditional with the appropriate generalization of Skyrms' evaluation as its probability of truth. Lewis has shown that this induces a function

$$P_{t[A]}(B) = P_t(A \ \Box\!\!\rightarrow B)$$

that counts as a kind of minimal revision of P_t required to have a belief function where A is accepted. (Lewis, 1976, p. 140). He calls this the result of imaging on A. All the probability weight is moved into A with the weight of each $\sim\!A$-world w going to $S_t^w(A)$ – its nearest A-world. The probability weight is moved into A with minimal shifting of weight. This is a rival to the kind of minimal revision involved in classical conditionalization where all the weight is shifted into A by normalizing so that probability ratios within A are undisturbed. Recently, James Fetzer and Donald Nute (1979) have proposed imaging as the appropriate formulation of a causally sensitive probability calculus. The core of Sobel's (1979) interesting proposals also consists in using a generalization of imaging to represent causally sensitive conditional chances.

BIBLIOGRAPHY

Adams, E.: 1965, 'On the Logic of Conditionals', *Inquiry* 8, 166–197.
Adams, E.: 1966, 'Probability and the Logic of Conditionals', in J. Hintikka and P. Suppes (eds.), *Aspects of Inductive Logic,* North-Holland, 1966, pp. 265–316.
Adams, E.: 1970, 'Subjunctive and Indicative Conditionals', *Foundations of Language* 6, 39–94.
Adams, E.: 1975, *The Logic of Conditionals,* D. Reidel, Dordrecht, Holland.
Adams, E.: 1976, 'Prior Probabilities and Counterfactual Conditionals', in W. L. Harper and C. A. Hooker (eds.), *Foundations of Probability Theory, Statistical Inference and Statistical Theories of Science,* Volume I, D. Reidel, Dordrecht, pp. 1–21.
Carnap, R.: 1945, 'The Two Concepts of Probability', *Philosophy and Phenomeno-logical Research* 5, 513–532.
Cramer, H.: 1955, *The Elements of Probability Theory,* John Wiley and Sons, New York.
De Finetti, B.: 1975, *Theory of Probability,* John Wiley and Sons, New York.
Ellis, B.: 1973, 'The Logic of Subjective Probability', *British Journal for the Philosophy of Science* 24 (2), 125–152.
Fetzer, J. and Nute, D.: 1979, 'Syntax, Semantics, and Ontology: A Probabilistic Causal Calculus', *Synthese* 40 (3), 453–495.
Gärdenfors, P.: 1978, 'Conditionals and Changes of Belief', *Acta Philosophica Femica* **XXVIII** (2-3), 381–404.
Gärdenfors, P.: 1979a, 'Even If', Manuscript.
Gärdenfors, P.: 1979b, 'An Epistemic Approach to Conditionals', Manuscript.
Gibbard, A.: 1979, 'Two Recent Theories of Conditionals', this volume, pp. 211–247.
Gibbard, A.: 1979, 'Reply to Pollock', this volume, pp. 253–256.
Gibbard, A. and Harper, W. L.: 1978, 'Counterfactuals and Two kinds of Expected Utility', in C. A. Hooker, J. J. Leach, and E. F. McClennen (eds.), *Foundations*

and Applications of Decision Theory, Volume I, D. Reidel, Dordrecht, Holland, pp. 125–162, also this volume, pp. 153–190.

Harper, W. L.: 1974, *Counterfactuals and Representations of Rational Belief*, Ph.D. Dissertation, Univeristy of Rochester.

Harper, W. L.: 1975, 'Rational Belief Change, Popper Functions and Counterfactuals', *Synthese* **30**, 221; also reprinted 1976 in W. L. Harper and C. A. Hooker (eds.), *Foundations of Probability Theory, Statistical Inference, and Statistical Theories of Science*, Volume I, D. Reidel, Dordrecht, pp. 73–113.

Harper, W. L.: 1976, 'Ramsey Test Conditionals and Iterated Belief Change', in W. L. Harper and C. A. Hooker (eds.), *Foundations of Probability Theory, Statistical Inference, and Statistical Theories of Science*, Volume I, D. Reidel, Dordrecht, pp. 117–136.

Harper, W. L.: 1978, 'Bayesian Learning Models with Revision of Evidence', *Philosophia*, June 1978.

Harper, W. L.: 1979, 'Conceptual Change, Incommensurability and Special Relativity Kinematics', *Acta Philosophical Fennica*.

Harper, W. L., Leblanc, H., and van Fraassen, B.: 1979, 'On Characterizing Popper and Carnap Probability Functions', forthcoming.

Herzberger, H.: 1979. 'Counterfactuals and Consistency', *Journal of Philosophy* **LXXVI** (2), 83–88.

Kyburg, H. E.: 1974, 'Propensities and Probabilities', *The British Journal for the Philosophy of Science* **25**, 358–375.

Kyburg, H. E.: 1976, 'Chance', *Journal of Philosophical Logic* **5**, 355–393.

Levi, I.: 1977, 'Subjunctive Dispositions and Chances', *Synthese* **34**, 423–455.

Lewis, D. K.: 1972, 'Counterfactuals and Comparative Possibility', *Journal of Philosophical Logic* **2**, 418; D. Hockney, W. Harper, and B. Preed (eds.), *Contemporary Research in Philosophical Logic and Linguistic Semantics*, D. Reidel, Dordrecht, Holland, pp. 1–29; and this volume, pp. 57–95.

Lewis, D. K.: 1973, *Counterfactuals*, Oxford.

Lewis, D. K.: 1975, 'Probabilities of Conditionals and Conditional Probabilities', *The Philosophical Review* **LXXXV** (3), 297–315; also this volume, pp. 129–147.

Lewis, D. K.: 1978, 'Subjectivist's Guide to Objective Chance', this volume, pp. 267–297.

Lewis, D. K.: 1979, 'Counterfactual Dependence and Time's Arrow', *Noûs* **XIII** (4).

Mackie, J. L.: 1962, 'Counterfactuals and Causal Forms', in R. J. Butler (ed.), *Analytical Philosophy*.

Mackie, J. L.: 1973, *Truth Probability and Paradox*, Oxford University Press, New York.

Mellor, H.: 1971, *The Matter of Chance*, Cambridge University Press, Cambridge, England.

Pollock, J.: 1976, *Subjunctive Reasoning*, D. Reidel, Dordrecht, Holland.

Pollock, J.: 1978, 'Indicative Conditionals and Conditional Probability', this volume, pp. 249–252.

Popper, K. R.: 1959, *The Logic of Scientific Discovery*, Harper Torchbook edition, New York.

Popper, K. R.: 1959, 'The Propensity Interpretation of Probability', *The British Journal of the Philosophy of Science* **10**, 25–92.

Reichenbach, H.: 1949, *The Theory of Probability,* University of California Press, Berkeley and Los Angeles.

Skyrms, B.: 1977, 'Resiliency, Propositions and Causal Necessity', *Journal of Philosophy* **74**, 704–713.

Skyrms, B.: 1980, *Causal Necessity,* forthcoming Yale Press.

Skyrms, B.: 1979, 'The Prior Propensity Account of Subjunctive Conditionals', this volume, pp. 259–265.

Sobel, J. H.: 1979, 'Probability, Chance and Choice', working paper.

Stalnaker, R. C.: 1968, 'A Theory of Conditionals', *Studies in Logical Theory,* American Philosophical Quarterly Monograph Series, No. 2, Blackwell, Oxford; in Ernest Sosa (ed.), *Causation and Conditionals,* Oxford, Oxford University Press, 1975, Oxford; and this volume, pp. 41–55.

Stalnaker, R. C.: 1970, 'Probability and Conditionals', *Philosophy of Science* **37**, 1970; and this volume, pp. 107–127.

Stalnaker, R. C.: 1975, 'Indicative Conditionals', *Philosophia* **5**, 1975; and this volume, pp. 193–210.

Stalnaker, R. C.: 1976, 'Letter to van Fraassen', in W. L. Harper and C. A. Hooker (eds.), *Foundations of Probability Theory, Statistical Inference, and Statistical Theories of Science,* Volume I, D. Reidel, Dordrecht, Holland, p. 302.

Stalnaker, R. C.: 1979, 'A Defense of Conditional Excluded Middle', this volume, pp. 87–104.

Stalnaker, R. C. and Thomason, R.: 1970, 'A Semantical Analysis of Conditional Logic', *Theoria.*

Thomason, R.: 1970, 'A Fitch-Style Formulation of Conditonal Logic', *Logique et Analyse* **52**, 397–412.

Thomason, R. and Gupta, A.: 1979, 'A Theory of Conditionals in the Context of Branching Time', this volume, pp. 299–322.

van Fraassen, B. C.: 1974, 'Hidden Variables and Conditional Logic', *Theoria* **40**.

van Fraassen, B. C.: 1976a, 'Probabilities of Conditionals', in W. L. Harper and C. A. Hooker (eds.), *Foundations of Probability Theory, Statistical Inference, and Statistical Theories of Science,* Volume I, D. Reidel, Dordrecht, Holland, pp. 262–300.

van Fraassen, B. C.: 1976b, 'Representations of Conditional Probabilities', *Journal of Philosophical Logic* **5**, 417–430.

van Fraassen, B. C.: 1977, 'Relative Frequencies', *Synthese* **34**, 133–166.

van Fraassen, B. C.: 'A Temporal Framework for Conditionals and Chance', this volume, pp. 232–340.

Von Mises, R.: 1957, *Probability Statistics and Truth,* Macmillan, New York.

PART 2

THE CLASSIC STALNAKER-LEWIS THEORY OF CONDITIONALS

ROBERT C. STALNAKER*

A THEORY OF CONDITIONALS

I. INTRODUCTION

A conditional sentence expresses a proposition which is a function of two
other propositions, yet not one which is a *truth* function of those propositions. I may know the truth values of "Willie Mays played in the American
League" and "Willie Mays hit four hundred" without knowing whether or not
Mays would have hit four hundred if he had played in the American League.
This fact has tended to puzzle, displease, or delight philosophers, and many
have felt that it is a fact that calls for some comment or explanation. It has
given rise to a number of philosophical problems; I shall discuss three of these.

My principal concern will be with what has been called the *logical problem
of conditionals*, a problem that frequently is ignored or dismissed by writers
on conditionals and counterfactuals. This is the task of describing the formal
properties of the *conditional function*: a function, usually represented in
English by the words "if ... then", taking ordered pairs of propositions into
propositions. I shall explain informally and defend a solution, presented
more rigorously elsewhere, to this problem.[1]

The second issue – the one that has dominated recent discussions of contrary-to-fact conditionals – is the *pragmatic problem of counterfactuals*. This
problem derives from the belief, which I share with most philosophers writing
about this topic, that the formal properties of the conditional function,
together with all of the *facts*, may not be sufficient for determining the truth
value of a counterfactual; that is, different truth values of conditional statements may be consistent with a single valuation of all nonconditional statements. The task set by the problem is to find and defend criteria for choosing
among these different valuations.

This problem is different from the first issue because these criteria are
pragmatic, and not semantic. The distinction between semantic and pragmatic
criteria, however, depends on the construction of a semantic theory. The
semantic theory that I shall defend will thus help to clarify the second problem by charting the boundary between the semantic and pragmatic components of the concept. The question of this boundary line is precisely what
Rescher, for example, avoids by couching his whole discussion in terms of

W. L. Harper, R. Stalnaker, and G. Pearce (eds.), Ifs, 41–55
Copyright © 1968 by Basil Blackwell Publisher.

conditions for belief, or justified belief, rather than truth conditions. Conditions for justified belief are pragmatic for any concept.[2]

The third issue is an epistemological problem that has bothered empiricist philosophers. It is based on the fact that many counterfactuals seem to be synthetic, and contingent, statements about unrealized possibilities. But contingent statements must be capable of confirmation by empirical evidence, and the investigator can gather evidence only in the actual world. How are conditionals which are both empirical and contrary-to-fact possible at all? How do we learn about possible worlds, and where are the facts (or counterfacts) which make counterfactuals true? Such questions have led philosophers to try to analyze the conditional in non-conditional terms[3] – to show that conditionals merely appears to be about unrealized possibilities. My approach, however, will be to accept the appearance as reality, and to argue that one can sometimes have evidence about nonactual situations.

In Sections II and III of this paper, I shall present and defend a theory of conditionals which has two parts, a formal system with a primitive conditional connective, and a semantical apparatus which provides general truth conditions for statements involving that connective. In Sections IV, V, and VI, I shall discuss in a general way the relation of the theory to the three problems outlined above.

II. THE INTERPRETATION

Eventually, I want to defend a hypothesis about the truth conditions for statements having conditional form, but I shall begin by asking a more practical question: how does one evaluate a conditional statement? How does one decide whether or not he believes it to be true? An answer to this question will not be a set of truth conditions, but it will serve as a heuristic aid in the search for such a set.

To make the question more concrete, consider the following situation: you are faced with a true-false political opinion survey. The statement is, "If the Chinese enter the Vietnam conflict, the United States will use nuclear weapons." How do you deliberate in choosing your response? What considerations of a logical sort are relevant? I shall first discuss two familiar answers to this question, and then defend a third answer which avoids some of the weaknesses of the first two.

The first answer is based on the simplest account of the conditional, the truth functional analysis. According to this account, you should reason as follows in responding to the true-false quiz: you ask yourself, first, will the

Chinese enter the conflict? and second, will the United States use nuclear weapons? If the answer to the first question is no, *or* if the answer to the second is yes, then you should place your X in the 'true' box. But this account is unacceptable since the following piece of reasoning is an obvious *non sequitur*: "I firmly believe that the Chinese will stay out of the conflict; *therefore* I believe that the statement is true." The falsity of the antecedent is never sufficient reason to affirm a conditional, even an indicative conditional.

A second answer is suggested by the shortcomings of the truth-functional account. The material implication analysis fails, critics have said, because it leaves out the idea of *connection* which is implicit in an if-then statement. According to this line of thought, a conditional is to be understood as a statement which affirms that some sort of logical or causal connection holds between the antecedent and the consequent. In responding to the true-false quiz, then, you should look, not at the truth values of the two clauses, but at the relation between the propositions expressed by them. If the 'connection' holds, you check the 'true' box. If not, you answer 'false'.

If the second hypothesis were accepted, then we would face the task of clarifying the idea of 'connection', but there are counter-examples even with this notion left as obscure as it is. Consider the following case: you firmly believe that the use of nuclear weapons by the United States in this war is inevitable because of the arrogance of power, the bellicosity of our president, rising pressure from congressional hawks, or other *domestic* causes. You have no opinion about future Chinese actions, but you do not think they will make much difference one way or another to nuclear escalation. Clearly, you believe the opinion survey statement to be true even though you believe the antecedent and consequent to be logically and causually independent of each other. It seems that the presence of a 'connection' is not a necessary condition for the truth of an if-then statement.

The third answer that I shall consider is based on a suggestion made some time ago by F. P. Ramsey.[4] Consider first the case where you have no opinion about the statement, "The Chinese will enter the Vietnam war." According to the suggestion, your deliberation about the survey statement should consist of a simple thought experiment: add the antecedent (hypothetically) to your stock of knowledge (or beliefs), and then consider whether or not the consequent is true. Your belief about the conditional should be the same as your hypothetical belief, under this condition, about the consequent.

What happens to the idea of connection on this hypothesis? It is sometimes relevant to the evaluation of a conditional, and sometimes not. If you believe that a causal or logical connection exists, then you will add the consequent to your stock of beliefs along with the antecedent, since the rational man accepts the consequences of his beliefs. On the other hand, if you already believe the consequent (and if you also believe it to be causally independent of the antecedent), then it will remain a part of your stock of beliefs when you add the antecedent, since the rational man does not change his beliefs without reason. In either case, you will affirm the conditional. Thus this answer accounts for the relevance of 'connection' when it is relevant without making it a necessary condition of the truth of a conditional.

Ramsey's suggestion covers only the situation in which you have no opinion about the truth value of the antecedent. Can it be generalized? We can of course extend it without problem to the case where you believe or know the antecedent to be true; in this case, no changes need be made in your stock of beliefs. If you already believe that the Chinese will enter the Vietnam conflict, then your belief about the conditional will be just the same as your belief about the statement that the U.S. will use the bomb.

What about the case in which you know or believe the antecedent to be false? In this situation, you cannot simply add it to your stock of beliefs without introducing a contradiction. You must make adjustments by deleting or changing those beliefs which conflict with the antecedent. Here, the familiar difficulties begin, of course, because there will be more than one way to make the required adjustments.[5] These difficulties point to the pragmatic problem of counterfactuals, but if we set them aside for a moment, we shall see a rough but general answer to the question we are asking. This is how to evaluate a conditional:

First, add the antecedent (hypothetically) to your stock of beliefs; second, make whatever adjustments are required to maintain consistency (without modifying the hypothetical belief in the antecedent); finally, consider whether or not the consequent is then true.

It is not particularly important that our answer is approximate – that it skirts the problem of adjustments – since we are using it only as a way of finding truth conditions. It is crucial, however, that the answer may not be restricted to some particular context of belief if it is to be helpful in finding a definition of the conditional function. If the conditional is to be understood as a function of the propositions expressed by its component clauses, then its truth value should not in general be dependent on the attitudes which anyone has toward those propositions.

Now that we have found an answer to the question, "How do we decide whether or not we believe a conditional statement?" the problem is to make the transition from belief conditions to truth conditions; that is, to find a set of truth conditions for statements having conditional form which explains why we use the method we do use to evaluate them. The concept of a *possible world* is just what we need to make this transition, since a possible world is the ontological analogue of a stock of hypothetical beliefs. The following set of truth conditions, using this notion, is a first approximation to the account that I shall propose:

Consider a possible world in which A is true, and which otherwise differs minimally from the actual world. *"If A, then B" is true (false) just in case B is true (false) in that possible world.*

An analysis in terms of possible worlds also has the advantage of providing a ready made apparatus on which to build a formal semantical theory. In making this account of the conditional precise, we use the semantical systems for modal logics developed by Saul Kripke.[6] Following Kripke, we first define a *model structure*. Let M be an ordered triple (K, R, λ). K is to be understood intuitively as the set of all possible worlds; R is the relation of relative possibility which defines the structure. If α and β are possible worlds (members of K), then $\alpha R \beta$ reads "β is possible with respect to α". This means that, where α is the actual world, β is a possible world. R is a reflexive relation; that is, every world is possible with respect to itself. If your modal intuitions so incline you, you may add that R must be transitive, or transitive and symmetrical.[7] The only element that is not a part of the standard modal semantics is λ, a member of K which is to be understood as the *absurd world* – the world in which contradictions and all their consequences are true. It is an isolated element under R; that is, no other world is possible with respect to it, and it is not possible with respect to any other world. The purpose of λ is to allow for an interpretation of "If A, then B" in the case where A is impossible; for this situation one needs an impossible world.

In addition to a model structure, our semantical apparatus includes a *selection function, f*, which takes a proposition and a possible world as arguments and a possible world as its value. The s-function selects, for each antecedent A, a particular possible world in which A is true. The *assertion* which the conditional makes, then, is that the consequent is true in the world selected. A conditional is true in the actual world when its consequent is true in the selected world.

Now we can state the semantical rule for the conditional more formally (using the corner, $>$, as the conditional connective):

$A > B$ is true in α if B is true in $f(A, \alpha)$;
$A > B$ is false in α if B is false in $f(A, \alpha)$.

The interpretation shows conditional logic to be an extension of modal logic. Modal logic provides a way of talking about what is true in the actual world, in all possible worlds, or in at least one, unspecified world. The addition of the selection function to the semantics and the conditional connective to the object language of modal logic provides a way of talking also about what is true in *particular* non-actual possible situations. This is what counterfactuals are: statements about particular counterfactual worlds.

But the world selected cannot be just any world. The s-function must meet at least the following conditions. I shall use the following terminology for talking about the arguments and values of s-functions: where $f(A, \alpha) = \beta$, A is the *antecedent*, α is the *base world*, and β is the *selected world*.

(1)   For all antecedents A and base worlds α, A must be true in $f(A, \alpha)$.

(2)   For all antecedents A and base worlds α, $f(A, \alpha) = \lambda$ only if there is no world possible with respect to α in which A is true.

The first condition requires that the antecedent be true in the selected world. This ensures that all statements like "if snow is white, then snow is white" are true. The second condition requires that the absurd world be selected only when the antecedent is impossible. Since everything is true in the absurd world, including contradictions, if the selection function were to choose it for the antecedent A, then "If A, then B and not B" would be true. But one cannot legitimately reach an impossible conclusion from a consistent assumption.

The informal truth conditions that were suggested above required that the world selected *differ minimally* from the actual world. This implies, first, that there are no differences between the actual world and the selected world except those that are required, implicitly or explicitly, by the antecedent. Further, it means that among the alternative ways of making the required changes, one must choose one that does the least violence to the correct description and explanation of the actual world. These are vague conditions which are largely dependent on pragmatic considerations for their application. They suggest, however, that the selection is based on an ordering of possible worlds with respect to their resemblance to the base world. If this is correct, then there are two further formal constraints which must be imposed on the s-function.

(3)   For all base worlds α and all antecedents A, if A is true in α, then $f(A, \alpha) = \alpha$.

(4)   For all base worlds α and all antecedents B and B', if B is true in $f(B', \alpha)$ and B' is true in $f(B, \alpha)$, then $f(B, \alpha) = f(B', \alpha)$.

The third condition requires that the base world be selected if it is among the worlds in which the antecedent is true. Whatever the criteria for evaluating resemblance among possible worlds, there is obviously no other possible world as much like the base world as the base world itself. The fourth condition ensures that the ordering among possible worlds is consistent in the following sense: if any selection established β as prior to β' in the ordering (with respect to a particular base world α), then no other selection (relative to that α) may establish β' as prior to β.[8] Conditions (3) and (4) together ensure that the s-function establishes a total ordering of all selected worlds with respect to each possible world, with the base world preceding all others in the order.

These conditions on the selection function are necessary in order that this account be recognizable as an explication of the conditional, but they are of course far from sufficient to determine the function uniquely. There may be further formal constraints that can plausibly be imposed on the selection principle, but we should not expect to find semantic conditions sufficient to guarantee that there will be a unique s-function for each valuation of non-conditional formulas on a model structure. The questions, "On what basis do we select a selection function from among the acceptable ones?" and "What are the criteria for ordering possible worlds?" are reformulations of the pragmatic problem of counterfactuals, which is a problem in the application of conditional logic. The conditions that I have mentioned above are sufficient, however, to define the semantical notions of validity and consequence for conditional logic.

III. THE FORMAL SYSTEM

The class of valid formulas of conditional logic according to the definitions sketched in the preceding section, is coextensive with the class of theorems of a formal system, C2. The primitive connectives of C2 are the usual $\supset$ and $\sim$ (with v, &, and $\equiv$ defined as usual), as well as a conditional connective, $>$ (called the corner). Other modal and conditional concepts can be defined in terms of the corner as follows:

$$\Box A \ =_{DF} \ \sim A > A$$

$$\Diamond A \ =_{DF} \ \sim(A > \sim A)$$

$$A \gtreqless B \ =_{DF} \ (A > B) \ \& \ (B > A)$$

The rules of inference of C2 are *modus ponens* (if A and $A \supset B$ are theorems, then B is a theorem) and the Gödel rule of necessitation (If A is a theorem, then $\Box A$ is a theorem). There are seven axiom schemata:

(a1)   Any tautologous wff (well-formed formula) is an axiom.

(a2) $\Box(A \supset B) \supset (\Box A \supset \Box B)$

(a3) $\Box(A \supset B) \supset (A > B)$

(a4) $\Diamond A \supset \cdot (A > B) \supset \sim(A > \sim B)$

(a5) $A > (B \vee C) \supset \cdot (A > B) \vee (A > C)$

(a6) $(A > B) \supset (A \supset B)$

(a7) $A \gtrless B \supset \cdot (A > C) \supset (B > C)$

The conditional connective, as characterized by this formal system, is intermediate between strict implication and the material conditional, in the sense that $\Box(A \supset B)$ entails $A > B$ by (a3) and $A > B$ entails $A \supset B$ by (a6). It cannot, however, be analyzed as a modal operation performed on a material conditional (like Burks's causal implication, for example).[9] The corner lacks certain properties shared by the two traditional implication concepts, and in fact these differences help to explain some peculiarities of counterfactuals. I shall point out three unusual features of the conditional connective.

(1) Unlike both material and strict implication, the conditional corner is a non-transitive connective. That is, from $A > B$ and $B > C$, one cannot infer $A > C$. While this may at first seem surprising, consider the following example: *Premisses.* "If J. Edgar Hoover were today a communist, then he would be a traitor." "If J. Edgar Hoover had been born a Russian, then he would today be a communist." *Conclusion.* "If J. Edgar Hoover had been born a Russian, he would be a traitor." It seems reasonable to affirm these premisses and deny the conclusion.

If this example is not sufficiently compelling, note that the following rule follows from the transitivity rule: From $A > B$ to infer $(A \,\&\, C) > B$. But it is obvious that the former rule is invalid; we cannot always strengthen the antecedent of a true conditional and have it remain true. Consider "If this match were struck, it would light," and "If this match had been soaked in water overnight *and* it were struck, it would light."[10]

(2) According to the formal system, the denial of a conditional is equivalent to a conditional with the same antecedent and opposite consequent (provided

that the antecedent is not impossible). That is, $\Diamond A - \sim(A > B) \equiv (A > \sim B)$. This explains the fact, noted by both Goodman and Chisholm in their early papers on counterfactuals, that the normal way to contradict a counterfactual is to contradict the consequent, keeping the same antecedent. To deny "If Kennedy were alive today, we wouldn't be in this Vietnam mess," we say, "If Kennedy were alive today, we would so be in this Vietnam mess."

(3) The inference of contraposition, valid for both the truth-functional horseshoe and the strict implication hook, is invalid for the conditional corner. $A > B$ may be true while $\sim B > \sim A$ is false. For an example in support of this conclusion, we take another item from the political opinion survey: "If the U.S. halts the bombing, then North Vietnam will not agree to negotiate." A person would believe that this statement is true if he thought that the North Vietnamese were determined to press for a complete withdrawal of U.S. troops. But he would surely deny the contrapositive, "If North Vietnam agrees to negotiate, then the U.S. will not have halted the bombing." He would believe that halt in the bombing, and much more, is required to bring the North Vietnamese to the negotiating table.[11]

Examples of these anomalies have been noted by philosophers in the past. For instance, Goodman pointed out that two counterfactuals with the same antecedent and contradictory consequents are "normally meant" as direct negations of each other. He also remarked that we may sometimes assert a conditional and yet reject its contrapositive. He accounted for these facts by arguing that semifactuals – conditionals with false antecedents and true consequents – are for the most part not to be taken literally. "In practice," he wrote, "full counterfactuals affirm, while semifactuals deny, that a certain connection obtains between antecedent and consequent ... The practical import of a semifactual is thus different from its literal import."[12] Chisholm also suggested paraphrasing semifactuals before analyzing them. "Even if you were to sleep all morning, you would be tired" is to be read "It is false that if you were to sleep all morning, you would not be tired."[13]

A separate and nonconditional analysis for semifactuals is necessary to save the 'connection' theory of counterfactuals in the face of the anomalies we have discussed, but it is a baldly *ad hoc* manoeuvre. Any analysis can be saved by paraphrasing the counter-examples. The theory presented in Section II avoids this difficulty by denying that the conditional can be said, in general, to assert a connection of any particular kind between antecedent and consequent. It is, of course, the structure of inductive relations and causal connections which make counterfactuals and semifactuals true or false, but they do this by determining the relationships among possible worlds, which in turn determine the truth values of conditionals. By treating the relation between connection and conditionals as an indirect relation in this way, the theory is able to give a unified account of conditionals which explains the variations in their behavior in different contexts.

IV. THE LOGICAL PROBLEM: GENERAL CONSIDERATIONS

The traditional strategy for attacking a problem like the logical problem of conditionals was to find an *analysis*, to show that the unclear or objectionable phrase was dispensable, or replaceable by something clear and harmless. Analysis was viewed by some as an *unpacking* – a making manifest of what was latent in the concept; by others it was seen as the *replacement* of a vague idea by a precise one, adequate to the same purposes as the old expression, but free of its problems. The semantic theory of conditionals can also be viewed either as the construction of a concept to replace an unclear notion of ordinary language, or as an *explanation* of a commonly used concept. I see the theory in the latter way: no recommendation or stipulation is intended. This does not imply, however, that the theory is meant as a description of linguistic usage. What is being explained is not the rules governing the use of an English word, but the structure of a concept. Linguistic facts – what we would say in this or that context, and what sounds odd to the native speaker – are relevant as evidence, since one may presume that concepts are to some extent mirrored in language.

The 'facts', taken singly, need not be decisive. A recalcitrant counterexample may be judged a deviant use or a different sense of the word. We can claim that a paraphrase is necessary, or even that ordinary language is systematically mistaken about the concept we are explaining. There are, of course, different senses and times when 'ordinary language' goes astray, but such *ad hoc* hypotheses and qualifications diminish both the plausibility and the explanatory force of a theory. While we are not irrevocably bound to the linguistic facts, there are no 'don't cares' – contexts of use with which we are not concerned, since any context can be relevant as evidence for or against an analysis. A general interpretation which avoids dividing senses and accounts for the behavior of a concept in many contexts fits the familiar pattern of scientific explanation in which diverse, seemingly unlike surface phenomena are seen as deriving from some common source. For these reasons, I take it as a strong point in favor of the semantic theory that it treats the conditional as a univocal concept.

V. PRAGMATIC AMBIGUITY

I have argued that the conditional connective is semantically unambiguous. It is obvious, however, that the context of utterance, the purpose of the assertion, and the beliefs of the speaker or his community may make a difference to the interpretation of a counterfactual. How do we reconcile the ambiguity of conditional sentences with the univocity of the conditional concept? Let us look more closely at the notion of ambiguity.

A sentence is ambiguous if there is more than one proposition which it may properly be interpreted to express. Ambiguity may be syntactic (if the sentence has more than one grammatical structure, semantic (if one of the words has more than one meaning), or pragmatic (if the interpretation depends directly on the context of use). The first two kinds of ambiguity are perhaps more familiar, but the third kind is probably the most common in natural languages. Any sentence involving pronouns, tensed verbs, articles or quantifiers is pragmatically ambiguous. For example, the proposition expressed by "L'état, c'est moi" depends on who says it; "Do it now" may be good or bad advice depending on when it is said; "Cherchez la femme" is ambiguous since it contains a definite description, and the truth conditions for "All's well that ends well" depends on the domain of discourse. If the theory presented above is correct, then we may add conditional sentences to this list. The truth conditions for "If wishes were horses, then beggers would ride" depend on the specification of an s-function.[14]

The grounds for treating the ambiguity of conditional sentences as pragmatic rather than semantic are the same as the grounds for treating the ambiguity of quantified sentences as pragmatic: simplicity and systematic coherence. The truth conditions for quantified statements vary with a change in the domain of discourse, but there is a single structure to these truth conditions which remains constant for every domain. The semantics for classical predicate logic brings out this common structure by giving the universal quantifier a single meaning and making the domain a parameter of the interpretation. In a similar fashion, the semantics for conditional logic brings out the common structure of the truth conditions for conditional statements by giving the connective a single meaning and making the selection function a parameter of the interpretation.

Just as we can communicate effectively using quantified sentences without explicitly specifying a domain, so we can communicate effectively using conditional sentences without explicitly specifying an s-function. This suggests that there are further rules beyond those set down in the semantics, governing the use of conditional sentences. Such rules are the subject matter of a *pragmatics* of conditionals. Very little can be said, at this point, about pragmatic rules for the use of conditionals since the logic has not advanced beyond the propositional stage, but I shall make a few speculative remarks about the kinds of research which may provide a framework for treatment of this problem, and related pragmatic problems in the philosophy of science.

(1) If we had a functional logic with a conditional connective, it is likely that $(\forall x)(Fx > Gx)$ would be a plausible candidate for the form of a law of nature. A law of nature says, not just that every actual F is a G, but further that for every possible F, if it were an F, it would be a G. If this is correct, then Hempel's confirmation paradox does not arise, since "All ravens are black" is not logically equivalent to "All non-black things are non-ravens." Also, the relation between counterfactuals and laws becomes clear: laws support counterfactuals because they entail them. "If this dove were a raven, it would be black" is simply an instantiation of "All ravens are black."[15]

(2) Goodman has argued that the pragmatic problem of counterfactuals is one of a cluster of closely related problems concerning induction and confirmation. He locates the source of these difficulties in the general problem of projectability, which can be stated roughly as follows: when can a predicate be validly projected from one set of cases to others? or when is a hypothesis confirmed by its positive instances? Some way of distinguishing between natural predicates and those which are artificially constructed is needed. If a theory of projection such as Goodman envisions were developed, it might find a natural place in a pragmatics of conditionals. Pragmatic criteria for measuring the inductive properties of predicates might provide pragmatic criteria for ordering possible worlds.[16]

(3) There are some striking structural parallels between conditional logic and conditional probability functions, which suggests the possibility of a connection between inductive logic and conditional logic. A probability assignment and an *s*-function are two quite different ways to describe the inductive relations among propositions; a theory which draws a connection between them might be illuminating for both.[17]

VI. CONCLUSION: EMPIRICISM AND POSSIBLE WORLDS

Writers of fiction and fantasy sometimes suggest that imaginary worlds have a life of their own beyond the control of their creators. Pirandello's six characters, for example, rebelled against their author and took the story out of his hands. The skeptic may be inclined to suspect that this suggestion is itself fantasy. He believes that nothing goes into a fictional world, or a possible world, unless it is put there by decision or convention; it is a creature of invention and not discovery. Even the fabulist Tolkien admits that Faërie is a land "full of wonder, but not of information."[18]

For similar reasons, the empiricist may be uncomfortable about a theory which treats counterfactuals as literal statements about non-actual situations. Counterfactuals are often contingent, and contingent statements must be supported by evidence. But evidence can be gathered, by us at least, only in this universe. To satisfy the empiricist, I must show how possible worlds, even if the product of convention, can be subjects of empirical investigation.

There is no mystery to the fact that I can partially define a possible world in such a way that I am ignorant of some of the determinate truths in that world. One way I can do this is to attribute to it features of the actual world which are unknown to me. Thus I can say, "I am thinking of a possible world in which the population of China is just the same, on each day, as it is in the actual world." *I* am making up this world – it is a pure product of my intentions – but there are already things true in it which I shall never know.

Conditionals do implicitly, and by convention, what is done explicitly by stipulation in this example. It is because counterfactuals are generally about possible worlds which are very much like the actual one, and defined in terms of it, that evidence is so often relevant to their truth. When I wonder, for example, what would have happened if I had asked my boss for a raise yesterday, I am wondering about a possible world that I have already roughly picked out. It has the same history, up to yesterday, as the actual world, the same boss with the same dispositions and habits. The main difference is that in that world, yesterday I asked the boss for a raise. Since I do not know everything about the boss's habits and dispositions in the actual world, there is a lot that I do not know about how he acts in the possible world that I have chosen, although I might find out by watching him respond to a similar request from another, or by asking his secetary about his mood yesterday. These bits of information about the actual world would not be decisive, of course, but they would be relevant, since they tell me more about the non-actual situation that I have selected.

If I make a conditional statement – subjunctive or otherwise – and the antecedent turns out to be true, then whether I know it or not, I have said something about the actual world, namely that the consequent is true in it. If the antecedent is false, then I have said something about a particular counterfactual world, even if I believe the antencedent to be true. The conditional provides a set of conventions for selecting possible situations which have a specified relation to what actually happens. This makes it possible for statements about unrealized possibilities to tell us, not just about the speaker's imagination, but about the world.

Yale University

NOTES

* I want to express appreciation to my colleague, Professor R. H. Thomason, for his collaboration in the formal development of the theory expounded in this paper, and for his helpful comments on its exposition and defense.

The preparation of this paper was supported in part by a National Science Foundation grant, GS-1567.

1 R. C. Stalnaker and R. H. Thomason, 'A Semantic Analysis of Conditional Logic', *(mimeo.,* 1967). In this paper, the formal system, C2, is proved sound and semantically complete with respect to the interpretation sketched in the present paper. That is, it is shown that a formula is a consequence of a class of formulas if and only if it is derivable from the class in the formal system, C2.

2 N. Rescher, *Hypothetical Reasoning*, Amsterdam, 1964.

3 Cf. R. Chisholm, 'The Contrary-to-fact Conditional', *Mind* 55 (1946), 289–307, reprinted in *Readings in Philosophical Analysis*, ed. by H. Feigl and W. Sellars, New York, 1949, pp. 482–497. The problem is sometimes posed (as it is here) as the task of analyzing the *subjunctive* conditional into an indicative statement, but I think it is a mistake to base very much on the distinction of mood. As far as I can tell, the mood tends to indicate something about the attitude of the speaker, but in no way effects the propositional content of the statement.

4 F. P. Ramsey, 'General Propositions and Causality', in Ramsey, *Foundations of Mathematics and other Logical Essays*, New York, 1950, pp. 237–257. The suggestion is made on p. 248. Chisholm, *op. cit.*, p. 489, quotes the suggestion and discusses the limitations of the "connection" thesis which it brings out, but he develops it somewhat differently.

5 Rescher, *op. cit.*, pp. 11–16, contains a very clear statement and discussion of this problem, which he calls the problem of the ambiguity of belief-contravening hypotheses. He argues that the resolution of this ambiguity depends on pragmatic consideration. Cf, also Goodman's problem of relevant conditions in N. Goodman, *Fact, Fiction, and Forecast*, Cambridge, Mass., 1955, pp. 17–24.

6 S. Kripke, 'Semantical Analysis of Modal Logics, I', *Zeitschrift für mathematische Logik und Grundlagen der Mathematik* 9 (1963), 67–96.

7 The different restrictions on the relation R provide interpretations for the different modal systems. The system we build on is von Wright's M. If we add the transitivity requirement, then the underlying modal logic of our system is Lewis's S4, and if we add both the transitivity and symmetry requirements, then the modal logic is S5. Cf. S. Kripke, *op. cit.*

8 If $f(A, \alpha) = \beta$, then β is established as prior to all worlds possible with respect to α in which A is true.

[9] A. W. Burks, 'The Logic of Causal Propositions', *Mind* **60** (1951), 363–382. The causal implication connective characterized in this article has the same structure as strict implication. For an interesting philosophical defense of this modal interpretation of conditionals, see B. Mayo, 'Conditional Statements', *The Philosophical Review* **66** (1957), 291–303.

[10] Although the transitivity inference fails, a related inference is of course valid. From $A > B$, $B > C$, and A, one can infer C. Also, note that the biconditional connective is transitive. From $A \gtrless B$ and $B \gtrless C$, one can infer $A \gtrless C$. Thus the biconditional is an equivalence relation, since it is also symmetrical and reflexive.

[11] Although contraposition fails, *modus tolens* is valid for the conditional: from $A > B$ and $\sim B$, one can infer $\sim A$.

[12] Goodman, *op. cit.*, pp. 15, 32.

[13] Chisholm, *op, cit.*, p. 492.

[14] I do not wish to pretend that the notions needed to define ambiguity and to make the distinction between pragmatic and semantic ambiguity (e.g., 'proposition', and 'meaning') are precise. They can be made precise only in the context of semantic and pragmatic theories. But even if it is unclear, in general, what pragmatic ambiguity is, it is clear, I hope, that my examples are cases of it.

[15] For a discussion of the relation of laws to counterfactuals, see E. Nagel, *Structure of Science*, New York, 1961, pp. 47–78. For a recent discussion of the paradoxes of confirmation by the man who discovered them, see C. G. Hempel, 'Recent Problems of Induction', in R. G. Colodny (ed.), *Mind and Cosmos*, Pittsburgh, 1966, pp. 112–134.

[16] Goodman, *op. cit.*, especially Ch. IV.

[17] Several philosophers have discussed the relation of conditional propositions to conditional probabilities. See R. C. Jeffrey, 'If', *The Journal of Philosophy* **61** (1964), 702–703; and E. W. Adams, 'Probability and the Logic of Conditionals', in J. Hintikka and P. Suppes (eds.), *Aspects of Inductive Logic*, Amsterdam, 1966, pp. 265–316. I hope to present elsewhere my method of drawing the connection between the two notions, which differs from both of these.

[18] J. R. Tolkien, 'On Fairy Stories', in *The Tolkien Reader*, New York, 1966, p. 3.

DAVID LEWIS

COUNTERFACTUALS AND COMPARATIVE POSSIBILITY*

In the last dozen years or so, our understanding of modality has been much improved by means of possible-world semantics: the project of analyzing modal language by systematically specifying the conditions under which a modal sentence is true at a possible world. I hope to do the same for counterfactual conditionals. I write $A \square \rightarrow C$ for the counterfactual conditional with antecedent A and consequent C. It may be read as 'If it were the case that A, then it would be the case that C' or some more idiomatic paraphrase thereof.

1. ANALYSES

I shall lead up by steps to an analysis I believe to be satisfactory.

ANALYSIS 0. $A \square \rightarrow C$ *is true at world i iff C holds at every A-world such that —.* 'A-world', of course, means 'world where A holds'.

The blank is to be filled in with some sort of condition restricting the A-worlds to be considered. The condition may depend on i but not on A. For instance, we might consider only those A-worlds that agree with i in certain specified respects. On this analysis, the counterfactual is some fixed strict conditional.

No matter what condition we put into the blank, Analysis 0 cannot be correct. For it says that if $A \square \rightarrow \bar{B}$ is true at i, $\bar{B}$ holds at every A-world such that —. In other words, there are no AB-worlds such that —. Then $AB \square \rightarrow \bar{C}$ and $AB \square \rightarrow C$ are alike vacuously true, and $-(AB \square \rightarrow C)$ and $-(AB \square \rightarrow \bar{C})$ are alike false, for any C whatever. On the contrary: it can perfectly well happen that $A \square \rightarrow \bar{B}$ is true, yet $AB \square \rightarrow \bar{C}$ is non-vacuous, and $AB \square \rightarrow C$ is false. In fact, we can have an arbitrarily long sequence like this of non-vacuously true counterfactuals and true denials of their opposites:

$$A \square \rightarrow \bar{B} \quad \text{and} \quad -(A \square \rightarrow B),$$
$$AB \square \rightarrow \bar{C} \quad \text{and} \quad -(AB \square \rightarrow C),$$

57

W. L. Harper, R. Stalnaker, and G. Pearce (eds.), Ifs, 57–85.

$$ABC \,\square\!\!\rightarrow \bar{D} \quad \text{and} \quad -(ABC \,\square\!\!\rightarrow D),$$
etc.

Example: if Albert had come to the party, he would not have brought Betty; for, as he knows, if he had come and had brought Betty, Carl would not have stayed; for, as Carl knows, if Albert had come and had brought Betty and Carl had stayed, Daisy would not have danced with him; ... Each step of the sequence is a counterexample to Analysis 0. The counterfactual is not any strict conditional whatever.

Analysis 0 also says that $A\,\square\!\!\rightarrow C$ implies $AB\,\square\!\!\rightarrow C$. If C holds at every A-world such that —, then C holds at such of those worlds as are B-worlds. On the contrary: we can have an arbitrarily long sequence like this of non-vacuously true counterfactuals and true denials of their opposites:

$$A \,\square\!\!\rightarrow \bar{Z} \quad \text{and} \quad -(A \,\square\!\!\rightarrow Z),$$
$$AB \,\square\!\!\rightarrow Z \quad \text{and} \quad -(AB \,\square\!\!\rightarrow \bar{Z}),$$
$$ABC \,\square\!\!\rightarrow \bar{Z} \quad \text{and} \quad -(ABC \,\square\!\!\rightarrow Z),$$
etc.

Example: if I had shirked my duty, no harm would have ensued; but if I had and you had too, harm would have ensued; but if I had and you had too and a third person had done far more than his duty, no harm would have ensued... For this reason also the counterfactual is not any strict conditional whatever.

More precisely, it is not any one, fixed strict conditional. But this much of Analysis 0 is correct: (1) to assess the truth of a counterfactual we must consider whether the consequent holds at certain antecedent-worlds; (2) we should not consider all antecedent-worlds, but only some of them. We may ignore antecedent-worlds that are gratuitously remote from actuality.

Rather than any fixed strict conditional, we need a *variably strict conditional*. Given a far-fetched antecedent, we look perforce at antecedent-worlds remote from actuality. There are no others to look at. But given a less far-fetched antecedent, we can afford to be more fastidious and ignore the very same worlds. In considering the supposition 'if I had just let go of my pen...' I will go wrong if I consider bizarre worlds where the law of gravity is otherwise than it actually is; whereas in considering the supposition 'if the planets traveled in spirals...' I will go just as wrong if I ignore such worlds.

It is this variable strictness that accounts for our counter-example sequences. It may happen that we can find an A-world that meets some stringent restriction; before we can find any AB-world we must relax the restriction; before we can find any ABC-world we must relax it still more; and so on. If so a counterexample sequence of the first kind definitely will appear, and one of the second kind will appear also if there is a suitable Z.

We dream of considering a world where the antecedent holds but everything else is just as it actually is, the truth of the antecedent being the one difference between that world and ours. No hope. Differences never come singly, but in infinite multitudes. Take, if you can, a world that differs from ours *only* in that Caesar did not cross the Rubicon. Are his predicament and ambitions there just as they actually are? The regularities of his character? The psychological laws exemplified by his decision? The orders of the day in his camp? The preparation of the boats? The sound of splashing oars? Hold *everything* else fixed after making one change, and you will not have a possible world at all.

If we cannot have an antecedent-world that is otherwise just like our world, what can we have? This, perhaps: an antecedent-world that does not differ gratuitously from ours; one that differs only as much as it must to permit the antecedent to hold; one that is closer to our world in similarity, all things considered, than any other antecedent world. Here is a first analysis of the counterfactual as a variably strict conditional.

ANALYSIS 1. $A \square\!\!\rightarrow C$ *is true at i iff C holds at the closest (accessible) A-world to i, if there is one.* This is Robert Stalnaker's proposal in 'A Theory of Conditionals', *Studies in Logical Theory* (*A.P.Q.* supplementary monograph series, 1968), and elsewhere.

It may be objected that Analysis 1 is founded on comparative similarity – 'closeness' – of worlds, and that comparative similarity is hopelessly imprecise unless some definite respect of comparison has been specified. Imprecise it may be; but that is all to the good. Counterfactuals are imprecise too. Two imprecise concepts may be rigidly fastened to one another, swaying together rather than separately, and we can hope to be precise about their connection. Imprecise though comparative similarity may be, we *do* judge the comparative similarity of complicated things like cities or people or philosophies – and we do it often without benefit of any definite respect of comparison stated in advance. We balance off various similarities and dissimilarities according to the importances we attach to various respects of comparison and according to the degrees of similarity in the various respects. Conversational context, of course, greatly affects our weighting of respects of comparison, and even in a fixed context we have plenty of latitude. Still, not anything goes. We have concordant mutual expectations, mutual expectations of expectations, etc., about the relative importances we will attach to respects of comparison. Often these are definite and accurate and firm enough to resolve the imprecision of comparative similarity to the point where we can converse without misunderstanding. Such imprecision we can live with. Still, I grant that a counterfactual based on comparative similarity has no place in the language of the exact sciences.

I imposed a restriction to A-worlds 'accessible' from i. In this I follow Stalnaker, who in turn is following the common practice in modal logic. We might think that there are some worlds so very remote from i that they should always be ignored (at i) even if some of them happen to be A-worlds and there are no closer A-worlds. If so, we have the wherewithal to ignore them by deeming them *inaccessible* from i. I can think of no very convincing cases, but I prefer to remain neutral on the point. If we have no need for accessibility restrictions, we can easily drop them by stipulating that all worlds are mutually interaccessible.

Unfortunately, Analysis 1 depends on a thoroughly implausible assumption: that there will never be more than one closest A-world. So fine are the gradations of comparative similarity that despite the infinite number and variety of worlds every tie is broken.

Example: A is 'Bizet and Verdi are compatriots', F is 'Bizet and Verdi are French', I is 'Bizet and Verdi are Italian'. Grant for the sake of argument that we have the closest F-world and the closest I-world; that these are distinct (dual citizenships would be a gratuitous difference from actuality); and that these are the two finalists in the competition for closest A-world. It might be that something favors one over the other – for all I know, Verdi narrowly escaped settling in France and Bizet did not narrowly escape settling in Italy. But we can count on no such luck. The case may be perfectly balanced between respects of comparison that favor the F-world and respects that favor the I-world. It is out of the question, on

Analysis 1, to leave the tie unbroken. That means there is no such thing as *the* closest A-world. Then anything you like holds at the closest A-world if there is one, because there isn't one. If Bizet and Verdi had been compatriots they would have been Ukranian.

ANALYSIS 2. *$A \square \rightarrow C$ is true at i iff C holds at every closest (accessible) A-world to i, if there are any.* This is the obvious revision of Stalnaker's analysis to permit a tie in comparative similarity between several equally close closest A-worlds.

Under Analysis 2 unbreakable ties are no problem. The case of Bizet and Verdi comes out as follows. $A \square \rightarrow F$, $A \square \rightarrow \bar{F}$, $A \square \rightarrow I$, and $A \square \rightarrow \bar{I}$ are all false. $A \square \rightarrow (F \vee I)$ and $A \square \rightarrow (\bar{F} \vee \bar{I})$ are both true. $A \square \rightarrow FI$ and $A \square \rightarrow \bar{F}\bar{I}$ are both false. These conclusions seem reasonable enough.

This reasonable settlement, however, does not sound so good in words. $A \square \rightarrow F$ and $A \square \rightarrow \bar{F}$ are both false, so we want to assert their negations. But negate their English readings in any straightforward and natural way, and we do not get $-(A \square \rightarrow F)$ and $-(A \square \rightarrow \bar{F})$ as desired. Rather the negation moves in and attaches only to the consequent, and we get sentences that seem to mean $A \square \rightarrow \bar{F}$ and $A \square \rightarrow \bar{\bar{F}}$ – a pair of falsehoods, together implying the further falsehood that Bizet and Verdi could not have been compatriots; and exactly the opposite of what we meant to say.

Why is it so hard to negate a whole counterfactual, as opposed to negating the consequent? The defender of Analysis 1 is ready with an explanation. Except when A is impossible, he says, there is a unique closest A-world. Either C is false there, making $-(A \square \rightarrow C)$ and $A \square \rightarrow \bar{C}$ alike true, or C is true there, making them alike false. Either way, the two agree. We have no need of a way to say $-(A \square \rightarrow C)$ because we might as well say $A \square \rightarrow \bar{C}$ instead (except when A is impossible, in which case we have no need of a way to say $-(A \square \rightarrow C)$ because it is false).

There is some appeal to the view that $-(A \square \rightarrow C)$ and $A \square \rightarrow \bar{C}$ are equivalent (except when A is impossible) and we might be tempted thereby to return to Analysis 1. We might do better to return only part way, using Bas van Fraassen's method of supervaluations to construct a compromise between Analyses 1 and 2.

ANALYSIS 1½. *$A \square \rightarrow C$ is true at i iff C holds at a certain arbitrarily chosen one of the closest (accessible) A-worlds to i, if there are any. A sen-*

tence is super-true iff it is true no matter how the arbitrary choices are made, super-false iff false no matter how the arbitrary choices are made. Otherwise it has no super-truth value. Unless a particular arbitrary choice is under discussion, we abbreviate 'super-true' as 'true', and so on. Something of this kind is mentioned at the end of Richmond Thomason, 'A Fitch-Style Formulation of Conditional Logic', *Logique et Analyse* 1970.

Analysis $1\frac{1}{2}$ agrees with Analysis 1 about the equivalence (except when A is impossible) of $-(A\,\square\!\!\rightarrow C)$ and $A\,\square\!\!\rightarrow \bar{C}$. If there are accessible A-worlds, the two agree in truth (i.e. super-truth) value, and further their biconditional is (super-)true. On the other hand, Analysis $1\frac{1}{2}$ tolerates ties in comparative similarity as happily as Analysis 2. Indeed a counterfactual is (super-)true under Analysis $1\frac{1}{2}$ iff it is true under Analysis 2. On the other hand, a counterfactual false under Analysis 2 may either be false or have no (super-)truth under Analysis $1\frac{1}{2}$. The case of Bizet and -Verdi comes out as follows: $A\,\square\!\!\rightarrow F, A\,\square\!\!\rightarrow \bar{F}, A\,\square\!\!\rightarrow I, A\,\square\!\!\rightarrow \bar{I}$, and their negations have no truth value. $A\,\square\!\!\rightarrow (F\vee I)$ and $A\,\square\!\!\rightarrow (\bar{F}\vee \bar{I})$ are (super-)true. $A\,\square\!\!\rightarrow FI$ and $A\,\square\!\!\rightarrow \bar{F}\bar{I}$ are (super-)false.

This seems good enough. For all I have said yet, Analysis $1\frac{1}{2}$ solves the problem of ties as well as Analysis 2, provided we're not too averse to (super-) truth value gaps. But now look again at the question how to deny a counterfactual. We have a way after all: to deny a 'would' counterfactual, use a 'might' counterfactual with the same antecedent and negated consequent. In reverse likewise: to deny a 'might' counterfactual, use a 'would' counterfactual with the same antecedent and negated consequent. Writing $A\,\Diamond\!\!\rightarrow C$ for 'If it were the case that A, then it might be the case that C' or some more idiomatic paraphrase, we have these valid-sounding equivalences:

(1) $-(A\,\square\!\!\rightarrow C)$ is equivalent to $A\,\Diamond\!\!\rightarrow \bar{C}$,
(2) $-(A\,\Diamond\!\!\rightarrow C)$ is equivalent to $A\,\square\!\!\rightarrow \bar{C}$.

The two equivalences yield an explicit definition of 'might' from 'would' counterfactuals:

$$A\,\Diamond\!\!\rightarrow C =^{\mathrm{df}} -(A\,\square\!\!\rightarrow \bar{C});$$

or, if we prefer, the dual definition of 'would' from 'might'. According to this definition and Analysis 2, $A\,\Diamond\!\!\rightarrow C$ is true at i iff C holds at some closest (accessible) A-world to i. In the case of Bizet and Verdi, $A\,\Diamond\!\!\rightarrow F$,

$A\diamondsuit\rightarrow F$, $A\diamondsuit\rightarrow I$, $A\diamondsuit\rightarrow \bar{I}$ are all true; so are $A\diamondsuit\rightarrow(F\vee I)$ and $A\diamondsuit\rightarrow(F\vee\bar{I})$; but $A\diamondsuit\rightarrow FI$ and $A\diamondsuit\rightarrow F\bar{I}$ are false.

According to the definition and Analysis 1 or $1\frac{1}{2}$, on the other hand, $A\diamondsuit\rightarrow C$ and $A\square\rightarrow C$ are equivalent except when A is impossible. That should put the defender of those analyses in an uncomfortable spot. He cannot very well claim that 'would' and 'might' counterfactuals do not differ except when the antecedent is impossible. He must therefore reject my definition of the 'might' counterfactual; and with it, the equivalences (1) and (2), uncontroversial though they sound. He then owes us some other account of the 'might' counterfactual, which I do not think he can easily find. Finally, once we see that we do have a way to negate a whole counterfactual, we no longer appreciate his explanation of why we don't need one. I conclude that he would be better off moving at least to Analysis 2.

Unfortunately, Analysis 2 is not yet satisfactory. Like Analysis 1, it depends on an implausible assumption. Given that some A-world is accessible from i, we no longer assume that there must be *exactly* one closest A-world to i; but we still assume that there must be *at least* one. I call this the *Limit Assumption*. It is the assumption that as we proceed to closer and closer A-worlds we eventually hit a limit and can go no farther. But why couldn't it happen that there are closer and closer A-worlds without end – for each one, another even closer to i? Example: A is 'I am over 7 feet tall'. If there are closest A-worlds to ours, pick one of them: how tall am I there? I must be $7+\varepsilon$ feet tall, for some positive ε, else it would not be an A-world. But there are A-worlds where I am only $7+\varepsilon/2$ feet tall. Since that is closer to my actual height, why isn't one of these worlds closer to ours than the purportedly closest A-world where I am $7+\varepsilon$ feet tall? And why isn't a suitable world where I am only $7+\varepsilon/4$ feet even closer to ours, and so ad infinitum? (In special cases, but not in general, there may be a good reason why not. Perhaps $7+\varepsilon$ could have been produced by a difference in one gene, whereas any height below that but still above 7 would have taken differences in many genes.) If there are A-worlds closer and closer to i without end, then any consequent you like holds at every closest A-world to i, because there aren't any. If I were over 7 feet tall I would bump my head on the sky.

ANALYSIS 3. *$A\square\rightarrow C$ is true at i iff some (accessible) AC-world is closer* *to i than any $A\bar{C}$-world, if there are any (accessible) A-worlds.* This is my final analysis.

Analysis 3 looks different from Analysis 1 or 2, but it is similar in principle. Whenever there are closest (accessible) A-worlds to a given world, Analyses 2 and 3 agree on the truth value there of $A\,\square\!\!\rightarrow C$. They agree also, of course, when there are no (accessible) A-worlds. When there are closer and closer A-worlds without end, $A\,\square\!\!\rightarrow C$ is true iff, as we proceed to closer and closer A-worlds, we eventually leave all the $A\bar{C}$-worlds behind and find only AC-worlds.

Using the definition of $A\,\Diamond\!\!\rightarrow C$ as $-(A\,\square\!\!\rightarrow\bar{C})$, we have this derived truth condition for the 'might' counterfactual: $A\,\Diamond\!\!\rightarrow C$ is true at i iff for every (accessible) $A\bar{C}$-world there is some AC-world at least as close to i, and there are (accessible) A-worlds.

We have discarded two assumptions about comparative similarity in going from Analysis 1 to Analysis 3: first Stalnaker's assumption of uniqueness, then the Limit Assumption. What assumptions remain?

First, the *Ordering Assumption*: that for each world i, comparative similarity to i yields a *weak ordering* of the worlds accessible from i. That is, writing $j\leqslant_i k$ to mean that k is not closer to i than j, each $\leqslant_i$ is *connected* and *transitive*. Whenever j and k are accessible from i either $j\leqslant_i k$ or $k\leqslant_i j$; whenever $h\leqslant_i j$ and $j\leqslant_i k$, then $h\leqslant_i k$. It is convenient, if somewhat artificial, to extend the comparative similarity orderings to encompass also the inaccessible worlds, if any: we stipulate that each $\leqslant_i$ is to be a weak ordering of *all* the worlds, and that j is closer to i than k whenever j is accessible from i and k is not. (Equivalently: whenever $j\leqslant_i k$, then if k is accessible from i so is j.)

Second, the *Centering Assumption*: that each world i is accessible from itself, and closer to itself than any other world is to it.

2. REFORMULATIONS

Analysis 3 can be given several superficially different, but equivalent, reformulations.

2.1. *Comparative Possibility*

Introduce a connective $\prec$. $A\prec B$ is read as 'It is less remote from actuality that A than that B' or 'It is more possible that A than that B' and is true at a world i iff some (accessible) A-world is closer to i than is any B-world. First a pair of modalities and then the counterfactual can be defined from this new connective of comparative possibility, as follows. (Let $\bot$ be a sentential constant false at every world, or an arbitrarily chosen contradiction; later, let $\top = {}^{df} - \bot$.)

$$\Diamond A = {}^{df} A \prec \bot; \quad \Box A = {}^{df} - \Diamond - A;$$
$$A \;\Box\!\!\rightarrow C = {}^{df} \Diamond A \supset (AC \prec A\bar{C}).$$

The modalities so defined are interpreted by means of accessibility in the usual way. $\Diamond A$ is true at i iff some A-world is accessible from i, and $\Box A$ is true at i iff A holds throughout all the worlds accessible from i. If accessibility restrictions are discarded, so that all worlds are mutually interaccessible, they became the ordinary 'logical' modalities. (We might rather have defined the two modalities and comparative possibility from the counterfactual.

$$\Box A = {}^{df} \bar{A} \;\Box\!\!\rightarrow \bot; \quad \Diamond A = {}^{df} - \Box - A;$$
$$A \prec B = {}^{df} \Diamond A \;\&\;((A \lor B) \;\Box\!\!\rightarrow A\bar{B}).$$

Either order of definitions is correct according to the given truth conditions.)

Not only is comparative possibility technically convenient as a primitive; it is of philosophical interest for its own sake. It sometimes seems true to say: It is possible that A but not that B, it is possible that B but not that C, C but not D, etc. Example: A is 'I speak English', B is 'I speak German' (a language I know), C is 'I speak Finnish', D is 'A dog speaks Finnish', E is 'A stone speaks Finnish', F is 'A number speaks Finnish'. Perhaps if I say all these things, as I would like to, I am equivocating – shifting to weaker and weaker noncomparative senses of 'possible' from clause to clause. It is by no means clear that there are enough distinct senses to go around. As an alternative hypothesis, perhaps the clauses are compatible comparsions of possibility without equivocation: $A \prec B \prec C \prec D \prec E \prec F$. (Here and elsewhere, I compress conjunctions in the obvious way.)

2.2. Cotenability

Call B *cotenable* at i with the supposition that A iff some A-world accessible from i is closer to i than any $\bar{B}$-world, or if there are no A-worlds accessible from i. In other words: iff, at i, the supposition that A is either more possible than the falsity of B, or else impossible. Then $A \,\square\!\!\rightarrow C$ is true at i iff C follows from A together with auxiliary premises $B_1, \ldots,$ each true at i and cotenable at i with the supposition that A.

There is less to this definition than meets the eye. A conjunction is cotenable with a supposition iff its conjuncts all are; so we need only consider the case of a single auxiliary premise B. That single premise may always be taken either as $\bar{A}$ (if A is impossible) or as $A \supset C$ (otherwise); so 'follows' may be glossed as 'follows by truth-functional logic'.

Common opinion has it that laws of nature are cotenable with any supposition unless they are downright inconsistent with it. What can we make of this? Whatever else laws may be, they are generalizations that we deem especially important. If so, then conformity to the prevailing laws of a world i should weigh heavily in the similarity of other worlds to i. Laws should therefore tend to be cotenable, unless inconsistent, with counterfactual suppositions. Yet I think this tendency may be overridden when conformity to laws carries too high a cost in differences of particular fact. Suppose, for instance, that i is a world governed (in all respects of the slightest interest to us) by deterministic laws. Let A pertain to matters of particular fact at time t; let A be false at i, and determined at all previous times to be false. There are some A-worlds where the laws of i are never violated; all of these differ from i in matters of particular fact at all times before t. (Nor can we count on the difference approaching zero as we go back in time.) There are other A-worlds exactly like i until very shortly before t when a small, local, temporary, imperceptible suspension of the laws permits A to come true. I find it highly plausible that one of the latter resembles i on balance more than any of the former.

2.3. *Degrees of Similarity*

Roughly, $A \,\square\!\!\rightarrow C$ is true at i iff either (1) there is some degree of similarity to i within which there are A-worlds and C holds at all of them, or (2) there are no A-worlds within any degree of similarity to i. To avoid the questionable assumption that similarity of worlds admits somehow of numerical measurement, it seems best to identify each 'degree of similarity to i' with a set of worlds regarded as the set of all worlds within that degree mof siilarity to i. Call a set S of worlds a *sphere* around i iff every S-world is accessible from i and is closer to i than is any S-world. Call a sphere A-*permitting* iff it contains some A-world. Letting spheres represent degrees of similarity, we have this reformulation: $A \,\square\!\rightarrow C$ is true at i iff $A \supset C$ holds throughout some A-permitting sphere around i, if such there be.

To review our other operators: $A \diamondsuit\!\rightarrow C$ is true at i iff AC holds somewhere in every A-permitting sphere around i, and there are such. $\square A$ is true at i iff A holds throughout every sphere around i. $\diamondsuit A$ is true at i iff A holds somewhere in some sphere around i. $A \prec B$ is true at i iff some sphere around i permits A but not B. Finally, B is cotenable at i with the supposition that A iff B holds throughout some A-permitting sphere around i, if such there be.

Restated in terms of spheres, the Limit Assumption says that if there is any A-permitting sphere around i, then there is a smallest one – the intersection of all A-permitting spheres is then itself an A-permitting sphere. We can therefore reformulate Analysis 2 as: $A \,\square\!\rightarrow C$ is true at i iff $A \supset C$ holds throughout the smallest A-permitting sphere around i, if such there be.

These systems of spheres may remind one of neighborhood systems in topology, but that would be a mistake. The topological concept of closeness captured by means of neighborhoods is purely local and qualitative, not comparative: adjacent vs. separated, no more. Neighborhoods do not capture comparative closeness to a point because arbitrary supersets of neighborhoods of the point are themselves neighborhoods of a point. The spheres around a world, on the other hand, are nested, wherefore they capture comparative closeness: j is closer to i than k is (according to the definition of spheres and the Ordering Assumption) iff some sphere around i includes j but excludes k.

2.4. *Higher-Order Quantification*

The formulation just given as a metalinguistic truth condition can also be stated, with the help of auxiliary apparatus, as an explicit definition in the object language.

$$A \,\square\!\rightarrow C =^{\mathrm{df}} \diamondsuit A \supset \exists S(\varPhi S \,\&\, \diamondsuit SA \,\&\, \square(SA \supset C)).$$

Here the modalities are as before; 'S' is an object-language variable over propositions; and $\varPhi$ is a higher-order predicate satisfied at a world i by a proposition iff the set of all worlds where that proposition holds is a sphere around i. I have assumed that every set of worlds is the truth-set of some – perhaps inexpressible – proposition.

We could even quantify over modalities, these being understood as certain properties of propositions. Call a modality *spherical* iff for every world i there is a sphere around i such that the modality belongs at i to all and only those propositions that hold throughout that sphere. Letting ■ be a variable over all spherical modalities, and letting ♦ abbreviate –■–, we have

$$A \,\square\!\!\rightarrow C =^{\mathrm{df}} \Diamond A \supset \exists \blacksquare (\blacklozenge A \,\&\, \blacksquare (A \supset C)).$$

This definition captures explicitly the idea that the counterfactual is a variably strict conditional.

To speak of variable strictness, we should be able to compare the strictness of different spherical modalities. Call one modality *(locally) stricter* than another at a world i iff the second but not the first belongs to some proposition at i. Call two modalities *comparable* iff it does not happen that one is stricter at one world and the other at another. Call one modality *stricter* than another iff they are comparable and the first is stricter at some world. Call one *uniformly stricter* than another iff it is stricter at every world. Comparative strictness is only a partial ordering of the spherical modalities: some pairs are incomparable. However, we can without loss restrict the range of our variable ■ to a suitable subset of the spherical modalities on which comparative strictness is a linear ordering. (Perhaps – iff the inclusion orderings of spheres around worlds all have the same order type – we can do better still, and use a subset linearly ordered by uniform comparative strictness.) Unfortunately, these linear sets are not uniquely determined.

Example: suppose that comparative similarity has only a few gradations. Suppose, for instance, that there are only five different (nonempty) spheres around each world. Let $\square_1 A$ be true at i iff A holds throughout the innermost (nonempty) sphere around i: let $\square_2 A$ be true at i iff A holds throughout the innermost-but-one; and likewise for $\square_3$, $\square_4$, and $\square_5$. Then the five spherical modalities expressed by these operators are a suitable linear set. Since we have only a finite range, we can replace quantification by disjunction:

$$A \, \square\!\!\rightarrow C =^{df} \Diamond A \supset . (\Diamond_1 A \ \& \ \square_1 (A \supset C))$$
$$\vee \cdots \vee (\Diamond_5 A \ \& \ \square_5 (A \supset C))$$

See Louis Goble, 'Grades of Modality', *Logique et Analyse* 1970.

2.5. *Impossible Limit-Worlds*

We were driven from Analysis 2 to Analysis 3 because we had reason to doubt the Limit Assumption. It seemed that sometimes there were closer and closer A-worlds to i without limit – that is, without any closest A-worlds. None, at least, among the *possible* worlds. But we can find the closest A-worlds instead among certain *impossible* worlds, if we are willing to look there. If we count these impossible worlds among the worlds to be considered, the Limit Assumption is rescued and we can safely return to Analysis 2.

There are various ways to introduce the impossible limits we need. The following method is simplest, but others can be made to seem a little less *ad hoc*. Suppose there are closer and closer (accessible, possible) A-worlds to i without limit; and suppose Σ is any maximal set of sentences such that, for any finite conjunction C of sentences in Σ, $A \Diamond\!\!\rightarrow C$ holds at i according to Analysis 3. (We can think of such a Σ as a full description of one – possible or impossible – way things might be if it were that A, from the standpoint of i.) Then we must posit an impossible limit-world where all of Σ holds. It should be accessible from i alone; it should be closer to i than all the possible A-worlds; but it should be no closer to i than any possible world that is itself clossr than all the possible A-worlds. (Accessibility from, and comparative similarity to, the impossible limit-worlds is undefined. Truth of sentences there is determined by the way in which these worlds were introduced as limits, not according to the ordinary truth conditions.) Obviously the Limit Assumption is satisfied once these impossible worlds have been added to the worlds under consideration. It is easy to verify that the truth values of counterfactuals at possible worlds afterwards according to Analyses 2 and 3 alike agrees with their original truth values according to Analysis 3.

The impossible worlds just posited are impossible in the least objectionable way. The sentences true there may be *incompatible*, in that not all of them hold together at any possible world; but there is no (correct) way to derive any contradiction from them. For a derivation proceeds from

finitely many premises; and any finite subset of the sentences true at one
of the limit-worlds *is* true together at some possible world. Example:
recall the failure of the Limit Assumption among possible worlds when
A is 'I am over 7 feet tall'. Our limit-worlds will be impossible worlds
where A is true but all of 'I am at least 7.1 feet tall', 'I am at least 7.01 feet
tall', 'I am at least 7.001 feet tall' etc. are false. (Do not confuse these with
possible worlds where I am infinitesimally more than 7 feet tall. For all I
know, there are such; but worlds where physical magnitudes can take
'non-standard' values differing infinitesimally from a real number presumably differ from ours in a very fundamental way, making them far
more remote from actuality than some of the standard worlds where I am,
say, 7.1 feet tall. If so, 'Physical magnitudes never take non-standard
values' is false at any possible world where I am infinitesimally more than
7 feet tall, but true at the impossible closest A-worlds to ours.)

How bad is it to believe in these impossible limit-worlds? Very bad, I
think; but there is no reason not to reduce them to something less objectionable, such as sets of propositions or even sentences. I do not like a
parallel reduction of possible worlds, chiefly because it is incredible in
the case of the possible world *we* happen to live in, and other possible
worlds do not differ in kind from ours. But this objection does not carry
over to the impossible worlds. We do not live in one of those, and possible
and impossible worlds do differ in kind.

2.6. *Selection Functions*

Analysis 2, vindicated either by trafficking in impossible worlds or by
faith in the Limit Assumption even for possible worlds, may conveniently
be reformulated by introducing a function f that selects, for any antecedent
A and possible world i, the set of all closest (accessible) A-worlds to i (the
empty set if there are none). $A \ \square\!\!\rightarrow C$ is true at a possible world i iff C
holds throughout the selected set $f(A, i)$. Stalnaker formulates Analysis 1
this way, except that his $f(A, i)$ is the unique member of the selected set,
if such there be, instead of the set itself.

If we like, we can put the selection function into the object language;
but to do this without forgetting that counterfactuals are in general contingent, we must have recourse to *double indexing*. That is, we must think
of some special sentences as being true or false at a world i not absolutely,
but in relation to a world j. An ordinary sentence is true or false at i, as the

case may be, in relation to any j; it will be enough to deal with ordinary counterfactuals compounded out of ordinary sentences. Let $\not{f}A$ (where A is ordinary) be a special sentence true at j in relation to i iff j belongs to $f(A, i)$. Then $\not{f}A \supset C$ (where C is ordinary) is true at j in relation to i iff, if j belongs to $f(A, i)$, C holds at j. Then $\square(\not{f}A \supset C)$ is true at j in relation to i iff C holds at every world in $f(A, i)$ that is accessible from j. It is therefore true at i in relation to i itself iff C holds throughout (fA, i) – that is, iff $A \square\!\!\rightarrow C$ holds at i. Introducing an operator $\dagger$ such that $\dagger B$ is true at i in relation to j iff B is true at i in relation to i itself, we can define the counterfactual:

$$A \square\!\!\rightarrow C = ^{\mathrm{df}}\dagger\square(\not{f}A \supset C).$$

An $\not{f}$-operator without double indexing is discussed in Lennart Åqvist, 'Modal Logic with Subjunctive Conditionals and Dispositional Predicates', *Filosofiska Studier* (Uppsala) 1971; the $\dagger$-operator was introduced in Frank Vlach, ' "Now" and "Then" '(in preparation).

2.7. *Ternary Accessibility*

If we like, we can reparse counterfactuals as $[A \square\!\!\rightarrow]C$, regarding $\square\!\!\rightarrow$ now not as a two-place operator but rather as taking one sentence A to make a one-place operator $[A \square\!\!\rightarrow]$. If we have closest A-worlds – possible or impossible – whenever A is possible, then each $[A \square\!\!\rightarrow]$ is a necessity operator interpretable in the normal way by means of an accessibility relation. Call j *A-accessible* from i (or *accessible from i relative to A*) iff j is a closest (accessible) A-world from i; then $[A \square\!\!\rightarrow]C$ is true at i iff C holds at every world A-accessible from i. See Brian F. Chellas, 'Basic Conditional Logic' (in preparation).

3. FALLACIES

Some familiar argument-forms, valid for certain other conditionals, are invalid for my counterfactuals.

Transitivity	Contraposition	Strengthening	Importation
$A \square\!\!\rightarrow B$			
$B \square\!\!\rightarrow C$	$A \square\!\!\rightarrow C$	$A \square\!\!\rightarrow C$	$A \square\!\!\rightarrow (B \supset C)$
$A \square\!\!\rightarrow C$	$\bar{C} \square\!\!\rightarrow \bar{A}$	$AB \square\!\!\rightarrow C$	$AB \square\!\!\rightarrow C$

However, there are related valid argument-forms that may often serve as substitutes for these.

$A \,\square\!\!\rightarrow B$

$AB \,\square\!\!\rightarrow C$

───

$A \,\square\!\!\rightarrow C$

$\bar{C}$

$A \,\square\!\!\rightarrow C$

───

$\bar{C} \,\square\!\!\rightarrow \bar{A}$

$A \,\Diamond\!\!\rightarrow B$

$A \,\square\!\!\rightarrow C$

───

$AB \,\square\!\!\rightarrow C$

$A \,\Diamond\!\!\rightarrow B$

$A \,\square\!\!\rightarrow (B \supset C)$

───

$AB \,\square\!\!\rightarrow C$

Further valid substitutes for transitivity are these.

$A \,\square\!\!\rightarrow B$

$\square\,(B \supset C)$

───

$A \,\square\!\!\rightarrow C$

$B \,\square\!\!\rightarrow A$

$A \,\square\!\!\rightarrow B$

$B \,\square\!\!\rightarrow C$

───

$A \,\square\!\!\rightarrow C$

$B \,\Diamond\!\!\rightarrow A$

$A \,\square\!\!\rightarrow B$

$B \,\square\!\!\rightarrow C$

───

$A \,\square\!\!\rightarrow C$

4. TRUE ANTECEDENTS

On my analysis, a counterfactual is so called because it is suitable for non-trivial use when the antecedent is presumed false; not because it implies the falsity of the antecedent. It is conversationally inappropriate, of course, to use the counterfactual construction unless one supposes the antecedent false; but this defect is not a matter of truth conditions. Rather, it turns out that a counterfactual with a true antecedent is true iff the consequent is true, as if it were a material conditional. In other words, these two arguments are valid.

$$(-)\frac{A, \ \bar{C}}{-(A \,\square\!\!\rightarrow C)} \qquad (+)\frac{A, \ C}{A \,\square\!\!\rightarrow C}.$$

It is hard to study the truth conditions of counterfactuals with true antecedents. Their inappropriateness eclipses the question whether they are true. However, suppose that someone has unwittingly asserted a counterfactual $A \,\square\!\!\rightarrow C$ with (what you take to be) a true antecedent A. Either of these replies would, I think, sound cogent.

(−) Wrong, since in fact A and yet not C.

(+) Right, since in fact A and indeed C.

The two replies depend for their cogency – for the appropriateness of the word 'since' – on the validity of the corresponding arguments.

I confess that the case for (−) seems more compelling than the case for (+). One who wants to invalidate (+) while keeping (−) can do so if he is prepared to imagine that another world may sometimes be just as similar to a given world as that world is to itself. He thereby weakens the Centering Assumption to this: each world is self-accessible, and at least as close to itself as any other world is to it. Making that change and keeping everything else the same, (−) is valid but (+) is not.

5. COUNTERPOSSIBLES

If A is impossible, $A \;\square\!\!\rightarrow C$ is vacuously true regardless of the consequent C. Clearly some counterfactuals with impossible antecedents are asserted with confidence, and should therefore come out true: 'If there were a decision procedure for logic, there would be one for the halting problem'. Others are not asserted by reason of the irrelevance of antecedent to consequent: 'If there were a decision procedure for logic, there would be a sixth regular solid' or '... the war would be over by now'. But would these be confidently *denied*? I think not; so I am content to let all of them alike be true. Relevance is welcome in the theory of conversation (which I leave to others) but not in the theory of truth conditions.

If you do insist on making discriminations of truth value among counterfactuals with impossible antecedents, you might try to do this by extending the comparative similarity orderings of possible worlds to encompass also certain impossible worlds where not-too-blatantly impossible antecedents come true. (These are worse than the impossible limit-worlds already considered, where impossible but consistent infinite combinations of possibly true sentences come true.) See recent work on impossible-world semantics for doxastic logic and for relevant implication; especially Richard Routley, 'Ultra-Modal Propositional Functors' (in preparation).

6. POTENTIALITIES

'Had the Emperor not crossed the Rubicon, he would never have become Emperor' does *not* mean that the closest worlds to ours where there is a unique emperor and he did not cross the Rubicon are worlds where there is a unique emperor and he never became Emperor. Rather, it is *de re* with respect to 'the Emperor', and means that he who actually is (or was at the time under discussion) Emperor has a counterfactual property, or *potentiality,* expressed by the formula: 'if x had not crossed the Rubicon, x

would never have become Emperor'. We speak of what would have befallen the actual Emperor, not of what would have befallen whoever would have been Emperor. Such potentialities may also appear when we quantify into counterfactuals: 'Any Emperor who would never have become Emperor had he not crossed the Rubicon ends up wishing he hadn't done it' or 'Any of these matches would light if it were scratched'. We need to know what it is for something to have a potentiality – that is, to satisfy a counterfactual formula $A(x) \square \rightarrow C(x)$.

As a first approximation, we might say that something x satisfies the formula $A(x) \square \rightarrow C(x)$ at a world i iff some (accessible) world where x satisfies $A(x)$ and $C(x)$ is closer to i than any world where x satisfies $A(x)$ and $\bar{C}(x)$, if there are (accessible) worlds where x satisfies $A(x)$.

The trouble is that this depends on the assumption that one and the same thing can exist – can be available to satisfy formulas – at various worlds. I reject this assumption, except in the case of certain abstract entities that inhabit no particular world, and think it better to say that concrete things are confined each to its own single world. He who actually is Emperor belongs to our world alone, and is not available to cross the Rubicon or not, become Emperor or not, or do anything else at any other world. But although he himself is not present elsewhere, he may have *counterparts* elsewhere: inhabitants of other worlds who resemble him closely, and more closely than do the other inhabitants of the same world. What he cannot do in person at other worlds he may do vicariously, through his counterparts there. So, for instance, I might have been a Republican not because I myself am a Republican at some other world than this – I am not – but because I have Republican counterparts at some worlds. See my 'Counterpart Theory and Quantified Modal Logic', *Journal of Philosophy* 1968.

Using the method of counterparts, we may say that something x satisfies the formula $A(x) \square \rightarrow C(x)$ at a world i iff some (accessible) world where some counterpart of x satisfies $A(x)$ and $C(x)$ is closer to i than any world where any counterpart of x satisfies $A(x)$ and $\bar{C}(x)$, if there are (accessible) worlds where a counterpart of x satisfies $A(x)$. This works also for abstract entities that inhabit no particular world but exist equally at all, if we say that for these things the counterpart relation is simply identity.

A complication: it seems that when we deal with relations expressed by counterfactual formulas with more than one free variable, we may need to mix different counterpart relations. 'It I were you I'd give up' seems to mean that some world where a character-counterpart of me is a predicament-counterpart of you and gives up is closer than any where a character-counterpart of me is a predicament-counterpart of you and does not give up. (I omit provision for vacuity and for accessibility restrictions.) The difference between Goodman's sentences

(1)   If New York City were in Georgia, New York City would be in the South.

(2)   If Georgia included New York City, Georgia would not be entirely in the South.

may be explained by the hypothesis that both are *de re* with respect to both 'New York City' and 'Georgia', and that a less stringent counterpart relation is used for the subject terms 'New York City' in (1) and 'Georgia' in (2) than for the object terms 'Georgia' in (1) and 'New York City' in (2). I cannot say in general how grammar and context control which counterpart relation is used where.

An independent complication: since closeness of worlds and counterpart relations among their inhabitants are alike matters of comparative similarity, the two are interdependent. At a world close to ours, the inhabitants of our world will mostly have close counterparts; at a world very different from ours, nothing can be a very close counterpart of anything at our world. We might therefore wish to fuse closeness of worlds and closeness of counterparts, allowing these to balance off. Working with comparative similarity among *pairs* of a concrete thing and the world it inhabits (and ignoring provision for vacuity and for accessibility restrictions), we could say that an inhabitant x of a world i satisfies $A(x) \square \rightarrow C(x)$ at i iff some such thing-world pair $\langle y, j \rangle$ such that y satisfies $A(x)$ and $C(x)$ at j is more similar to the pair $\langle x, i \rangle$ than is any pair $\langle z, k \rangle$ such that z satisfies $A(x)$ and $\bar{C}(x)$ at k. To combine this complication and the previous one seems laborious but routine.

7. COUNTERCOMPARATIVES

'If my yacht were longer than it is, I would be happier than I am' might be handled by quantifying into a counterfactual formula: $\exists x, y$ (my yacht is

x feet long & I enjoy y hedons & (my yacht is more than x feet long $\square\!\!\rightarrow$ I enjoy more than y hedons)). But sometimes, perhaps in this very example, comparison makes sense when numerical measurement does not. An alternative treatment of countercomparatives is available using double indexing. (Double indexing has already been mentioned in connection with the f-operator; but if we wanted it both for that purpose and for this, we would need triple indexing.) Let A be true at j in relation to i iff my yacht is longer at j than at i (more precisely: if my counterpart at j has a longer yacht than my counterpart at i (to be still more precise, decide what to do when there are multiple counterparts or multiple yachts)); let C be true at j in relation to i iff I am happier at j than at i (more precisely: if my counterpart...). Then $A\,\square\!\!\rightarrow C$ is true at j in relation to i iff some world (accessible from j) where A and C both hold in relation to i is closer to j than any world where A and $\bar{C}$ both hold in relation to i. So far, the relativity to i just tags along. Our countercomparative is therefore true at i (in relation to any world) iff $A\,\square\!\!\rightarrow C$ is true at i in relation to i itself. It is therefore $\dagger(A\,\square\!\!\rightarrow C)$.

8. COUNTERFACTUAL PROBABILITY

'The probability that C, if it were the case that A, would be r' cannot be understood to mean any of:

(1) $\quad$ Prob $(A\,\square\!\!\rightarrow C) = r$,
(2) $\quad$ Prob $(C\mid A) = r$, or
(3) $\quad A\,\square\!\!\rightarrow \text{Prob}(C) = r$.

Rather, it is true at a world i (with respect to a given probability measure) iff for any positive ε there exists an A-permitting sphere T around i such that for any A-permitting sphere S around i within T, $\text{Prob}(C\mid AS)$, unless undefined, is within ε of r.

Example. A is 'The sample contained abracadabrene', C is 'The test for abracadabrene was positive', Prob is my present subjective probability measure after watching the test come out negative and tentatively concluding that abracadabrene was absent. I consider that the probability of a positive result, had abracadabrene been present, would have been 97%. (1) I know that false negatives occur because of the inherently indeterministic character of the radioactive decay of the tracer used in the test, so I am convinced that no matter what the actual conditions were, there might have been a false negative even if abracadabrene had been present. $\text{Prob}(A \Diamond \to \bar{C}) \approx 1$; $\text{Prob}(A \Box \to C) \approx 0$. (2) Having seen that the test was negative, I disbelieve C much more strongly than I disbelieve A; $\text{Prob}(AC)$ is much less than $\text{Prob}(A)$; $\text{Prob}(C \mid A) \approx 0$. (3) Unknown to me, the sample was from my own blood, and abracadabrene is a powerful hallucinogen that makes white things look purple. Positive tests are white, negatives are purple. So had abracadabrene been present, I would have strongly disbelieved C no matter what the outcome of the test really was. $A \Box \to \text{Prob}(C) \approx 0$. (Taking (3) *de re* with respect to 'Prob' is just as bad: since actually $\text{Prob}(C) \approx 0$, $A \Box \to \text{Prob}(C) \approx 0$ also.) My suggested definition seems to work, however, provided that the outcome of the test at a close A-world does not influence the closeness of that world to ours.

9. ANALOGIES

The counterfactual as I have analyzed it is parallel in its semantics to operators in other branches of intensional logic, based on other comparative relations. There is one difference: in the case of these analogous operators, it seems best to omit the provision for vacuous truth. They correspond to a doctored counterfactual $\Box \Rightarrow$ that is automatically false instead of automatically true when the antecedent is impossible: $A \Box \Rightarrow C = ^{\text{df}} \Diamond A \mathbin{\&} (A \Box \to C)$.

Deontic: We have the operator $A \Box \Rightarrow_d C$, read as 'Given that A, it ought to be that C', true at a world i iff some AC-world evaluable from the standpoint of i is better, from the standpoint of i, than any $A\bar{C}$-world. Roughly (under a Limit Assumption), iff C holds at the best A-worlds. See the operator of 'conditional obligation' discussed in Bengt Hansson, 'An Analysis of Some Deontic Logics', *Noûs* 1969.

Temporal: We have $A \Box \Rightarrow_f C$, read as 'When next A, it will be that C', true at a time t iff some AC-time after t comes sooner after t than any $A\bar{C}$-time; roughly, iff C holds at the next A-time. We have also the past mirror image: $A \Box \Rightarrow_p C$, read as 'When last A, it was that C'.

Egocentric (in the sense of A. N. Prior, 'Egocentric Logic', *Noûs* 1968): We have $A \Box \Rightarrow_e C$, read as 'The A is C', true for a thing x iff some AC-thing in x's ken is more salient to x than any $A\bar{C}$-thing; roughly, iff the most salient A-thing is C.

To motivate the given truth conditions, we may note that these operators all permit sequences of truths of the two forms:

$$A \, \Box\!\Rightarrow \bar{B}, \qquad\qquad A \, \Box\!\Rightarrow Z,$$
$$AB \, \Box\!\Rightarrow \bar{C}, \quad \text{and} \quad AB \, \Box\!\Rightarrow Z,$$
$$ABC \, \Box\!\Rightarrow \bar{D}, \qquad\qquad ABC \, \Box\!\Rightarrow Z,$$
$$\text{etc.;} \qquad\qquad\qquad \text{etc.}$$

It is such sequences that led us to treat the counterfactual as a variably strict conditional. The analogous operators here are likewise variably strict conditionals. Each is based on a binary relation and a family of comparative relations in just the way that the (doctored) counterfactual is based on accessibility and the family of comparative similarity orderings. In each case, the Ordering Assumption holds. The Centering Assumption, however, holds only in the counterfactual case. New assumptions hold in some of the other cases.

In the deontic case, we may or may not have different comparative orderings from the standpoint of different worlds. If we evaluate worlds according to their conformity to the edicts of the god who reigns at a given world, then we will get different orderings; and no worlds will be evaluable from the standpoint of a godless world. If rather we evaluate worlds according to their total yield of hedons, then evaluability and comparative goodness of worlds will be absolute.

In the temporal case, both the binary relation and the families of comparative relations, both for 'when next' and for 'when last', are based on the single underlying linear order of time.

The sentence $(A \vee \bar{B}) \Box\!\Rightarrow_f AB$ is true at time t iff some A-time after t precedes any $\bar{B}$-time after t. It thus approximates the sentence 'Until A, B', understood as being true at t iff some A-time after t is not preceded by any $\bar{B}$-time after t. Likewise $(A \vee \bar{B}) \Box\!\Rightarrow_p AB$ approximates 'Since A, B', with 'since' understood as the past mirror image of 'until'. Hans Kamp has shown that 'since' and 'until' suffice to define all possible tense operators, provided that the order of time is a complete linear order; see his *Tense Logic and the Theory of Order* (U.C.L.A. dissertation, 1968). Do my approximations have the same power? No; consider 'Until $\top$, $\bot$', true at t iff there is a next moment after t. This sentence cannot be translated using my operators. For if the order of time is a complete linear order with discrete stretches and dense stretches, then the given sentence will vary in truth value; but if in addition there is no beginning or end of time, and if there are no atomic sentences that vary in truth value, then no sentences that vary in truth value can be built up by means of truth-functional connectives, $\Box\Rightarrow_f$, and $\Box\Rightarrow_p$.

Starting from any of our various $\Box\Rightarrow$-operators, we can introduce one-place operators I shall call the *inner modalities*:

$$\boxdot A =^{\mathrm{df}} \mathbf{T} \,\Box\Rightarrow A,$$
$$\diamondsuit A =^{\mathrm{df}} -\boxdot-A,$$

and likewise in the analogous cases. The inner modalities in the counter-factual case are of no interest (unless Centering is weakened), since $\boxdot A$ and $\diamondsuit A$ are both equivalent to A itself. Nor are they anything noteworthy in the egocentric case. In the deontic case, however, they turn out to be slightly improved versions of the usual so-called obligation and permission operators. $\boxdot_d A$ is true at i iff some (evaluable) A-world is better, from the standpoint of i, than any $\bar{A}$-world; that is, iff either (1) there are best (evaluable) worlds, and A holds throughout them, or (2) there are better and better (evaluable) worlds without end, and A holds throughout all sufficiently good ones. In the temporal case, $\boxdot_f A$ is true at t iff some A-time after t comes sooner than any $\bar{A}$-time; that is, iff either (1) there is a next moment, and A holds then, or (2) there is no next moment, and A holds throughout some interval beginning immediately and extending into the future. $\boxdot_f A$ may thus be read 'Immediately, A'; as may $\diamondsuit_f A$, but in a somewhat different sense.

If no worlds are evaluable from the standpoint of a given world – say, because no god reigns there – it turns out that $\boxdot_d A$ is false and $\diamondsuit_d A$ is true for any A whatever. Nothing is obligatory, everything is permitted. Similarly for $\boxdot_f A$ and $\diamondsuit_f A$ at the end of time, if such there be; and for $\boxdot_p A$ and $\diamondsuit_p A$ at its beginning. Modalities that behave in this way are called *abnormal*, and it is interesting to find these moderately natural examples of abnormality.

10. Axiomatics

The set of all sentences valid under my analysis may be axiomatised taking the counterfactual connective as primitive. One such axiom system – not the neatest – is the system **C1** of my paper 'Completeness and Decidability of Three Logics of Counterfactual Conditionals', *Theoria* 1971, essentially as follows.

Rules:

> If A and $A \supset B$ are theorems, so is B.
> If $(B_1 \ \& \ \cdots) \supset C$ is a theorem, so is
> $$((A \ \Box\!\!\rightarrow B_1) \ \& \ \cdots) \supset (A \ \Box\!\!\rightarrow C).$$

Axioms:

> All truth-functional tautologies are axioms.
> $A \ \Box\!\!\rightarrow A$
> $(A \ \Box\!\!\rightarrow B) \ \& \ (B \ \Box\!\!\rightarrow A) . \supset . (A \ \Box\!\!\rightarrow C) \equiv (B \ \Box\!\!\rightarrow C)$
> $((A \lor B) \ \Box\!\!\rightarrow A) \lor ((A \lor B) \ \Box\!\!\rightarrow B) \lor (((A \lor B) \ \Box\!\!\rightarrow C) \equiv$
> $$(A \ \Box\!\!\rightarrow C) \ \& \ (B \ \Box\!\!\rightarrow C))$$
> $A \ \Box\!\!\rightarrow B . \supset . A \supset B$
> $AB \supset . A \ \Box\!\!\rightarrow B$

(Rules and axioms here and henceforth should be taken as schematic.) Recall that modalities and comparative possibility may be introduced via the following definitions: $\Box A =^{df} \bar{A} \Box\!\!\rightarrow \perp$; $\Diamond A =^{df} -\Box -A$; $A \prec B =^{df} \Diamond A \ \& \ ((A \lor B) \Box\!\!\rightarrow A\bar{B})$.

A more intuitive axiom system, called **VC**, is obtained if we take comparative possibility instead of the counterfactual as primitive. Let $A \preccurlyeq B =^{df} -(B \prec A)$.

Rules:

> If A and $A \supset B$ are theorems, so is B.
> If $A \supset B$ is a theorem, so is $B \preccurlyeq A$.

Basic Axioms:

> All truth-functional tautologies are basic axioms.
> $A \preccurlyeq B \preccurlyeq C . \supset . A \preccurlyeq C$
> $A \preccurlyeq B . \lor . B \preccurlyeq A$
> $A \preccurlyeq (A \lor B) . \lor . B \preccurlyeq (A \lor B)$

Axiom **C**:

> $A\bar{B} \supset . A \prec B$

Recall that modalities and the counterfactual may be introduced via the following definitions: $\Diamond A =^{\mathrm{df}} A \prec \perp$; $\Box A =^{\mathrm{df}} - \Diamond - A$; $A \Box\!\!\rightarrow C =^{\mathrm{df}}$ $\Diamond A \supset (AC \prec A\bar{C})$.

VC and **C1** turn out to be definitionally equivalent. That is, their respective definitional extensions (via the indicated definitions) yield exactly the same theorems. It may now be verified that these theorems are exactly the ones we ought to have. Since the definitions are correct (under my truth conditions) it is sufficient to consider sentences in the primitive notation of **VC**.

In general, we may define a *model* as any quadruple $\langle I, R, \leqslant, [\![\ ]\!] \rangle$ such that

(1) I is a nonempty set (regarded as playing the role of the set of worlds);

(2) R is a binary relation over I (regarded as the accessibility relation);

(3) $\leqslant$ assigns to each i in I a weak ordering $\leqslant_i$ of I (regarded as the comparative similarity ordering of worlds from the standpoint of i) such that whenever $j \leqslant_i k$, if iRk then iRj;

(4) $[\![\ ]\!]$ assigns to each sentence A a subset $[\![A]\!]$ of I (regarded as the set of worlds where A is true);

(5) $[\![-A]\!]$ is $I - [\![A]\!]$, $[\![A \& B]\!]$ is $[\![A]\!] \cap [\![B]\!]$, and so on;

(6) $[\![A \prec B]\!]$ is $\{i\varepsilon I$: for some j in $[\![A]\!]$ such that iRj, there is no k in $[\![B]\!]$ such that $k \leqslant_i j\}$.

The *intended models*, for the counterfactual case, are those in which I, R, $\leqslant$, and $[\![\ ]\!]$ really are what we regarded them as being: the set of worlds, some reasonable accessibility relation, some reasonable family of comparative similarity orderings, and an appropriate assignment to sentences of truth sets. The Ordering Assumption has been written into the very definition of a model (clause 3) since it is common to the counterfactual case and the analogous cases as well. As for the Centering Assumption, we must impose it on the intended models as a further condition:

(C) R is reflexive on I: and $j \leqslant_i i$ only if $j = i$.

It seems impossible to impose other purely mathematical conditions on the intended models (with the possible exception of (U), discussed below). We therefore hope that **VC** yields as theorems exactly the sentences valid – true at all worlds – in all models that meet condition (C). This is the case.

VC is sound for models meeting (C); for the basic axioms are valid, and the rules preserve validity, in all models; and Axiom **C** is valid in any model meeting (C).

VC is complete for models meeting (C): for there is a certain such model in which only theorems of **VC** are valid. This model is called the *canonical model* for **VC**, and is as follows:

(1) I is the set of all maximal **VC**-consistent sets of sentences;

(2) iRj iff, for every sentence A in j, $\Diamond A$ is in i;

(3) $j \leqslant_i k$ iff there is no set Σ of sentences that overlaps j but not k, such that whenever $A \leqslant B$ is in i and A is in Σ then B also is in Σ;

(4) i is in $[\![A]\!]$ iff A is in i.

In the same way, we can prove that the system consisting of the rules, the basic axioms, and *any* combination of the axioms listed below is sound and complete for models meeting the corresponding combination of conditions. Nomenclature: the system generated by the rules, the basic axioms, and the listed axioms — is called **V—**. (Note that the conditions are not independent. (C) implies (W), which implies (T), which implies (N). (S) implies (L). (A—) implies (U—). (W) and (S) together imply (C). (C) and (A—) together imply (S) by implying the stronger, trivializing condition that no world is accessible from any other. Accordingly, many combinations of the listed axioms are redundant.)

Axioms

N: $\Box^{\mathsf{T}}$

T: $\Box A \supset A$

W: $AB \supset . \Diamond A \; \& \; A \leqslant B$

C: $A\bar{B} \supset A \prec B$

L: (no further axiom, or some tautology)

S: $A \,\Box\!\!\rightarrow C . \vee . A \,\Box\!\!\rightarrow \bar{C}$

U: $\Box A \supset \Box\Box A$ and $\Diamond A \supset \Box\Diamond A$

A: $A \leqslant B \supset \Box(A \leqslant B)$ and $A \prec B \supset \Box(A \prec B)$.

Conditions

(N)   (normality): For any i in I there is some j in i such that iRj.

(T)   (total reflexivity): R is reflexive on I.

(W)   (weak centering): R is reflexive on I; for any i and j in I, $i \leqslant_i j$.

(C)   (centering): R is reflexive on I; and $j \leqslant_i i$ only if $j = i$.

(L)   (Limit Assumption): Whenever iRj for some j in $[\![A]\!]$, $[\![A]\!]$ has at least one $\leqslant_i$-minimal element.

(S)   (Stalnaker's Assumption): Whenever iRj for some j in $[\![A]\!]$, $[\![A]\!]$ has exactly one $\leqslant_i$-minimal element.

(U−)   (local uniformity): If iRj, then jRk iff iRk.

(A−)   (local absoluteness): If iRj, then jRk iff iRk and $h \leqslant_j k$ iff $h \leqslant_i k$.

The Limit Assumption (L) corresponds to no special axiom. Any one of our systems is sound and complete both for a combination of conditions without (L) and for that combination plus (L). The reason is that our canonical models always are rich enough to satisfy the Limit Assumption, but our axioms are sound without it. (Except **S**, for which the issue does not arise because (S) implies (L).) Moral: the Limit Assumption is irrelevant to the logical properties of the counterfactual. Had our interest been confined to logic, we might as well have stopped with Analysis 2.

Omitting redundant combinations of axioms, we have the 26 distinct systems shown in the diagram.

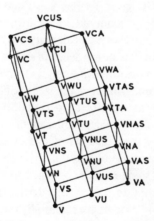

The general soundness and completeness result still holds if we replace the local conditions (U−) and (A−) by the stronger global conditions (U) and (A).

(U)   (uniformity): For any i, j, k in I, jRk iff iRk.
(A)   (absoluteness): For any h, i, j, k in I, jRk iff iRk and $h \leqslant_j k$ iff $h \leqslant_i k$.

Any model meeting $(U-)$ or $(A-)$ can be divided up into models meeting (U) or (A). The other listed conditions hold in the models produced by the division if they held in the original model. Therefore a sentence is valid under a combination of conditions including (U) or (A) iff it is valid under the combination that results from weakening (U) to $(U-)$, or (A) to $(A-)$.

In the presence of (C), (W), or (T), condition (U) is equivalent to the condition: for any i and j in I, iRj. **VCU** is thus the correct system to use if we want to drop accessibility restrictions. **VW**, or perhaps **VWU**, is the correct system for anyone who wants to invalidate the implication from A and C to $A \Box \rightarrow C$ by allowing that another world might be just as close to a given world as that world is to itself. **VCS**, or **VCUS** if we drop accessibility restrictions, is the system corresponding to Analysis 1 or $1\frac{1}{2}$. **VCS** is definitionally equivalent to Stalnaker's system **C2**.

The systems given by various combinations of **N**, **T**, **U**, and **A** apply, under various assumptions, to the deontic case. **VN** is definitionally equivalent to a system **CD** given by Bas van Fraassen in 'The Logic of Conditional Obligation' (forthcoming), and shown there to be sound and complete for the class of what we may call *multi-positional models* meeting (N). These differ from models in my sense in that a world may occur at more than one position in an ordering $\leqslant_i$. (Motivation: different positions may be assigned to one world *qua* realizer of different kinds of value.) Technically, we no longer have a direct ordering of the worlds themselves; rather, we have for each i in I a linear ordering of some set V_i and an assignment to each world j such that iRj of one or more members of V_i, regarded as giving the positions of j in the ordering from the standpoint of i. $A \prec B$ is true at i iff some position assigned to some A-world j (such that iRj) is better according to the given ordering than any position assigned to any B-world. My models are essentially the same as those multi-positional models in which no world does have more than one assigned position in any of the orderings. Hence **CO** is at least as strong as **VN**; but no stronger, since **VN** is already sound for all multi-positional models meeting (N).

All the systems are decidable. To decide whether a given sentence A is a theorem of a given system, it is enough to decide whether the validity of A under the corresponding combination of conditions can be refuted by a *small* countermodel – one with at most 2^n worlds, where n is the number of subsentences of A. (Take (U) and (A), rather than (U−) and (A−), as the conditions corresponding to U and A.) That can be decided by examining finitely many cases, since it is unnecessary to consider two models separately if they are isomorphic, or if they have the same I, R, $\leqslant$, and the same $[\![P]\!]$ whenever P is a sentence letter of A. If A is a theorem, then by soundness there is no countermodel and *a fortiori* no small countermodel. If A is not a theorem, then by completeness there is a countermodel $\langle I, R, \leqslant, [\![\ ]\!]\rangle$. We derive thence a small countermodel, called a *filtration* of the original countermodel, as follows. Let D_i, for each i in I, be the conjunction in some definite arbitrary order of all the subsentences of A that are true at i in the original countermodel, together with the negations of all the subsentences of A that are false at i in the original countermodel. Now let $\langle I^*, R^*, \leqslant^*, [\![\ ]\!]^*\rangle$ be as follows:

(1) I^* is a subset of I containing exactly one member of each nonempty $[\![D_i]\!]$;

(2)   for any i and j in I^*, iR^*j iff i is in $[\![\Diamond D_j]\!]$;

(3)   for any i, j, k in I^*, $j \leqslant^*_i k$ iff i is in $[\![D_j \leqslant D_k]\!]$;

(4)   for any sentence letter P, $[\![P]\!]^*$ is $[\![P]\!] \cap I^*$; for any compound sentence B, $[\![B]\!]^*$ is such that $\langle I^*, R^*, \leqslant^*, [\![\ ]\!]^*\rangle$ meets conditions (5) and (6) in the definition of a model.

Then it may easily be shown that $\langle I^*, R^*, \leqslant^*, [\![\ ]\!]^*\rangle$ is a small countermodel to the validity of A under the appropriate combination of conditions, and thereby to the theoremhood of A in the given system.

Princeton University

NOTE

* The theory presented in this paper is discussed more fully in my book *Counterfactuals* (Blackwell and Harvard University Press). My research on counterfactuals was supported by a fellowship from the American Council of Learned Societies.

ROBERT C. STALNAKER

A DEFENSE OF CONDITIONAL EXCLUDED MIDDLE[*]

This paper is a polemic about a detail in the semantics for conditionals. It takes for granted what is common to semantic theories proposed by David Lewis,[1] John Pollock,[2] Brian Chellas,[3] and myself and Richmond Thomason[4] in order to focus on some small points of difference between the theory I favor and the others. I will sketch quickly and roughly the general ideas which lie behind all of these theories, and the common semantical framework in which these ideas are developed. Then I will describe the divergences between my theory and the others – I will focus on the difference between my theory and the one favored by Lewis – and argue that my theory gives a better account of the way conditionals work in natural language.

The differences between the theories I will be comparing may seem small and unimportant. Does it really matter very much whether we conclude that conditionals like *If Bizet and Verdi had been compatriots, Bizet would have been Italian* are false or (as I will suggest) neither true nor false? This judgment may not be important in itself, but it is not an isolated judgment. Conditionals interact with negation, quantifiers, modal auxiliaries like *may* and *might*, adverbs like *even*, *only* and *probably*. Small differences among analyses of conditionals may have consequences for many complex constructions involving conditionals. A small distortion in the analysis of the conditional may create spurious problems with the analysis of other concepts. So if the facts about usage favor one among a number of subtly different theories, it may be important to determine which one it is.

All of the semantic analyses of conditional logic that I have in mind are given within the possible worlds framework. All begin with the general idea that a counterfactual conditional is true in the actual world if and only if the consequent is true in some possibly different possible world or worlds. The world or set of worlds in which the consequent is said to be true is determined by the antecedent. These must be possible worlds in which the antecedent is true, and which are otherwise minimally different from the actual world. To make sense of this idea, the semantic theory needs a semantic determinant which selects the minimally different world or worlds, or which orders the possible worlds with respect to their comparative similarity

W. L. Harper, R. Stalnaker, and G. Pearce (eds.), Ifs, 87–104

to the actual world. Truth conditions for conditionals are given relative to such a semantic determinant.

The semantic determinant for the account I prefer is a world selection function: a function f which takes a proposition and a possible world into a possible world.[5] A conditional, if A *then* B, is then said to be true in a world i if and only if B is true in $f(A, i)$ – the possible world which is the value of the function for arguments A and i. Various constraints are placed on the selection function, constraints which are motivated by the intuitive idea that the nearest, or least different, world in which antecedent is true is the one that should be selected. For example, it is required that the world selected relative to proposition A be an A-world – a possible world in which A is true $(f(A, i) \in A)$. And if the actual world meets this condition, it is required that it be selected. (If $i \in A$, then $f(A, i) = i$).

David Lewis's theory of conditionals is formulated in terms of a different semantic determinant. It states truth conditions for conditionals in terms of a three place comparative similarity relation instead of a selection function. Let $C_i(j, k)$ mean that j is more similar to i than k is to i. For any fixed i, the relation is assumed to be transitive and connected, and so to determine a weak total ordering of all possible worlds with respect to each possible world. A counterfactual, *if A, then B*, is then said by Lewis's theory to be true if and only if there is an A-world j such that B is true in it, and in all A-worlds which are at least as similar to i as j.[6]

Part of the difference between the two theories of conditionals that I have sketched is superficial. The theory I favor could have been formulated in terms of a comparative similarity relation instead of a selection function, and a comparative similarity relation is definable in terms of the selection function as follows: $C_i(j, k)$ if and only if for some proposition A such that both j and k are members of A, $f(A, i) = j$. It can be shown, first, that this defined relation meets all the conditions Lewis imposes on his comparative similarity relation, and second, that Lewis's truth conditions, applied to the defined relation, coincide with the truth conditions given in my theory. The theories are not equivalent since the defined comparative similarity relation necessarily has properties beyond those imposed by Lewis's theory; specifically, the comparative similarity relation defined in terms of a selection function determines not just a weak total ordering, but a well ordering of all possible worlds with respect to each possible world. So my theory, formulated in terms of a comparative similarity relation, is a special case of Lewis's.

Lewis identifies two assumptions about the comparative similarity relation which my theory makes and this does not: he calls them the limit assumption and the uniqueness assumption. The first is the assumption that for every possible world i and non-empty proposition A, there is at least one A-world minimally different from i. The second is the assumption that for every world i and proposition A there is at most one A-world minimally different from i. Lewis's theory, with the addition of these two assumptions, is essentially equivalent to mine.

Each of the two assumptions about the comparative similarity relation corresponds to an entailment principle in the semantics for conditionals. To accept the limit assumption is to accept the following consequence condition for conditionals: for any class of propositions Γ and propositions A and C, if Γ semantically entails C, then $\{A > B : B \in \Gamma\}$ semantically entails $A > C$. To accept in addition the uniqueness assumption is to accept the validity of the principle of conditional excluded middle:

$$\Vdash (A > C) \vee (A > {\sim}C).$$

Lewis argues that it is not reasonable to make the two assumptions which distinguish his theory from mine. I will argue that one of the assumptions is reasonable to make, and that the other need not be made in application. I will also discuss a number of examples which I think tend to show that the analysis I have proposed gives a better account of the phenomena.

Let me look first at the uniqueness assumption. This is the assumption which rules out ties in similarity. It says that no distinct possible worlds are ever equally similar to any given possible world. That is, without a doubt, a grossly implausible assumption to make about the kind of similarity relation we use to interpret conditionals, and it is an assumption which the abstract semantic theory that I want to defend does make. But like many idealized assumptions made in abstract semantic theory, it may be relaxed in the application of the theory. In general, to apply a semantic theory to the interpretation of language as it is used, one need not assume that every semantic determinant is completely and precisely defined. In application, domains of individuals relative to which quantifiers are interpreted, sets of possible worlds relative to which modal auxiliaries are interpreted, propositional functions used to interpret predicates all may admit borderline cases even though the abstract semantic theory assumes well defined sets with sharp boundaries. To reconcile the determinacy of abstract semantic theory with the indeterminacy of realistic application, we need a general theory of vagueness. But given such a theory, we can reconcile the uniqueness assumption, as an assumption of the abstract semantics for conditionals, with the

fact that it is unrealistic to assume that our conceptual resources are capable
of well ordering the possible worlds.

The theory of vagueness that I will recommend is the theory of supervaluations first developed by Bas Van Fraassen.[7] The main idea of this theory
is this: any partially defined semantic interpretation will correspond to a class
of completely defined interpretations – the class of all ways of arbitrarily
completing it. For example, a partial ordering will correspond to a class of
total orderings, and a domain with fuzzy borders will correspond to a class
of domains with sharp borders. The theory of supervaluations defines the
truth values assigned by partial interpretations in terms of the corresponding
class of complete, two-valued, *classical valuations*. In this way, it explains the
values under partial interpretations in terms of the kind of valuations assumed
by idealized abstract semantic theories. A sentence is *true* according to a
supervaluation if and only if it is true on *all* corresponding classical valuations,
false if and only if it is false on *all* corresponding classical valuations and
neither true nor false it it is true on some of the classical valuations and false
on others.

Using the method of supervaluations, we may acknowledge, without
modifying the abstract semantic theory of conditionals, that the selection
functions that are actually used in making and interpreting counterfactual
conditional statements correspond to orderings of possible worlds that admit
ties and incomparabilities. In doing this, we are not resorting to an *ad hoc*
device to save a theory, since the method of supervaluations, or some account
of semantic indeterminacy, is necessary anyway to account for pervasive
semantic underdetermination in natural language. Whatever theory of conditionals one favors, one must admit that vagueness is particularly prevalent
in the use of conditional sentences.[8]

What effect does the recognition of indeterminacy by the introduction of
supervaluations have on the *logic* of conditionals? None at all: it is one of the
virtues of this method of treating semantic indeterminacy that it leaves
classical two-valued logic virtually untouched.[9] Classical logical truths are
true in *all* classical valuations, and so will be true in all classical valuations
defined by any partial interpretation. Therefore, they will be true in all
supervaluations. Also, since classical valuations are themselves special cases
of supervaluations, any sentence true in all supervaluations will be true on all
classical valuations. So, whatever the details of the particular classical semantic theory, the concept of logical truth defined by it will not be changed by
the introduction of supervaluations.

For example, in the conditional logic C2 (the logic of the theory I am defending), the principle of conditional excluded middle, $(A > B) \vee (A > {\sim}B)$, remains valid when supervaluations are added, even though there may be cases where neither $(A > B)$ nor $(A > {\sim}B)$ is true. It may be that neither disjunct is made true by every arbitrary extension of a given partial interpretation, but it will always be that each arbitrary extension makes true one disjunct or the other. The theory of supervaluations, applied to this logic of conditionals, gives the principle of conditional excluded middle the same status as it gives the simple principle of excluded middle. $(B \vee {\sim}B)$ is logically true even though sometimes neither B nor ${\sim}B$ is true.

My aim so far, in this defense of my analysis of conditionals against its close relatives, has been simply to neutralize one important objection to the analysis: that it makes an implausible assumption about our conceptual resources, the assumption that we need a well ordering of all possible worlds with respect to each possible world in order to interpret conditional statements. I have argued that in the context of a general recognition of semantic indeterminacy, the dispute over the uniqueness assumption should be regarded not as a dispute about how much and what kind of structure there is in the actual contextual parameter we use to interpret conditionals, but rather a dispute about what degree and kind of structure that parameter is aiming at: about what would count as a determinate complete interpretation. In practice, what the issue comes down to is a disagreement about whether certain counterfactual conditionals are false or neither true nor false, and about whether certain inferences involving conditionals are valid.

Before looking at some examples of inferences and judgments which I think support the analysis I have proposed, I should point out, as Lewis does, that the limit assumption cannot be neutralized by the introduction of supervaluations in the same way as the uniqueness assumption. In my defense of the principle of conditional excluded middle, I shall take for granted that this assumption is a reasonable assumption to make. Later, I will explain and defend this decision.

Let us look at some examples. I will begin with a familiar pair of counterfactual conditionals first discussed in 1950 by W. V. Quine:[11]

> If Bizet and Verdi had been compatriots, Bizet would have been Italian.
>
> If Bizet and Verdi had been compatriots, Verdi would have been French.

These examples have been taken, in the context of possible worlds analyses of

conditionals, to illustrate the possibility of virtual ties in closeness of counterfactual possible worlds to the actual world. Worlds in which Bizet and Verdi
are both French or both Italian, it seems plausible to assume, are more like
the actual world than worlds in which both are Argentinian or Japanese. But
there is no apparent reason to favor a world in which both are French over
one in which both are Italian, or vice versa. This seems right; it would be
arbitrary to require a choice of one of the above counterfactuals over the
other, but as we have seen, this is not at issue. What is at issue is what conclusion about the truth values of the counterfactuals should be drawn from
the fact that such a choice would be arbitrary. On Lewis's and Pollock's
analyses, both counterfactuals are false. On the analysis I am defending, both
are indeterminate – neither true nor false. It seems to me that the latter conclusion is clearly the more natural one. I think most speakers would be as
hesitant to deny as to affirm either of the conditionals, and it seems as clear
that one cannot deny them both as it is that one cannot affirm them both.
Lewis seems to agree that unreflective linguistic intuition favors this conclusion. He writes:

> Given Conditional Excluded Middle, we cannot truly say such things as this:
>
> *It is not the case that if Bizet and Verdi were compatriots, Bizet would be Italian; and it
> is not the case that if Bizet and Verdi were compatriots, Bizet would not be Italian;
> nevertheless, if Bizet and Verdi were compatriots, Bizet either would or would not be
> Italian . . .*
>
> I want to say this, and think it is probably true; my own theory was designed to make it
> true. But offhand, I must admit, it does sound like a contradiction. Stalnaker's theory
> does, and mine does not, respect the opinion of any ordinary language speaker who cares
> to insist that it is a contradiction.[12]

Lewis goes on to say that the cost of respecting this 'offhand opinion' is too
great, but as I have argued, the introduction of supervaluations avoid the need
to pay the main cost that he has in mind.

Quine originally presented this example, not to defend one analysis of
counterfactuals against another, but to create doubt about the possibility of
any acceptable analysis. "It may be wondered, indeed," he writes introducing
the two Bizet–Verdi counterfactuals, "whether any really coherent theory of
the contrafactual conditional of ordinary usage is possible at all, particularly
when we imagine trying to adjudicate between such examples as these."[13]
There is a problem, Quine suggests, because we are required to *adjudicate*
between the two. But why are we required to adjudicate? The argument is
implicit, but I suspect that what Quine had in mind might be reconstructed
as follows: "It is clear that if Bizet and Verdi had been compatriots, then

either Bizet would have been Italian, or Verdi French. But then one (and only one) of the two counterfactuals, *If Bizet and Verdi had been compatriots, Verdi would have been French*, or *If Bizet and Verdi had been compatriots, Bizet would have been Italian* must be true. How are we to adjudicate between them?" The crucial inference in this reconstructed argument relies on the distribution principle, $(A > (B \lor C))$, therefore $(A > B) \lor (A > C)$, a rule of inference that is equivalent, in the context of conditional logic, to the principle of conditional excluded middle. Quine takes for granted, by tacitly using this principle of inference, that a counterfactual antecedent purports to represent a unique, determinate counterfactual situation. It is because counterfactual antecedents *purport* to represent unique possible situations that examples which show that they may fail to do so are a problem. One should respond to the problem, I think, not by revising the truth conditions for conditionals so that it does not arise, but rather by recognizing what we must recognize anyway: that in application there is great potential for indeterminacy in the truth conditions for counterfactuals.

The failure of the distribution principle we have been discussing is a symptom of the fact that, on Lewis's analysis, the antecedents of conditionals act like necessity operators on their consequents. To assert *if A, then B* is to assert that B is true in every one of a set of possible worlds defined relative to A. Therefore, if this kind of analysis is correct, we should expect to find, when conditionals are combined with quantifiers, all the same scope distinctions as we find in quantified modal logic. In particular, corresponding to the distinction between $(A > (B \lor C)$ and $((A > B) \lor (A > C))$ is the quantifier scope distinction between $(A > (\exists x)Fx)$ and $(\exists x)(A > Fx)$. On Lewis's account, even when the domain of the quantifier remains fixed across possible worlds, there is a semantically significant difference between these two formulas of conditional logic, and we should expect to find scope ambiguities in English sentences that might be formalized in either way.

Before seeing if such ambiguities are found in conditional statements, let us look at a case where the ambiguity is uncontroversial. The following dialogue illustrates a quantifier scope ambiguity in a necessity statement:

> X: President Carter has to appoint a woman to the Supreme Court.
>
> Y: Who do you think he has to appoint?
>
> X: He doesn't have to appoint any particular woman; he just has to appoint some woman or other.

Y, perversely, gives the quantified expression, *a woman*, wide scope in interpreting X's statement. X, in his response to Y's question, shows that he meant the quantifier to have narrow scope. The difference is, of course, not a matter of whether the speaker *knows* who the woman is. X might have meant the wide scope reading – the reading Y took it to have – and still not have known who the woman is. In that case, his response to Y's question would have been something like this:

> X: I don't know; I just know it's a woman that he has to appoint.

In this alternative response, the appropriateness of the question is not challenged. X just confesses inability to answer it. This alternative reply is appropriate only if the speaker intended the wide scope reading.

Now compare a parallel dialogue beginning with a statement that is clearly unambiguous:

> X: President Carter will appoint a woman to the Supreme Court.
>
> Y: Who do you think he will appoint?
>
> X: He won't appoint any particular woman; he just will appoint some woman or other.

X's response here is obviously nonsense.[14] There must be a particular person that he will appoint, although the speaker need not know who it is. If he does not know, the analogue of the alternative response is the one he will give:

> X: I don't know; I just know it's a woman that he will appoint.

Now look at a corresponding example with a counterfactual conditional and consider which of the above examples it most resembles.

> X: President Carter would have appointed a woman to the Supreme Court last year if there had been a vacancy.
>
> Y: Who do you think he would have appointed?
>
> X: He wouldn't have appointed any particular woman; he just would have appointed some woman or other.

Or, the alternative response:

> X: I don't know; I just know it's a woman that he would have appointed.

If Lewis's analysis is correct, you should perceive a clear scope ambiguity in X's original statement. Y's question, and X's alternative response, should seem appropriate only when the strong, wide scope reading was the intended one. I do not see an ambiguity; X's first response seems as bad, or almost as bad, as the analogous response in the future tense case. And I do not think there is any interpretation for which Y's question shows a misreading of the statement.

There is still, on the analysis I am defending, a relevant difference between the future tense example and the counterfactual example – a way in which the latter is more like the necessity example than the former. In the future tense case, if X's initial statement is true, then it follows that Y's question has a correct answer, even if no one knows what it is.[15] But in the counterfactual case, this need not be true. There may be no particular woman of whom it is true to say, President Carter would have appointed *her* if a vacancy had occurred. This is possible because of the possibility of under-determination, but it does not imply that there is any scope ambiguity in the original statement. The situation is analogous to familiar examples of under-determination in fiction. The question, *exactly how many sisters and cousins did Sir Joseph Porter have*? may have no correct answer, but one who asks it in response to the statement that his sisters and cousins numbered in the dozens does not exhibit a misunderstanding of the semantic structure of the statement.[16] It is not surprising, from the point of view of the analysis I am defending, that the possible situations determined by the antecedents of counterfactual conditionals are like the imaginary worlds created by writers of fiction. In both cases, one purports to represent and describe a unique determinate possible world, even though one never really succeeds in doing so.

As we have seen, Lewis agrees that the analysis I am defending respects, as his does not, certain "offhand" opinions of ordinary language speakers. He argues that the cost of respecting these opinions is too high. But Lewis also recognizes – in fact emphasizes – that counterfactual conditionals are frequently vague, and he adopts the same account of vagueness that allows the analysis I am defending to avoid implausible assumptions about our conceptual resources. Why, then, does Lewis still reject this analysis? "Two major problems remain," he writes. "First, the revised version [C2, revised by the introduction of supervaluations] still depends for its success on the Limit Assumption . . . Second, the revised version still gives us no 'might' counterfactual."[17] I will conclude the defense of my analysis by responding to these two further problems. I will first argue that the limit assumption, unlike the uniqueness assumption, is a plausible assumption to make about the orderings of possible worlds that are determined by our conceptual resources, and that the rejection of this assumption has some bizarre consequences. Second, I will say how I think *might* conditional should be understood, and argue that Lewis's analysis fails to give a satisfactory account of the relation between *might* and *would* conditionals.

When the uniqueness assumption fails to hold for a comparative similarity relation among possible worlds, then the selection function in terms of which conditionals are interpreted in C2 is left underdetermined by that relation. Many selection functions may be compatible with the comparative similarity relation, and it would be arbitrary to choose one over the others. But if the limit assumption were to fail, there would be too few candidates to be the selection function rather than too many. Any selection function would be forced to choose worlds which were less similar to the actual world than other eligible worlds. This is why the supervaluation method does not provide a way to avoid making the limit assumption.

The limit assumption implies that for any proposition A which is possibly true, there is a non-empty set of *closest* worlds in which A is true. Is this a plausible assumption to make about the orderings of possible worlds which are relevant to the interpretation of conditionals? If one were to begin with a concept of overall similarity among possible worlds which is understood independently of its application to the interpretation of conditionals, this clearly would be an arbitrary and unjustified assumption. Nothing that I can think of in the concept of similarity, or in the respects of similarity that are relevant, would motivate imposing this restrictive formal structure on the ordering determined by a similarity relation. But, on the other hand, if one begins with a selection function and thinks of the similarity orderings as induced by the selection function, the assumption will not be arbitrary or unmotivated: the fact that it holds will be explained by the way in which the orderings are determined. To the extent that an intuitive notion of similarity among possible worlds plays a role, it is a device used for the purpose of selecting possible worlds. Given this rule, it is not unreasonable to require that the way respects of similarity are weighed should be such as to make selection possible.

Even if we take the selection function as the basic primitive semantic determinant in the analysis of conditionals, we still must rely on some more or less independently understood notion of similarity or closeness of worlds to describe the intuitive basis on which the section is made. The intuitive idea is something like this: the function selects a possible world in which the antecedent is true, but which otherwise is as much like the actual world, in relevant respects, as possible. So, one might argue, we still need to give some justification for the limit assumption. How can we be sure that it will be possible to select a world, or a set of worlds, on this basis? Consider one of Lewis's examples: suppose that this line, ———, were more than one inch long. (The line is actually a little less than one inch long.) Every possible world in which the line is more than one inch long is one in which it is longer than it needs to be in order to make the antecedent true. It appears that the intuitive rule to select a world that makes the minimal change in the actual world necessary to make the antecedent true is one that cannot be followed.

The qualification in the intuitive rule that is crucial for answering this objection is the phrase, 'in relevant respects'. The selection function may ignore respects of similarity which are not relevant to the context in which the conditional statement is made. Even if, in terms of some general notion of overall similarity, i is clearly more similar to the actual world than j, if the ways in which it is more similar are irrelevant, than j may be as good a candidate for selection as i.[18] In the example, it may be that what matters is that the line is more than one inch long, and still short enough to fit on the page. In this case, all lengths over one inch, but less than four or five inches will be equally good.

But what about a context in which every millimeter matters? If relative to the issue under discussion, every difference in length is important, then it is just inappropriate to use the antecedent, *if the line were more than an inch long*. This would, in such a context, be like using the definite description, *the shortest line longer than one inch*. The selection function will be undefined for such antecedents in such contexts.

To summarize: from a naive point of view, nothing seems more obvious than that a conditional antecedent asks one to imagine a possible situation in which the antecedent is true. To say *if pigs could fly* is to envision a situation, or a kind of situation, in which pigs can fly. This is the motivation for making a selection function the basic semantic determinant. But it is equally obvious that the basis for the selection is some notion of similarity or minimal difference between worlds. The situations in which pigs can fly that you are asked to envision are ones which are as much as possible like the actual situation. The problem is that it is theoretically possible for these two intuitions to clash. There could be similarity relations and antecedents relative to which selection would be impossible. Lewis's response to this problem is to generalize the analysis of conditionals so that selection is no longer essential. The alternative response, which seems to me more natural, is to exclude as inappropriate antecedents and contexts in which the relevant similarity relation fails to make selection possible. Given that the appropriate similarity notion is one that may ignore irrelevant respects of similarity altogether, this exclusion should not be unreasonably restrictive.

I think a closer look at our example will support the conclusion that Lewis's response to the problem is intuitively less satisfactory than the one I am suggesting. On Lewis's analysis, every conditional of the following form is true: *If the line had been more than one inch long, it would not have been x inches long*, where *x* is any real number. This implies (given Lewis's analysis of *might* conditionals) that there is *no* length such that the line might have had that length if it had been more than one inch long. Yet, the line might have been more than an inch long, and if it had been, it would have had some length or other. The point is not just that there is no particular length that the line *would* have had. More than this, there is not even any length that it *might* have had. That conclusion seems, intuitively, to contradict the assumption that the line might have been more than one inch long, yet on Lewis's account, both the conclusion and the assumption may be true.[19]

The second problem that Lewis finds with the analysis I am defending is that it gives us no account of the *might* conditional. Lewis analyzes this kind of conditional in terms of his *would* conditional as follows: the *might* conditional, *if A, it might be that B*, is true if and only if the *would* conditional, *if A, it would be that not-B*, is false. In Lewis's notation, $(A \diamondsuit \!\!\!\rightarrow B) =_{df} \sim (A \;\square\!\!\!\rightarrow \sim B)$. Ordinary counterfactuals express a kind of variable necessity on the consequent, according to Lewis. *Might* counterfactuals express the corresponding kind of possibility.

It is clear that this definition conflicts with the analysis of conditionals I am proposing, since the principle of excluded middle, together with Lewis's definition of *might* conditionals, implies that a *might* conditional is equivalent to the corresponding *would* conditional. This is obviously an unacceptable conclusion, so if Lewis's definition is supported by the facts, this counts against an analysis that validates the principle of conditional excluded middle. But I will argue that Lewis's definition has unacceptable consequences, and that a more satisfactory analysis, compatible with the principle of excluded middle, can be given.

Note that Lewis's definition treats the apparently complex construction, *if ... might*, as an idiom instead of analyzing it in terms of the meanings of *if* and *might*. This is not a serious defect, but it would be methodologically preferable – less *ad hoc* – to explain the complex construction in terms of its parts. So I will begin by looking at uses of *might* outside of conditional contexts, and then consider what the result would be of combining the account of *might* suggested by those uses with our analysis of *if*.

Might, of course, expresses possibility. *John might come to the party* and *John might have come to the party* each say that it is possible, in some sense, that John come, or have come, to the party. I think the most common kind of possibility which this word is used to express is epistemic possibility. Normally, a speaker using one of the above sentences will be saying that John's coming, or having come, to the party is compatible with the speaker's knowledge. But *might* sometimes expresses some kind of non-epistemic possibility. *John might have come to the party* could be used to say that it was within John's power to come, or that it was not inevitable that he not come. The fact that the sentence, *John might come to the party, although he won't*, is somewhat strange indicates that the epistemic sense is the dominant one for this example. There is no strangeness in *John could come to the party, although he won't*. The epistemic interpretation seems less dominant in the past tense example: *John might have come to the party, although he didn't* is not so strange.

What I want to suggest is that *might*, when it occurs in conditional contexts, has the same range of senses as it has outside of conditional contexts. Normally, but not always, it expresses epistemic possibility. The scope of the *might*, when it occurs in conditional contexts is normally the whole conditional, and not just the consequent. This claim may seem *ad hoc*, since the surface form of English sentences such as *If John had been invited, he might have come to the party* certainly suggests that the antecedent is outside the scope of the *might*. But there are parallel constructions where the wide scope analysis is uncontroversial. For example, *If he is a bachelor, he must be unmarried*. Also, the wide scope interpretation is supported by the fact that *might* conditionals can be paraphrased with the *might* preceding the antecedent: *It might be that if John had been invited, he would have come to the party*.

The main evidence that *might* conditionals are epistemic is that it is unacceptable to conjoin a *might* conditional with the denial of the corresponding *would* conditional. This fact is also strong evidence against Lewis's account, according to which such conjunctions should be perfectly normal. On Lewis's account, *might* conditionals stand to *would* conditionals as ordinary *might* stands to *must*. There is no oddity in denying the categorical claim, *John must come to the party*, while affirming that he might come. But it would sound strange to deny that he would have come if he had been invited, while affirming that he might have come.

Consider a variation on the Supreme Court appointment dialogues discussed above:

> X: Does President Carter have to appoint a woman to the Supreme Court?
>
> Y: No, certainly not, although he might appoint a woman.

This is perfectly okay. Now compare:

> X: Would President Carter have appointed a woman to the Supreme Court last year if a vacancy had occurred?
>
> Y: No, certainly not, although he might have appointed a woman.

On Lewis's analysis, one should expect Y's second response to be as acceptable as his first.

One should not conclude from the conflict between the denial of the *would* conditional and the affirmation of the *might* conditional that these two statements contradict each other. To draw that conclusion would be to confuse pragmatic with semantic anomaly. On the epistemic interpretation, what Y does is to represent himself as knowing something by asserting it, and then to deny that he knows it. The conflict is thus like Moore's paradox, rather than like a contradictory assertion.

My account predicts, while Lewis's does not, that the example given above should seem Moore-paradoxical. I think it is clear that the evidence supports this prediction. Rich Thomason has pointed out that there are also examples of the reverse: cases for which Lewis's account predicts a Moore's paradox, while mine does not. Here too, I think it is clear that the evidence supports my account. Consider any statement of the form *If A, it might be that not-B, although I believe that if A then it would be that B*. Lewis's definition implies that such a statement is equivalent to a statement of the form *Not-C, although I believe that C*, and so implies that such a statement should seem Moore-paradoxical. But there is nothing wrong with saying *John might not have come to the party if he had been invited, but I believe he would have come.* As my account predicts, this statement is as acceptable as the parallel statement with non-conditional *might*: *John might not come to the party, although I believe that he will.*[20]

Lewis considers and rejects a number of alternatives to his analysis of *might* counterfactuals, including an analysis which treats them as *would* counterfactuals prefixed by an epistemic possibility operator. Here is his counterexample: Suppose there is in fact no penny in my pocket, although I do not know it since I did not look. "Then *'If I had looked, I might have found a penny'* is plainly false." But it is true that it might be, for all I know, that I would have found a penny if I had looked.[21]

I do not think that Lewis's example is *plainly* false since the epistemic reading, according to which it is true, seems to be one perfectly reasonable interpretation of it. I can also see the non-epistemic sense that Lewis has in mind, but I think that this sense can also be captured by treating the *might* as a possibility operator on the conditional. Consider not what is, in fact, compatible with my knowledge, but what would be compatible with it if I knew all the relevant facts. This will yield a kind of quasi-epistemic possibility – possibility relative to an idealized state of knowledge. If there is some indeterminacy in the language, there will still remain some different possibilities, even after all the facts are in, and so this kind of possibility will not collapse into truth. Propositions that are neither true nor false because of the indeterminacy will still be possibly true in this sense. Because *if Bizet and Verdi had been compatriots, Verdi would have been French* is neither true nor false, *If Bizet and Verdi had been compatriots, Verdi might have been French* will be true in this sense of *might*.

Now this interpretation of *might* conditionals is very close to Lewis's. It agrees with Lewis's account that *If A, it might be that B* is true if and only if *If A it would be that not-B* is not true. But my explanation has the following three advantages over Lewis's: First, it treats the *might* as a kind of possibility operator on the conditional – an operator that can also operate on other kinds of propositions – rather than treating *if . . . might* as a semantically unanalyzed unit. With Lewis's analysis of the *would* conditional, this cannot be done. Second, it treats this particular kind of *might* as a special case of a more general analysis – one that includes the ordinary epistemic interpretation as another special case. Third, it explains, as Lewis's analysis cannot, why it is anomalous to deny the *would* conditional while affirming the corresponding *might*.

It may seem strange that I have called the use of *might* which expresses semantic indeterminacy a quasi-epistemic use, but I think that there is a general tendency to use epistemic terminology to describe indeterminacy, and to think of indeterminacy as a limiting case of ignorance – the ignorance that remains after all the facts that are in. I will conclude with an example that illustrates this tendency, as well, I think, as the general point that we tend to think of counterfactual suppositions as determining a unique possible situation.

If President Kennedy had not been assassinated in 1963, would the United States have avoided the Vietnam debacle? It is a controversial question. We will probably never know for sure. If we could look back into the minds of President Kennedy and his advisors, if we could learn all there is to learn about their policy plans and priorities, their expectations and perceptions, then maybe we could settle the question. But on the other hand, it could be that the answer turns on possible actions and events which are not determined by facts about the actual situation. In that case, we *could* never know, no matter how much we learned. In that case, even an omniscient God wouldn't know.[22] If this is true, then our failure to answer the question is not really an epistemic limitation, but we still use the language of knowledge and ignorance to characterize it. Even when we recognize that such a question really has no answer, we continue to talk and think as if there were an answer that we cannot know. This is, I think, because we tend to think of the counterfactual situations determined by suppositions as being as complete and determinate as our own actual world.

Cornell University

NOTES

* This paper was completed while I was a National Endowment for the Humanities Fellow at the Center for Advanced Study in the Behavioral Sciences. I am grateful both to NEH and to the Center for providing time off from teaching and an idyllic setting in which to work. I am also grateful to David Lewis and Richmond Thomason for valuable conversation and correspondence over the years on the topic of the paper.

[1] [7].

[2] [11].

[3] [2].

[4] [13] and [14].

[5] Propositions, in the formal semantic theory, are identified with sets of possible worlds, or equivalently, with functions from possible worlds into truth values. If A is a proposition and i is a possible world, then '$i \in A$' means that A is true at i.

In [13] and [14], the selection function was a function from sentences rather than propositions, but it was assumed that this function assigned the same values to sentences which expressed the same proposition.

[6] This is the truth condition only on the assumption that there are some A-worlds, that is, that the antecedent is not necessarily false. Both Lewis's theory and mine stipulate that a counterfactual is vacuously true when its antecedent is necessarily false.

[7] See [8]. This theory, and theories based on the same general idea, have been widely applied. See, for example, [3], [4], [6], [9], [15], and [16].

[8] Lewis emphasizes the vagueness and context dependence of counterfactuals in his discussion of resemblance. See [7], pp. 91–95.

[9] This statement needs qualification when the language has a truth operator or other resources capable of expressing the fact that a statement lacks a truth value. In such cases, there will be a divergence between the validity of an argument from A to B and the logical truth of a statement $A \supset B$. See [18].

[10] The relationship between the semantics for C2 with supervaluations and Lewis's semantics is explored by Bas Van Fraassen in [17]. A C2 model with supervaluations is defined as a family of determinate C2 models. Van Fraassen shows that Lewis models which satisfy the limit assumption are equivalent to what he calls *regular* families of C2 models — families meeting a certain restriction. I believe that if one drops the regularity restriction, then the same equivalence holds between families of C2 models and the models of the semantic theory of simple subjunctive conditionals favored by John Pollock. For an exposition of Pollock's theory, see [11].

[11] [12], p. 14.

[12] [7], p. 80.

[13] [12], p. 14.

[14] Philosophers who deny truth to statements about future contingents may disagree. They may want to say that X's reply makes sense, and might be true. Readers who are inclined to this view may substitute a past tense example.

[15] I am assuming that statements about future contingents may be true. For one who treats future contingent statements as truth-valueless, and uses supervaluations to interpret truth-value gaps, the contrast between future tense and counterfactual examples that I am pointing to will disappear. See [16].

[16] This example from Gilbert and Sullivan's *H.M.S. Pinafore* is borrowed from David Lewis, [10].

[17] [7], pp. 82–83.

[18] David Lewis, in [8], suggests that the similarity ordering relevant to interpreting counterfactuals may, in some cases, give zero weight to some respects of similarity.

[19] See [5] and [11] for arguments against the limit assumption.

[20] This argument was given to me by Rich Thomason, private communication.

[21] [7], p. 80.

[22] Cf. [1].

BIBLIOGRAPHY

[1]   Adams, Robert, 'Middle Knowledge and the Problem of Evil', *American Philosophical Quarterly* **14** (1977), 109–117.

[2]   Chellas, Brian, 'Basic Conditional Logic', *Journal of Philosophical Logic* **4** (1975), 133–228.

[3]   Field, Hartry, 'Quine and the Correspondence Hypothesis', *Philosophical Review* **83** (1974), 200–228.

[4]   Fine, Kit, 'Vagueness, Truth and Logic', *Synthese* **30** (1975), 265–300.

[5]   Herzberger, Hans, 'Counterfactuals and Consistency', *Journal of Philosophy* **76** (1979), 83–88.

[6]   Kamp, Hans, 'Two Theories About Adjectives', in Edward L. Keenan (ed.), *Formal Semantics of Natural Language,* Cambridge U. Press, Cambridge, 1975, pp. 123–155.

[7]   Lewis, David, *Counterfactuals*, Basil Blackwell, Oxford, 1973.

[8] Lewis, David, 'Counterfactual Dependence and Time's Arrow', *Noûs* **13** (1979), 455–476.

[9] Lewis, David, 'General Semantics', in Donald Davidson and Gilbert Harman (eds.), *Semantics of Natural Language*, D. Reidel, Dordrecht, Holland, 1972, pp. 169–218.

[10] Lewis, David, 'Truth in Fiction', *American Philosophical Quarterly* **15** (1978), 37–46.

[11] Pollock, John, *Subjunctive Reasoning*, D. Reidel, Dordrecht, Holland, 1976.

[12] Quine, W. V., *Methods of Logic,* Holt, Rinehart, and Winston, New York, 1950.

[13] Stalnaker, Robert, 'A Theory of Conditionals', in N. Rescher (ed.), *Studies in Logical Theory,* Basil Blackwell, Oxford, 1968, pp. 98–112.

[14] Stalnaker, Robert and Richmond Thomason, 'A Semantic Analysis of Conditional Logic', *Theoria* **36** (1970), 23–42.

[15] Thomason, Richmond, 'A Semantic Theory of Sortal Incorrectness', *Journal of Philosophical Logic* **1** (1972), 209–258.

[16] Thomason, Richmond, 'Indeterminist Time and Truth-value Gaps', *Theoria* **36** (1970), 246–281.

[17] Van Fraassen, Bas, 'Hidden Variables in Conditional Logic', *Theoria* **40** (1974), 176–190.

[18] Van Fraassen, Bas, 'Singular Terms, Truth-value Gaps, and Free Logic', *Journal of Philosophy* **63** (1966), 481–495.

PART 3

CONDITIONALS AND SUBJECTIVE CONDITIONAL PROBABILITY

(The Ramsey Test Paradigm)

ROBERT C. STALNAKER[1]

PROBABILITY AND CONDITIONALS

ABSTRACT. The aim of the paper is to draw a connection between a semantical theory of conditional statements and the theory of conditional probability. First, the probability calculus is interpreted as a semantics for truth functional logic. Absolute probabilities are treated as degrees of rational belief. Conditional probabilities are explicitly defined in terms of absolute probabilities in the familiar way. Second the probability calculus is extended in order to provide an interpretation for counterfactual probabilities – conditional probabilities where the condition has zero probability. Third, conditional propositions are introduced as propositions whose absolute probability is equal to the conditional probability of the consequent on the antecedent. An axiom system for this conditional connective is recovered from the probabilistic definition. Finally, the primary semantics for this axiom system, presented elsewhere, is related to the probabilistic interpretation.

According to some interpretations of probability theory, a conditional probability statement represents a semantic or pragmatic relation between two propositions. An if–then statement in English, or an analogue in some formal language, also represents a relation between two propositions – the antecedent and the consequent. A lot of philosophical effort has been devoted to the clarification of these two conditional relations, and recently a few philosophers have tried to draw a connection between them.[2] There are at least two reasons motivating the attempts to bring these two problems together. First, although the interpretation of probability is controversial, the abstract calculus is a relatively well defined and well established mathematical theory. In contrast to this, there is little agreement about the logic of conditional sentences. Diverse systems of strict implication, conditional logic, entailment, connexive implication, and causal implication have been proposed and defended on the basis of the vague set of linguistic and methodological intuitions about conditionality, which is all we have to go on. Probability theory could be a source of insight into the formal structure of conditional sentences. Second, one approach to the philosophical problems of induction and confirmation has linked these problems to the analysis of counterfactual conditionals. Other approaches have discussed the problem in the context of interpretations of probability. A connection between the semantics of conditionals and the interpretation of probability might help to bring together the different treatments of these philosophical problems.

W. L. Harper, R. Stalnaker, and G. Pearce (eds.), Ifs, 107–128.

In this paper, I shall use probability theory to defend an analysis of conditional propositions which was proposed in another context. My argument has three steps; each step consists of the construction of a probability system. By analogy with Quine's grades of modal involvement, I might call the systems three grades of conditional involvement, since each is an extension of the preceding one, and with each, conditionality plays a more central role.

In the first system, an absolute probability function is interpreted as an autonomous semantics for propositional calculus, based on the concept of knowledge rather than truth. Conditional probabilities are introduced by definition in the usual way, but are left undefined for some pairs of wffs. The fact that conditional probabilities are sometimes undefined proves a crucial limitation to the system.

The second system provides an interpretation for an extension of the probability calculus in which conditional probabilities are primitive. This system is also an autonomous semantics for propositional calculus based on a concept of *conditional* knowledge.

The third system introduces conditional propositions by adding a primitive conditional connective to the object language and a requirement to the definition of the conditional probability function. The leading idea of the added requirement is that the probability of a conditional statement should equal the conditional probability of the consequent on the antecedent. An axiom system for the conditional connective is then recovered from this probabilistic definition. This system is the formal system of conditional logic, **C2**, which was developed and interpreted independently. I shall conclude the paper by discussing briefly the relation between the probabilistic interpretation of conditional logic and the standard semantics.

1. ABSOLUTE PROBABILITY FUNCTIONS

The first system that I shall discuss, P_1, consists of two semantical functions, an absolute probability function and a truth valuation function. I shall first characterize the syntax of the object language, and define these functions. Second, I shall discuss the intuitive content of the functions, and show how one can justify the definition of the probability function in terms of the definition of the truth valuation function. Finally, I shall introduce conditional probabilities as abbreviations, and discuss their interpretation and their limitations.

The primitive symbols of the object language consist of an infinite set of propositional variables, $\{P, Q, R, P', \ldots\}$, two primitive connectives, $\wedge$ and $\sim$

(conjunction and negation, respectively), and parentheses. Any variable is a wff. also, if A and B are wffs, then $\sim A$ and $(A \wedge B)$ are wffs. The additional connectives, $\supset$, $\vee$, $\equiv$ (material conditional, disjunction and material equivalence, respectively) may be defined in terms of the primitives. In this exposition, we shall abbreviate wffs in the usual way.

(1)   A *truth valuation function* (tvf) is any function v taking wffs into $\{1, 0\}$ which meets the following two conditions for all wffs A and B:

        (a) $v(\sim A) = 1 - v(A)$

        (b) $v((A \wedge B)) = v(A) \times v(B)$

(2)   An *absolute probability function* (apf) is any function, Pr, taking wffs into real numbers which meets the following six conditions for all wffs A, B, and C:

        (a) $1 \geqslant \Pr(A) \geqslant 0$

        (b) $\Pr(A) = \Pr(A \wedge A)$

        (c) $\Pr(A \wedge B) = \Pr(B \wedge A)$

        (d) $\Pr(A \wedge (B \wedge C)) = \Pr((A \wedge B) \wedge C)$

        (e) $\Pr(A) + \Pr(\sim A) = 1$

        (f) $\Pr(A) = \Pr(A \wedge B) + \Pr(A \wedge \sim B)$

(3)   A P_1 interpretation is an ordered pair, $\langle v, \Pr \rangle$ where v is a tvf and Pr is an apf, and where for all wffs A, if $\Pr(A) = 1$, then $v(A) = 1$.

A tvf and an apf are two ways to provide an interpretation for wffs, the first in terms of truth and falsity, the second in terms of knowledge and degrees of rational belief. A tvf provides a representation of a *possible world*. Wffs receiving a value of one correspond to propositions which are true in that world, and those with value zero correspond to propositions which are false. An apf provides a representation of a *state of knowledge*.[3] A state of knowledge is here understood to include not only a specification of those propositions known to be true and false, but also a measure of the degree to which the knower has a right to believe propositions which are neither known true nor known false. Values of the function between zero and one exclusive represent the degrees assigned to propositions whose truth value is unknown.

Wffs having values of one and zero represent propositions known to be true, and false, respectively.[4]

These two modes of interpretation are not exclusive alternatives, but complementary. A P_1 interpretation combines the two: it provides a representation of a possible world and of the state of knowledge of a knower in that world. The two components of a P_1 interpretation are not completely independent, since a knower cannot know something that is not true. But any probability value between the extremes is compatible with any truth value. Therefore, there is a wide range of apfs which are compatible with any given tvf; this is to say, there may be a diversity of knowers in a single possible world. Also, most apfs are compatible with a variety of tvfs, which is to say that knowledge need not be omniscient: a single state of knowledge may be compatible with many possible states of the world.

For any given state of knowledge, there is a class K of possible worlds which are compatible with that state of knowledge. If the relevant state of knowledge is represented by the apf, Pr, and possible worlds are represented by tvfs, then the class can be defined as follows:

(4) $K =_{df} \{v/\langle v, \text{Pr} \rangle \text{ is a } P_1 \text{ interpretation}\}$.

The class K is the class of *epistematically possible worlds.*

Because an apf is compatible with a range of possible state of the world, it is not possible to define 'degree of rational belief' in terms of truth. We can, however, justify all of the constraints on the belief function given in definition (2) in terms of the definition (1) of the tvf. This is accomplished by linking the general concept of degree of rational belief to the general concept of a logically possible world, or a model. This connection is drawn independently of any particular language. In terms of it, the specification of the models for a particular language can be used to evaluate the specific definition of the belief function for that language.

A degree of rational belief in a given proposition for a given subject is interpreted as a number determining the minimum odds which the subject should be willing to accept were he to bet on the truth of that proposition. If $\text{Pr}(A) = r$, then the subject should be willing to bet on A at odds $r/(1-r)$, and he should be unwilling to accept a bet at odds less favourable than this. The ratio, $r/(1-r)$ is the ratio of the probability that the proposition is true (that he wins the bet) to the probability that it is false (that he loses the bet). This characterization seems reasonable, since it is reasonable to act on one's beliefs. If you find gambling games a narrow and unsuitable basis on which to build the interpretation of a belief function, consider a 'bet' as any action in the face of uncertainty, and the 'odds' as the ratio of the value of what you risk by taking the action to the value of what you hope to gain, should the uncertain even turn out in your favor.

A probability assignment to a set of propositions is defined to be *incoherent* if there exists a set of bets for or against those propositions that should be accepted by the subject (according to the assignment), but are such that the subject would sustain a net loss from the set of bets *in every possible outcome*. A probability function is *coherent* if it is not incoherent. If *possible outcomes* are identified with *models of the language*, then we have a general condition of adequacy, stated in terms of the notion of a model, for any belief function. It is obviously reasonable to require that any function determining odds be coherent. If you are willing to accept bets which you are *logically certain* to lose, then you are as irrational as if you had beliefs which are logically certain to be false.

We may use the general definition of coherence to evaluate the system P_1. It can be shown that the conditions defining apf in (2) above are necessary and sufficient to ensure coherence, relative to the class of all models, or tvfs, defined in (1). Every apf is coherent, and every coherent probability function of propositional logic is an apf.[5]

In so far as coherence is our only constraint, the definition of apf is demonstrably correct. But we may still ask, are there further purely logical conditions which should be used to evaluate the adequacy of a definition of belief function? One stronger condition – strict coherence – has been suggested.[6] Strict coherence appears to be a simple and natural strengthening of coherence, and has generally been treated as such. It turns out, however, to require the introduction of some rather different considerations. Strict coherence is not a *logical* constraint on the belief function, but rather a constraint on the intuitive interpretation of the function, as defined.

A function determining reasonable betting odds is coherent if there is no set of bets consistent with it such that the bettor is certain to suffer a net loss. A function determining betting odds is *strictly* coherent if it is coherent, and also, there is no set of bets consistent with it such that the bettor cannot possibly win, and might lose. The first criterion rules out bets that *must* lose; the second rules out those that *might* lose, and cannot win. This strengthening of coherence seems perfectly reasonable. It is surely irrational to take a risk with no hope of gain, even if there is *some* hope of breaking even.

Kemeny showed, in his paper on fair betting odds, that to ensure that a coherent probability function be strictly coherent, it is necessary and sufficient to add the following requirement:

(5)   If $\Pr(A) = 1$, then A is true in all possible outcomes.

The application of this condition depends not only on the truth semantics for the language, but also on an *independent* specification of a class of possible outcomes, or models.[7] Any $\mathbf{P}_1$ interpretation can be shown to be strictly coherent if we take the possible outcomes to be the *epistemically* possible worlds: the situations consistent with the subject's knowledge. This seems reasonable; I take no risk if I bet on the truth of a statement that I know to be true, so I should be willing to accept any odds. And no matter what the odds, I would not bet on something that I know to be false. The set of epistemically possible outcomes is the set K defined in terms of a given apf in (4) above. With K as the set of all possible outcomes, Kemeny's condition (5) follows from the definition of $\mathbf{P}_1$ interpretation. Therefore, we may conclude that every apf is strictly coherent, relative to the set of possible outcomes defined in this way, and that every probability function which is strictly coherent relative to some set of possible outcomes is an apf.

To characterize conditional probabilities in terms of absolute probabilities, we use the familiar definition:

(6) $$\Pr(A, B) =_{df} \frac{\Pr(A \wedge B)}{\Pr(B)} \text{ (provided } \Pr(B) \neq 0)$$

$\Pr(A, B)$ is undefined when $\Pr(B) = 0$.

Since a conditional probability is simply an abbreviation for a ratio of two absolute probabilities, it is already fully interpreted. We do require, however, a justification for calling this ratio a conditional probability. We can get this justification by giving a separate interpretation to conditional probabilities in terms of odds for conditional bets, and showing that the definition is appropriate to this interpretation.

A conditional bet is a bet that is called off unless a specified condition is met. A bet that P on the condition Q is a bet that is won if P and Q are both true, lost if P is false, and Q is true, and called off is Q is false. A conditional probability is taken as representing reasonable odds for a conditional bet. Where $\Pr(P, Q) = r$, the fair odds for a bet that P on the condition Q are $r/(1 - r)$.

We can justify the definition by showing that such a conditional bet is equivalent to a *pair* of simple bets in the sense that the outcome of the conditional bet (win, lose, or draw) is the same in each possible world as the net outcome of the pair of simple bets. Rather than betting X dollars on the truth of P, conditional on Q, I can achieve the same result by dividing my X dollars in a specifiable way between two bets, one that both P and Q are true, and the other that Q is false. I can always divide the money in such a way that I break even in case I win the second bet (and thus lose the first). For any coherent belief function, if I do divide the money in this way, then I will obtain a net gain or at least break even, should I win the first bet, and lose the second.

Since the two betting situations are equivalent no matter what the outcome, I can determine the fair odds for conditional bets by calculating the ratio of the net gain (in case P and Q are both true) to the net loss (in case P is false and Q true) in the simple betting situation. This calculation gives the same result in every case as the above definition of conditional probability.

Under the intuitive interpretation that we have given to the system P_1, conditionality is given a meaning only when the condition is consistent with the subject's knowledge. In terms of conditional bets, this restriction makes sense: there can be no rational criteria for determining the odds on conditional bets where it is *known* that the condition will remain unfulfilled, and the bet neither won nor lost. This restriction also fits in with some interpretations of conditional *assertions*. Quine, for example, argues that an

affirmation of the form 'if p then q' is commonly felt less as an affirmation of a conditional than as a conditional affirmation of the consequent. If, after we have made such an affirmation, the antecedent turns out true, then we consider ourselves commited to the consequent, and are ready to acknowledge error if it proves false. If, on the other hand, the antecedent turns out to have been false, our conditional affirmation is as if it had never been made ([10], p. 12).

On this view of conditional assertions, to affirm something on a condition known to be false is to commit oneself to nothing at all, since in such a case it is already known that the affirmation is "as if it had never been made."

Completely excluded by this concept of conditionality, however, is *counterfactual* knowledge, and partial belief. I may believe that *if* Kennedy had not been assassinated, it is highly probable that he would have won the 1964 presidential election. I *know* that the condition is false, but that does not prevent me from speculating – and perhaps speculating rationally – about what would have happened contrary-to-fact. Perhaps I could not place a bet on my counterfactual belief, but this is only because there would be no decisive way of telling who wins. For the same reason, I would not normally bet, say, that no woman will ever run a four-minute mile, or that Moses was actually an Egyptian. For a bet to be practical, there must be an operational decision procedure for determining the truth or falsity of the proposition in question. There must be some expected future event which both I and my gambling opponent would regard as decisively and unambiguously settling the issue. But this is a fact about bets, not about degrees of rational belief or knowledge. The lack of an operational procedure for settling disagreements about what would have been true contrary-to-fact shows not that counter-factual conditional probabilities should not be interpreted, but rather that their interpretation requires an extension of the idea of coherence. Counter-factual assertions are the most controversial and interesting conditional state-ments. If we are to use probability theory to throw light on these cases, we must first extend the theory to cover counterfactual probabilities. Section 2 presents a generalization of the system P_1 which attempts to do this.

2. COUNTERFACTUAL PROBABILITIES

The second system P_2, again provides a pair of complementary interpretations to a formulation of classical propositional calculus. The object language, and its primary semantics given by the truth valuation function, are the same as before, but the second semantical function is a conditional probability func-tion. I shall first characterize this function, and P_2 interpretation, and then discuss their intuitive rationale.

(7)   An *extended probability function* (*epf*) is any function, Pr, taking ordered pairs of wffs into real numbers which meets the following six conditions for all wffs, A, B, C, and D:

(a) $\Pr(A, B) \geqslant 0$

(b) $\Pr(A, A) = 1$

(c)   If $\Pr(\sim C, C) \neq 1$, then $\Pr(\sim A, C) = 1 - \Pr(A, C)$.

(d)   If $\Pr(A, B) = \Pr(B, A) = 1$, then $\Pr(C, A) = \Pr(C, B)$

(e) $\Pr(A \wedge B, C) = \Pr(B \wedge A, C)$

(f) $\Pr(A \wedge B, C) = \Pr(A, C) \times \Pr(B, A \wedge C)$[8]

(8)   A P_2 *interpretation* is an ordered pair, $\langle v, \Pr \rangle$, where v is a tvf and Pr is an epf, such that for all wffs A and B, if $\Pr(A, B) = 1$, then $v(B \supset A) = 1$.

An epf represents an extended state of knowledge. An extended state of knowledge includes, not only a measure of the degree to which the knower has a right to believe certain propositions, but also the degree to which he *would* have a right to believe certain propositions *if* he knew something which in fact he does not know. An epf represents, not just one state of knowledge, but a set of hypothetical states of knowledge, one for each condition. For example, the set of values of $Pr(A, B)$ for all wffs, A and for a fixed wff B represents the state of knowledge that the knower would be in if he knew B.

Absolute probabilities are not represented by a primitive function, but they may be defined as a special case of conditional probabilities as follows:

$$(9) \qquad Pr(A) =_{df} Pr(A, t),$$

where t is some arbitrarily specified tautology.

In the case where the condition is a tautology, conditional knowledge coincides with knowledge *tout court*. It can also be shown that if the condition is known to be true, then the conditional probability is equal to the absolute probability defined in this way. Where $Pr(B, t) = 1$, $Pr(A, B) = Pr(A, t)$ for all A.

Where the condition is itself not known to be true, but also not known to be false, then the conditional state of knowledge will be a function of the actual state of knowledge, exactly as in the classical probability system. An analogue of definition (6), will be a simple consequence of the characterization of epf, (7) together with the above definition of absolute probabilities, (9). In this case, the set of epistemically possible worlds relative to the hypothetical state of knowledge will be a proper subset of the set of epistemically possible worlds, relative to the actual state of knowledge.

When the condition has an absolute probability value of zero, however, the conditional probability values are logically independent of the absolute probability values. Where $Pr(B) = 0$, $Pr(A, B)$ may equal zero, one, or anything in between, whatever the absolute probability value of A. In this case, the set of epistemically possible worlds relative to the hypothetical state of knowledge, will be disjoint from the set of epistemically possible worlds relative to the actual state of knowledge.

In the case where the selected state of knowledge is independent of the given one, we require only two things: first, that the resulting hypothetical state of knowledge contain the supposition as an item of knowledge, and second, that the state of knowledge be itself consistent and coherent. For some suppostions, however, it is impossible to meet even these modest requirements. For the supposition may be itself inconsistent or impossible, in which case no coherent state of knowledge can suppose it.

A proposition is an *impossible proposition* if its negation is known true no matter what. A represents an impossible proposition just in case $Pr(\sim A, A) = 1$. A state of knowledge obtained by assuming an impossible proposition to be true, I shall call an *absurd state of knowledge*. For reasons of determinateness and formal convenience, it is stipulated that where B is impossible, $\Pr(A, B) = 1$ for all A. In the absurd state of knowledge, everything is 'known'.

An epf, then, is intended to represent an actual state of knowledge and a set of hypothetical state of knowledge, related in a certain way. To show that it succeeds in this intention, I must prove that the constraints set down in the formal definition of an epf are necessary and sufficient for this representation. A few more definitions are needed to make this criterion of adequacy precise.

(10) A bet that A at odds $r/(1 - r)$ is *acceptable under condition C* if and only if $\Pr(A, C) \geqslant r$.

(11) A *conditional* bet that A on condition B at odds $r/(1 - r)$ *is acceptable under condition C* if and only if $\Pr(A, B \wedge C) \geqslant r$.

(12) $K_C^{\mathrm{Pr}} =_{\mathrm{df}} \{v / \text{for all } A, \text{ if } \Pr(A, C) = 1, \text{ then } v(A) = 1\}$.

(13) A knower's probability function, Pr is *strictly coherent with respect to condition C* if and only if there does not exist a set of bets and/or conditional bets acceptable to the knower under condition C such that the knower suffers a net loss in *some $v \in K_C^{\mathrm{Pr}}$* and a net gain in *no $v \in K_C^{\mathrm{Pr}}$*.

(14) A function Pr is *admissible as an extended belief function* if and only if it is a function taking ordered pairs of wffs into real numbers which meets the following three conditions:

(a) For all wffs C, $v(C) = 1$ for every $v \in K_C^{\mathrm{Pr}}$

(b) Pr is strictly coherent with respect to every C

(c) If K_C^{Pr} is empty, then $\Pr(A, C) = 1$ for every A.

Definition (10) interprets conditional probabilities not as the odds for a *conditional* bet which are *actually* fair, but rather as the odds for an *unconditional* bet which would hypothetically be fair if the knower were in a different state of knowledge. Definition (11), however, requires that conditional probabilities also represent fair odds for conditional bets – both actual and hypothetical conditional bets. This seems reasonable: the odds that I *would* accept if I knew C to be true for a bet that A should be the same as the odds that I will now accept for a conditional bet that A on condition C.

Definition (12) defines a set of tvfs relative to a belief function Pr and a condition C. This set represents the set of possible worlds that are epistemically possible with respect to the proposition represented by C, or the set of worlds consistent with the hypothetical state of knowledge selected by condition C. Note that where C is a tautology, K_C^{Pr} represents the set of worlds which are in fact epistemically possible to the knower, and where C is an impossible proposition, K_C^{Pr} is empty.

Definition (13) is the obvious generalization of the standard definition of strict coherence, and definition (14) states the criterion of adequacy for an extended belief function. Requirement (a) ensures that each hypothetical state of knowledge be the right one – namely one in which the condition C is known to be true. Requirement (b) ensures that each hypothetical state of knowledge meet the same standard of strict coherence that a simple state of knowledge, represented by an apf, must meet. Requirement (c) isolates the absurd state of knowledge and gives the probabilities definite values for it.

Using the results discussed in the first section, I shall sketch a proof of the following theorem:

(15)   A function is admissible as an extended belief function if and only if it is an epf.

First, the reader can easily verify that for each condition, (a)–(f) of (7), if it is violated, then one of the conditions, (a)–(c) of (14) will be violated. This suffices to prove the first half of the theorem: If a function is admissible as an extended belief function, then it is an epf. To prove the converse, we shall assume that the function, Pr, is an epf and show that each of the three conditions, (a) to (c) of (14) holds.

(a) By (7b), $\Pr(C, C) = 1$ for all C. Therefore, by definition of K_C^{Pr}, for all $C, v(C) = 1$ for every $v \in K_C^{\text{Pr}}$.

(b) Let a function taking single wffs into real numbers be defined for any given C as follows: $\Pr_C(A) =_{\text{df}} \Pr(A, C)$. The function $\Pr_C$ will either be an apf, or else it will be a constant function: $\Pr_C(A) = 1$ for all A. If $\Pr_C$ is an apf, then it will be strictly coherent with respect to the class of tvfs, K_C^{Pr}. Therefore, in this case, the strict coherence condition is met. If $\Pr_C$ is the constant function, then the strict coherence condition is trivially met, since K_C^{Pr} is empty.

(c) Finally, if K_C^{Pr} is empty, then there must be some class of wffs, Γ, such that (i) for all $A \in \Gamma$, $\Pr(A, C) = 1$, and (ii) for every tvf v, there is some $A \in \Gamma$ such that $v(A) = 0$. That is, there is a class of wffs all having probability values of one on the condition C, which is not simultaneously satisfiable. Therefore, by the semantical completeness of propositional calculus, $\Gamma \vdash B$ for all wffs B, from which it follows that for some finite set of wffs,

$\{A_1, A_2, \ldots, A_n\}$, all members of Γ, $A_1 \wedge A_2 \wedge \ldots \wedge A_n \vdash B$. But if $\Pr(A_1, C) = \Pr(A_2, C) = \ldots = \Pr(A_n, C) = 1$, then $\Pr(A_1 \wedge A_2 \wedge \ldots \wedge A_n, C) = 1$. Therefore, since the probability of a proposition is always equal to or greater than the probability of something that entails it, $\Pr(B, C) = 1$ for all wffs B. This completes the proof.

To conclude this section I wish to contrast the intuitive content of the extended probability system with that of the standard system. What is the nature of the additional information which would be contained in an extended system? A classical probability function as I have interpreted it, provided a measure of the simple epistemological status of propositions. Things are better or less well known according as their probability values are greater or less. The standard function does not, however, make any distinctions among propositions which are known to be true, and it can say nothing about the relations between propositions which are known to be true. Mathematical theorems may be ranked with empirical hypotheses. Simple facts are not distinguished from basic scientific principles. And one statement may be evidence for another, or independent of it, without this difference being reflected in the probability values. An extended function, on the other hand, contains information which is relevant to these differences in at least three ways:

First, an epf distinguishes between items of knowledge which are contingent and items of knowledge which are necessary. The former are *merely* known, while the latter *would* be known in all states of knowledge, or under every supposition. That is, A is a necessary truth if $\Pr(A, C) = 1$ for all C. What would be known under any condition is the same as what is true in all possible worlds, where the set, K of possible worlds is defined as the union of the sets K_C^{Pr} for all C. A world is ontologically possible if it is epistemologically possible relative to some supposition.

Second, an epf allows for a distinction between superficial facts – things we just happen to know – and items of empirical knowledge which have profound systematic interconnections with other parts of our knowledge. A superficial bit of information is an item of knowledge which would easily be called into question by counterfactual suppositions, and which could be hypothetically denied with only minor changes in the state of knowledge. An entrenched systematically important truth, on the other hand, would remain an item of knowledge under diverse counterfactual suppositions, and its hypothetical denial would force a radical change in the state of knowledge.

Third, an epf contains some information about the inductive relations among propositions known to be true. If there is a strong correlation between the rise and fall of the probability values of A and B under different counter-factual assumptions, for example, then one could conclude that the events described by A and B were causally connected in some way. By looking at the values of $\Pr(A, C)$ and $\Pr(B, C)$ for various particular C's, one might deter-mine *how* they were causally connected.

In general, counterfactual suppositions allow us to go beneath the surface of our knowledge in order to get at both the inductive and the conceptual relations among the things that we know, or believe to various degrees. The rules defining an epf do not, of course, provide any procedures for answering questions about these underlying relations, any more than logic provides criteria for truth. They do, however, offer a framework in which the counter-factual beliefs, which we undoubtedly have and use, can be represented.

In the final section, I shall extend this system by introducing conditional *propositions*. This will make it possible for the inductive and conceptual relations reflected in an extended probability system to be represented as explicit beliefs and items of knowledge.

3. CONDITIONAL PROPOSITIONS

The third system, P_3, involves not only an extension of the probability func-tion defined in Section 2, but also a change in the object language, and the truth semantics. In defining this system, I shall proceed somewhat differently than in the first two cases. First, I shall describe the syntax of the new object language, C2. Second, I shall add a requirement to the definition of prob-ability function which establishes a connection between conditional prop-ositions and conditional probabilities. Third, I shall ask what logical proper-ties conditional propositions must have in order that the probability function have the form that it does have. Thus, our procedure is here the reverse of what it was in Sections 1 and 2. In those sections, the established primary semantics was used, in conjunction with an idea of coherence, to justify the probability semantics. In this section, a natural extension of the probability semantics in conjunction with the idea of coherence will be used to discover and justify the rules of truth for conditional propositions.

The object language, C2, is as before except that one connective, $>$ (called the corner) is added to the list of primitive symbols, and one clause is added to the definition of wff as follows: if A and B are wffs, then $(A > B)$ is a wff.

A C2-epf is defined as a function taking ordered pairs of wffs of C2 into real numbers. The function must meet all of the requirements of an ordinary epf, as set down in definition (7), Section 2 above. It must also meet one additional requirement. Our first problem is to determine exactly what that should be.

The absolute probability of a conditional proposition – a proposition of the form $A > B$ – must be equal to the conditional probability of the consequent on the condition of the antecedent.

$$(16) \qquad \Pr(A > B) = \Pr(B, A)$$

The probability of the proposition, if Nixon is nominated then Johnson will win, should be the same as the probability that Johnson will win, on the condition that Nixon is nominated. This is the basic requirement, but by itself it is too weak, since it sets no limits on the *conditional* probability of conditional propositions. On the basis of the requirement (16), we could draw certain conclusions about the *absolute* probabilities of conditional propositions – for example that for all wffs A and B, $\Pr(A > B) = 1 - \Pr(A > {\sim}B)$ whenever $\Pr(A > {\sim}A) = 0$. But we could draw no conclusion at all about conditional probabilities of conditionals. For example, for any wffs C such that $\Pr(C) < 1$, the relation between $\Pr(A > B, C)$ and $\Pr(A > {\sim}B, C)$ would be completely open. Thus no real constraints would be placed on the logic of conditionals since any set of conditional formulas would be simultaneously satisfiable in the sense that there would exist a **C2**-epf which assigned each formula the value one on some consistent condition.

The following generalization of the proposed requirement suggests itself:

$$(17) \qquad \Pr(A > B, C) = \Pr(B, A \wedge C)$$

This condition, however, is clearly too strong.

The antecedent A may be a *counterfactual* assumption with respect to the condition C. That is, the antecedent A may be incompatible with the state of knowledge selected by the condition C. In this case the antecedent A cannot simply be added to the set of things known in that state of knowledge. Some deletions and adjustments will have to be made, and the condition C may be one of the things that gets deleted. In fact, the adoption of the strong requirement, (17) would trivially give all counterfactual propositions a probability of one, collapsing the distinction between knowledge and necessity, and reducing the probability system, $\mathbf{P_3}$ to one roughly equivalent to $\mathbf{P_1}$. This can be seen by the following argument: suppose $A > B$ represents a counterfactual – that is a conditional proposition whose antecedent is known to be false. Then $\Pr(A) = 0$, so $\Pr({\sim}A) = 1$. But for all C such that $\Pr(C) = 1$, and for all D, $\Pr(D, C) = \Pr(D)$. Therefore $\Pr(A > B) = \Pr(A > B, {\sim}A)$. But by requirement (17), $\Pr(A > B, {\sim}A) = \Pr(B, A \wedge {\sim}A)$, which always equals one.

Therefore $\Pr(A > B) = 1$. But all we assumed was that $A > B$ was counter-factual.

In order to steer a course between the unacceptably weak condition (16) and the unacceptably strong (17), we must generalize (16) in a different way. In order to carry out this generalization, we need a few more definitions.

(18)   A function, $\Pr_C$, taking ordered pairs of wffs of **C2** into real numbers is a *subfunction of Pr with respect to C* iff Pr is also a function taking pairs of wffs into real numbers, C is a wff of **C2**, and for all wffs A and B, $\Pr_C(B, A) = \Pr(A > B, C)$.

(19)   A function Pr taking ordered pairs of wffs of **C2** into real numbers is *acceptable on the first level* iff it is an epf and for all wffs A and B, $\Pr(A > B) = \Pr(B, A)$.

(20)   A function Pr taking pairs of wffs of **C2** into real numbers is *acceptable on the $(n + 1)th$ level* if for every wff C, the subfunction of Pr with respect to C is acceptable on the nth level.

(21)   A function Pr taking pairs of wffs of **C2** into real numbers is a **C2**-*epf* if it is acceptable on the nth level for every n.

The introduction of subfunctions is simply a device to allow the weak requirement (16) to be applied more generally without collapsing conditions as does the rejected requirement (17).

Definition (21) gives a complete semantical characterization of a conditional concept, not in terms of its truth relations, but in terms of its probability relations. The next step in the investigation is to define notions of satisfiability and validity for the wffs of the language **C2**, relative to this probability semantics. Then I shall present an axiom system which implicitly defines syntactical notions of consistency and theoremhood for conditional logic. This system, will then be proved semantically sound and complete relative to the probability semantics.

The final step of the argument – the construction of an appropriate truth semantics – has already been taken. The axiom system for **C2** has elsewhere been shown to be semantically sound and complete relative to a primary semantics which was given an independent philosophical justification.[9] In the conclusion to this paper, I shall discuss the relation between the two semantical systems.

(22)   A class Γ of wffs of **C2** is *p-simultaneously satisfiable* if there exists a **C2**-epf Pr and a wff C such that $\Pr(\sim C, C) \neq 1$, and for all $A \in \Gamma, \Pr(A, C) = 1$.

(23)   A wff A is *p-valid* if $(\sim A)$ is not *p*-simultaneously satisfiable.

A simultaneously satisfiable class, by this definition, represents a class of propositions, all of whose members might be known to be true. A valid formula represents a proposition whose negation could not possible be known to be true.

To specify the formal system, I shall use two nonprimitive modal operators, defined as follows:

(24)   Definition schemata:

   (a) $\Box A =_{df} {\sim}A > A$

   (b) $\Diamond A =_{df} {\sim}\Box{\sim}A$

These definitions bring out the fact that by moving from conditional probabilities to conditional propositions, we have also implicitly moved from a modal predicate of propositions, in the meta-language, to a modal operator, in the object language. In Quine's terminology, we have moved from the first to the second grade of modal involvement. In the system, P_2, $\Pr(A, {\sim}A) = 1$ just in case A is a necessary truth. Therefore, in P_3, we have a *proposition* which states that A is a necessary truth.

The following two rules and seven axiom schemata determine the formal system **C2**:

(25)   Rules:

   (a) If $A \supset B$ and A are theorems, then B is a theorem

   (b) If A is a theorem then $\Box A$ is a theorem

(26)   Axiom schemata:

   (a) Any tautologous wff is an axiom

   (b) $\Box(A \supset B) \supset \cdot \Box A \supset \Box B$

   (c) $\Box(A \supset B) \supset \cdot A > B$

   (d) $\Diamond A \supset \cdot A > B \supset {\sim}(A > {\sim}B)$

   (e) $A > (B \vee C) \supset \cdot (A > B) \vee (A > C)$

   (f) $A > B \supset \cdot A \supset B$

   (g) $(A > B) \wedge (B > A) \supset \cdot (A > C) \supset \cdot (B > C)$

In the usual way, these rules and axioms determine the syntactical notions, **C2**-provability, **C2**-derivability and **C2**-consistency.

Before stating the semantical completeness theorem, I shall list some object language theorem schemata which will be useful in the metaproof.

(27)   Theorem schemata:

(a) $\vdash (t > A) \equiv A$ (where t is any tautology)

(b) $\vdash A > A$

(c) $\vdash \Diamond C \supset \cdot (C > A) \equiv \sim (C > \sim A)$

(d) $\vdash C > (A \wedge B) \equiv (C > (B \wedge A))$

(e) $\vdash C > (A \wedge B) \equiv ((C > A) \wedge ((A \wedge C) > B))$

We are now equipped to sketch a proof of the following semantical completeness theorem:

(28)   A class Γ of wffs of **C2** is p-simultaneously satisfiable if and only if it is **C2**-consistent.

The first half of the proof consists of validating the axioms and showing that the rules preserve validity. First, note that for any wffs A and C, if there exists a **C2**-epf Pr which satisfies A on condition C then there exists a **C2**-epf which satisfies A on condition t (where t is a tautology), namely the subfunction of Pr, Pr_C. Therefore, to validate an axiom, it suffices to show that it is not satisfiable on condition t; Second, note that every axiom, in unabbreviated form, is the negation of a conjunction. For each axiom, assume that this conjunction has an absolute probability value of one (that is, assume that the negation of the axiom is satisfiable on condition t). In each case, a contradiction will fall out relatively easily. To show that modus ponens, (25a), preserves validity, assume that $Pr(A \supset B) = 1$ and $Pr(A) = 1$ for all **C2**-epfs. Then $Pr(A \wedge \sim B) = Pr(A) \times Pr(\sim B, A) = Pr(\sim B, A) = 0$. So $Pr(B, A) = 1$. But since $Pr(A) = 1$, $Pr(B, A) = Pr(B)$, so $Pr(B) = 1$ in all **C2**-epfs. To show that the necessitation rule, (25b) preserves validity, assume A is valid. Then $\{\sim A\}$ is not p-satisfiable, so for all **C2**-epfs Pr and wffs C such that $Pr(\sim C, C) \neq 1$, $Pr(\sim A, C) < 1$. But for all **C2**-epf's, Pr, $Pr(\sim A, \sim A) = 1$, so to avoid contradiction we must conclude that $Pr(\sim \sim A, \sim A) = 1$, and hence that $Pr(A, \sim A) = 1$ for all **C2**-epfs Pr. Therefore $Pr(\sim A > A)$, which is the same as $Pr(\Box A)$, must be equal to one. So both rules preserve validity.

To prove the converse, I shall show that given any **C2**-consistent class of wffs, Γ, it is possible to construct a **C2**-epf and a wff C such that

$\Pr(\sim C, C) \neq 1$ and $\Pr(A, C) = 1$ for all $A \in \Gamma$. The argument follows the familiar method developed by Henkin. First, in the usual manner, construct a maximally consistent class, Γ^* which contains Γ. Then let a function, Pr, taking ordered pairs of wffs into real numbers are defined as follows: For all wffs A and B, $\Pr(A, B) = 1$ if $(B > A) \in \Gamma^*$, and $\Pr(A, B) = 0$ otherwise. Let C be an arbitrarily selected tautology, t. Substituting $\sim$t for A in theorem (27a), we get $\vdash (\text{t} > \sim \text{t}) \equiv \sim \text{t}$. Since Γ^* is consistent, $\sim \text{t} \notin \Gamma^*$, and therefore $\text{t} > \sim \text{t} \notin \Gamma^*$, so $\Pr(\sim \text{t}, \text{t}) = 0$. Also by theorem, (27a) and the consistency of Γ^*, it is evident that $\Pr(A, \text{t}) = 1$ iff $A \in \Gamma^*$. Since $\Gamma \subseteq \Gamma^*$, $\Pr(A, \text{t}) = 1$ for all $A \in \Gamma$. Therefore, the function, Pr and the wff C that we have constructed meet the conditions of definition (22). It remains only to show that the function Pr is a **C2**-epf. This we shall do by going through the six defining requirements for epf given in (7), and the added requirement for **C2**-epf given in (21).

(a) $\Pr(A, B) = 0$ or 1 for all A and B, so $\Pr(A, B) \geqslant 0$.

(b) $\vdash A > A$ by (27b), so $\Pr(A, A) = 1$ for all A.

(c) Assume $\Pr(\sim C, C) \neq 1$. Then $\Pr(\sim (C > \sim C), \text{t}) = \Pr(\Diamond C, \text{t}) = 1$, so $\Diamond C \in \Gamma^*$. Then by (27c), $(C > A) \equiv \sim (C > \sim A) \in \Gamma^*$. Therefore $\Pr(A, C) = 1(0)$ iff $\Pr(\sim A, C) = 0(1)$. Hence provided $\Pr(\sim C, C) \neq 1$, $\Pr(\sim A, C) = 1 - \Pr(A, C)$.

(d) Assume $\Pr(A, B) = \Pr(B, A) = 1$. In this case, $A > B \in \Gamma^*$ and $B > A \in \Gamma^*$. Therefore, by an axiom, (26g), $(A > C) \equiv (B > C) \in \Gamma^*$, so $\Pr(C, A) = \Pr(C, B)$, provided $\Pr(A, B) = \Pr(B, A) = 1$.

(e) By (27d), $C > (A \land B) \in \Gamma^*$ iff $C > (B \land A) \in \Gamma^*$, so $\Pr(A \land B, C) = \Pr(B \land A, C)$.

(f) By (27e), $C > (A \land B) \in \Gamma^*$ iff $C > A \in \Gamma^*$ and $(A \land C) > B \in \Gamma^*$. Therefore, $\Pr(A \land B, C) = 1$ iff $\Pr(A, C) = 1$ and $\Pr(B, A \land C) = 1$. Therefore, $\Pr(A \land B, C) = \Pr(A, C) \times \Pr(B, A \land C)$.

(g) That the function Pr is acceptable on the first level follows from a special case of (27a) $\vdash \text{t} > (A > B) \equiv (A > B)$.

(h) To show the function acceptable on the n-th level, in general, it suffices to show that every subfunction of Pr, and subfunction of a subfunction of Pr, etc. meets the first six conditions, and that each is acceptable on the first level. We do this by generalizing each of the above seven arguments. Using the following derived rules and distribution principles, the generalizations are quite straightforward, although in a few cases tedious.

(29)   Derived rules and theorems schemata

(a) If $\vdash A$, then $\vdash C_1 > (C_2 > \ldots > (C_n > A))$.

(b) If $\vdash A \supset B$, then $\vdash (C > A) \supset (C > B)$

(c) $\vdash C > (A \equiv B) \equiv ((C > A) \equiv (C > B))$

(d) $\vdash C > (A \wedge B) \equiv (C > A) \wedge (C > B)$

These generalizations complete the argument. The function Pr is a **C2**-epf, and thus the arbitrary consistent class Γ is p-satisfiable.

4. POSSIBLE WORLDS AND KNOWLEDGE

In conclusion, I shall explain briefly the intuitive idea behind the primary semantics for **C2** and consider the relation between this system and the one based on probability that I have been discussing.

A conditional statement, according to a theory of conditionals that I have defended elsewhere, is a statement about a particular possible world. *Which* possible world it is about is a function of the antecedent. Which statement is made about that world is a function of the consequent. The particular possible world selected by the antecedent cannot be just any world. First, it must be one in which the antecedent is true; when we say "if A ...". We are supposing A to be *true*. Second, it must resemble the actual world as closely as possible, given the first requirement. This latter restriction means that, where the antecedent is true in the actual world, the actual world is the world I am talking about. That is why when one asserts a conditional which turns out to have a true antecedent, he is committed to the consequent. The latter restriction also means that the world selected carries over as much of the explanatory and descriptive structure of the actual world as is consistent with the antecedent. That is why causal laws and well entrenched empirical relations are relevant to the evaluation of a counterfactual.

These intuitive ideas can be represented in a semantic theory for a formal language which includes a primitive conditional connective. An interpretation of a set of formulas is defined on a *model structure* which consists of a structure of possible worlds. The interpretation on the structure is relative to a *selection function, f* – a function that selects, for each formula A and possible world α a possible world in which A is true. The truth rule for conditional formulas – formulas of the form $(A > B)$ – can be stated as follows:

(30)   For all wffs A and B, and all possible worlds α, $(A > B)$ is true in α iff B is true in $f(A, \alpha)$.

These truth conditions, together with constraints on the selection function which are appropriate to the intuitive picture sketched above, give rise to semantical concepts of satisfiability and validity for the formulas of **C2**. The axiom system given in Section 3 is sound and complete with respect to these concepts.

According to this semantical theory, the evaluation of a conditional statement involves, implicitly, the weighing of possible worlds against each other. To decide about a conditional, I must answer a hypothetical question about how I would revise my beliefs in the face of a particular potential discovery. We are all, of course, continually making such revisions, both actual and hypothetical, and this process of change reflects methodological patterns and principles. There are always alternative ways to patch up our structure of beliefs, as Quine has persuasively argued, but the choice among the alternatives is not arbitrary. Some opinions acquire a healthy immunity to contrary evidence and become the core of our conceptual system, while others remain near the surface, vulnerable to slight shifts in the phenomena. The policies by which we make distinctions like this lend some stability to the changing process of inquiry.

A selection function, selecting and ordering possible worlds, is intended as a representation of these methodological policies. A probability system is also a representation of them, since the same policies would be involved in the determination of degrees of belief. The difference is that a probability system represents in addition the limited perspective of an individual knower. The move through the various grades of conditional involvement – P_1 to P_3 – is an attempt to sort out the general principles from the factors that depend on the particular part of the actual world a knower has experienced, or learned about. The primary semantics for **C2** is the final step in this sorting out.

My intention in developing these formal and intuitive parallels between the theory of conditional probability and the semantics for conditional logic has been to give some additional support to the analysis of conditional statements sketched above. Beyond this, it is hoped that with the further development of the theories (for example the addition of quantifiers), this approach may provide some tools for the philosophical analysis of induction and confirmation.

University of Illinois

NOTES

[1] The preparation of this paper was supported under National Science Foundation Grant, GS-1567.

[2] For some of these discussions, see [4], [1], [2], and [12].

[3] More properly, I should say that an apf represents an *idealized* state of knowledge, or a state of *virtual* knowledge, or *implicit* knowledge. I assume that a knower knows implicitly all of the consequences of his knowledge, and more generally, that where A entails B, the degree of rational belief in B is at least as great as that in A. Cf. [3], pp. 31–39.

[4] Some will perhaps be tempted to argue that the identification of knowledge with probability of one is too stringent a condition on knowledge. This temptation should be resisted, since it misses the point of this identification. I am not using a well-established interpretation of probability to provide an analysis of knowledge. Rather, I am using the intuitive notion of knowledge to place constraints on the less clear intuitive notion of probability. No claims about the nature of knowledge are implied by the identification except that knowledge entails truth, and that a state of knowledge is, ideally, deductively closed.

[5] The notion of coherence was developed by the subjective probability theorists, F. P. Ramsey and Bruno de Finetti. See [6] for the classic papers. For proofs that the probability calculus provides necessary and sufficient conditions for coherence, see [5] and [8].

[6] Strict coherence was first discussed by Abner Shimony in [11].

[7] The requirement, "If A is true in all possible outcomes, then $Pr(A) = 1$" may be treated as a purely logical constraint, with all possible outcomes interpreted as all tvfs. Then the requirement comes down to "If A is a tautology, $Pr(A) = 1$, which is entailed by the coherence condition. The converse requirement, however, cannot be treated in the same way without making a host of untenable assumptions. To interpret (5) to mean "If $Pr(A) = 1$, then A is a tautology," is to confuse a formula with the proposition it represents. Under this interpretation, we should have to accept that every necessary truth – in fact, everything that is known – is a tautology, and that all atomic formulas represent contingent propositions, each of which is logically independent of all the others. If we wish to accept the strict coherence condition without accepting logical atomism, we must allow for an independent specification of a class of models, representing the possible outcomes.

[8] The extended probability function is based on one constructed by Sir Karl Popper. Cf. [9], appendix iv. Popper presents his system as an abstract calculus rather than as a semantics. Also, his system has the additional postulate that there must be elements, A, B, C and D such that $Pr(A, B) \neq Pr(C, D)$. This has the effect of ruling out the limiting case where $Pr(A, B) = 1$ for all A and B. In [7], Hughes Leblanc presents two formulations of Popper's system without the added postulate, as a measure on formulas of propositional logic. One of his two formulations is equivalent to the definition of epf.

It should be noted here that Leblanc confuses validity with necessity in the above mentioned article, defining validity so that it is a function of the probability assignment to the variables. Also, the proof that he offers for the equivalence of his two formulations is defective and the equivalence claim is false.

[9] The completeneess proof is presented in [14]; [13] is an informal exposition and philosophical defense of the theory.

BIBLIOGRAPHY

[1]   Adams, E. W., 'The Logic of Conditionals', *Inquiry* 8 (1965), 166–97.
[2]   Adams, E. W., 'Probability and the Logic of Conditionals', in Jaakko Hintikka and Patrick Suppes (eds.), *Aspects of Inductive Logic*, North-Holland Publ. Co., Amsterdam, 1966, pp. 265–316.
[3]   Hintikka, J., *Knowledge and Belief*, Cornell U. Press, Ithaca, New York, 1962.
[4]   Jeffrey, R. C., 'If' (abstract), *Journal of Philosophy* 61 (1964), 702–3.
[5]   Kemeny, J., 'Fair Bets and Inductive Probabilities', *Journal of Symbolic Logic* 20 (1955), 263–73.
[6]   Kyburg, H. and Smokler, H., *Studies in Subjective Probability*, John Wiley & Sons, Inc., New York, 1964.
[7]   Leblanc, H., 'On Requirements for Conditional Probability Functions', *Journal of Symbolic Logic* 25 (1960), 238–42.
[8]   Lehman, R., 'On Confirmation and Rational Betting', *Journal of Symbolic Logic* 20 (1955), 255–62.
[9]   Popper, K. R., *The Logic of Scientific Discovery*, Science Editions, Inc., New York, 1961.
[10] Quine, W. V., *Methods of Logic* (revised edition), Holt, Rinehart, and Winston, New York, 1959.
[11] Shimony, A., 'Coherence and the Axioms of Confirmation', *Journal of Symbolic Logic* 20 (1955), 1–28.
[12] Skyrms, B., 'Nomological Necessity and the Paradoxes of Confirmation', *Philosophy of Science* 33 (1966), 230–349.
[13] Stalnaker, R. C., 'A Theory of Conditionals', *Studies in Logical Theory* (supplementary monograph to the *American Philosophical Quarterly*) (1968), 98–112.
[14] Stalnaker, R. C. and Thomason, R. H., 'A Semantic Analysis of Conditional Logic', mimeo., 1967.

DAVID LEWIS

PROBABILITIES OF CONDITIONALS AND
CONDITIONAL PROBABILITIES

The truthful speaker wants not to assert falsehoods, wherefore he is willing to assert only what he takes to be very probably true. He deems it permissible to assert that A only if $P(A)$ is sufficiently close to 1, where P is the probability function that represents his system of degrees of belief at the time. Assertability goes by subjective probability.

At least, it does in most cases. But Ernest Adams has pointed out an apparent exception.[1] In the case of ordinary indicative conditionals, it seems that assertability goes instead by the conditional subjective probability of the consequent, given the antecedent. We define the conditional probability function $P(-/-)$ by a quotient of absolute probabilities, as usual:

(1) $P(C/A) = \mathrm{df}\, P(CA)/P(A)$, if $P(A)$ is positive.

(If the denominator $P(A)$ is zero, we let $P(C/A)$ remain undefined.) The truthful speaker evidently deems it permissible to assert the indicative conditional that if A, then C (for short, $A \to C$) only if $P(C/A)$ is sufficiently close to 1. Equivalently: only if $P(CA)$ is sufficiently much greater than $P(\bar{C}A)$.

Adams offers two sorts of evidence. There is direct evidence, obtained by contrasting cases in which we would be willing or unwilling to assert various indicative conditionals. There also is indirect evidence, obtained by considering various inferences with indicative conditional premises or conclusions. The ones that seem valid turn out to be just the ones that preserve assertability, if assertability goes by conditional probabilities for conditionals and by absolute probabilities otherwise.[2] Our judgements of validity are not so neatly explained by various rival hypotheses. In particular, they do not fit the hypothesis that the inferences that seem valid are just the ones that preserve truth if we take the conditionals as truth-functional.

Adams has convinced me. I shall take it as established that the assertability of an ordinary indicative conditional $A \to C$ does indeed go by the conditional subjective probability $P(C/A)$. But why? Why not rather by the absolute probability $P(A \to C)$?

The most pleasing explanation would be as follows: The assertability of $A \to C$ does go by $P(A \to C)$ after all; indicative conditionals are not

129

W. L. Harper, R. Stalnaker, and G. Pearce (eds.), Ifs, 129–147.
Copyright © 1976 by David Lewis.

exceptional. But also it goes by $P(C/A)$, as Adams says; for the meaning of $\rightarrow$ is such as to guarantee that $P(A \rightarrow C)$ and $P(C/A)$ are always equal (if the latter is defined). For short: *probabilities of conditionals are conditional probabilities.* This thesis has been proposed by various authors.[3]

If this is so, then of course the ordinary indicative conditional $A \rightarrow C$ cannot be the truth-functional conditional $A \supset C$. $P(A \supset C)$ and $P(C/A)$ are equal only in certain extreme cases. The indicative conditional must be something else: call it a *probability conditional.* We may or may not be able to give truth conditions for probability conditionals, but at least we may discover a good deal about their meaning and their logic just by using what we know about conditional probabilities.

Alas, this most pleasing explanation cannot be right. We shall see that there is no way to interpret a conditional connective so that, with sufficient generality, the probabilities of conditionals will equal the appropriate conditional probabilities. If there were, probabilities of conditionals could serve as links to establish relationships between the probabilities of non-conditionals, but the relationships thus established turn out to be incorrect. The quest for a probability conditional is futile, and we must admit that assertability does not go by absolute probability in the case of indicative conditionals.

PRELIMINARIES

Suppose we are given an interpreted formal language equipped at least with the usual truth-functional connectives and with the further connective $\rightarrow$. These connectives may be used to compound any sentences in the language. We think of the interpretation as giving the truth value of every sentence at every possible world. Two sentences are *equivalent* iff they are true at exactly the same worlds, and *incompatible* iff there is no world where both are true. One sentence *implies* another iff the second is true at every world where the first is true. A sentence is *necessary, possible* or *impossible* iff it is true at all worlds, at some, or at none. We may think of a probability function P as an assignment of numerical values to all sentences of this language, obeying these standard laws of probability:

(2) $1 \geqslant P(A) \geqslant 0$,

(3)   if A and B are equivalent, then $P(A) = P(B)$,

(4)   if A and B are incompatible, then $P(A \vee B) = P(A) + P(B)$,

(5)   if A is necessary, then $P(A) = 1$.

The definition (1) gives us the multiplication law for conjunctions.

Whenever $P(B)$ is positive, there is a probability function P' such that $P'(A)$ always equals $P(A/B)$; we say that P' *comes from P by conditionalizing on B.* A class of probability functions is *closed under conditionalizing* iff any probability function that comes by conditionalizing from one in the class is itself in the class.

Suppose that $\rightarrow$ is interpreted in such a way that, for some particular probability function P, and for any sentences A and C.

(6) $P(A \rightarrow C) = P(C/A)$, if $P(A)$ is positive;

iff so, let us call $\rightarrow$ a *probability conditional for P.* Iff $\rightarrow$ is a probability conditional for every probability function in some class of probability functions, then let us call $\rightarrow$ a *probability conditional for* the class. And iff $\rightarrow$ is a probability conditional for all probability functions, so that (6) holds for any P, A, and C, then let us call $\rightarrow$ a *universal probability conditional,* or simply a *probability conditional.*

Observe that if $\rightarrow$ is a universal probability conditional, so that (6) holds always, then (7) also holds always:

(7) $P(A \rightarrow C/B) = P(C/AB)$, if $P(AB)$ is positive.

To derive (7), apply (6) to the probability function P' that comes from P by conditionalizing on B; such a P' exists if $P(AB)$ and hence also $P(B)$ are positive. Then (7) follows by several applications of (1) and the equality between $P'(-)$ and $P(-/B)$. In the same way, if $\rightarrow$ is a probability conditional for a class of probability functions, and if that class is closed under conditionalizing, then (7) holds for any probability function P in the class, and for any A and C. (It does not follow, however, that if (6) holds for a particular probability function P, then (7) holds for the same P.)

FIRST TRIVIALITY RESULT

Suppose by way of *reductio* that $\rightarrow$ is a universal probability conditional. Take any probability function P and any sentences A and C such that $P(AC)$ and $P(A\bar{C})$ both are positive. Then $P(A)$, $P(C)$, and $P(\bar{C})$ also are positive. By (6) we have:

(8) $P(A \rightarrow C) = P(C/A)$.

By (7), taking B as C or as $\bar{C}$ and simplifying the right-hand side, we have:

(9) $P(A \to C/C) = P(C/AC) = 1$,

(10) $P(A \to C/\bar{C}) = P(C/A\bar{C}) = 0$.

For any sentence D, we have the familiar expansion by cases:

(11) $P(D) = P(D/C) \cdot P(C) + P(D/\bar{C}) \cdot P(\bar{C})$.

In particular, take D as $A \to C$. Then we may substitute (8), (9), and (10) into (11) to obtain:

(12) $P(C/A) = 1 \cdot P(C) + 0 \cdot P(\bar{C}) = P(C)$.

With the aid of the supposed probability conditional, we have reached the conclusion that if only $P(AC)$ and $P(A\bar{C})$ both are positive, then A and C are probabilistically independent under P. That is absurd. For instance, let P be the subjective probability function of someone about to throw what he takes to be a fair die, let A mean that an even number comes up, and let C mean that the six comes up. $P(AC)$ and $P(A\bar{C})$ are positive. But, *contra* (12), $P(C/A)$ is $\frac{1}{3}$ and $P(C)$ is $\frac{1}{6}$; A and C are not independent. More generally, let C, D, and E be possible but pairwise incompatible. There are probability functions that assign positive probability to all three: let P be any such. Let A be the disjunction $C \vee D$. Then $P(AC)$ and $P(A\bar{C})$ are positive but $P(C/A)$ and $P(C)$ are unequal.

Our supposition that $\to$ is a universal probability conditional has led to absurdity, but not quite to contradiction. If the given language were sufficiently weak in expressive power, then our conclusion might be unobjectionable. There might not exist any three possible but pairwise incompatible sentences to provide a counterexample to it. For all I have said, such a weak language might be equipped with a universal probability conditional. Indeed, consider the extreme case of a language in which there are none but necessary sentences and impossible ones. For this very trivial language, the truth-functional conditional itself is a universal probability conditional.

If an interpreted language cannot provide three possible but pairwise incompatible sentences, then we may justly call it a *trivial language*. We have proved this theorem: *any language having a universal probability conditional is a trivial language.*

SECOND TRIVIALITY RESULT

Since our language is not a trivial one, our indicative conditional must not be a universal probability conditional. But all is not yet lost for the thesis that probabilities of conditionals are conditional probabilities. A much less than universal probability conditional might be good enough. Our task, after all, concerns subjective probability: probability functions used to represent people's systems of beliefs. We need not assume, and indeed it seems rather implausible, that any probability function whatever represents a system of beliefs that it is possible for someone to have. We might set aside those probability functions that do not. If our indicative conditional were a probability conditional for a limited class of probability functions, and if that class were inclusive enough to contain any probability function that might ever represent a speaker's system of beliefs, that would suffice to explain why assertability of indicative conditionals goes by conditional subjective probability.

Once we give up on universality, it may be encouraging to find that probability conditionals for particular probability functions, at least, commonly do exist. Given a probability function P, we may be able to tailor the interpretation of $\rightarrow$ to fit.[4] Suppose that for any A and C there is some B such that $P(B/\overline{A})$ and $P(C/A)$ are equal if both defined; this should be a safe assumption when P is a probability function rich enough to represent someone's system of beliefs. If for any A and C we arbitrarily choose such a B and let $A \rightarrow C$ be interpreted as equivalent to $AC \vee \overline{A}B$, then $\rightarrow$ is a probability conditional for P. But such piecemeal tailoring does not yet provide all that we want. Even if there is a probability conditional for each probability function in a class, it does not follow that there is one probability conditional for the entire class. Different members of the class might require different interpretations of $\rightarrow$ to make the probabilities of conditionals and the conditional probabilities come out equal. But presumably our indicative conditional has a fixed interpretation, the same for speakers with different beliefs, and for one speaker before and after a change in his beliefs. Else how are disagreements about a conditional possible, or changes of mind? Our question, therefore, is whether the indicative conditional might have one fixed interpretation that makes it a probability conditional for the entire class of all those probability functions that represent possible systems of beliefs.

This class, we may reasonably assume, is closed under conditionalizing. Rational change of belief never can take anyone to a subjective probability function outside the class; and there are good reasons why the change of belief that results from coming to know an item of new evidence should take place by conditionalizing on what was learned.[5]

Suppose by way of *reductio* that $\rightarrow$ is a probability conditional for a class of probability functions, and that the class is closed under conditional-izing. The argument proceeds much as before. Take any probability function P in the class and any sentences A and C such that $P(AC)$ and $P(A\bar{C})$ are positive. Again we have (6) and hence (8); (7) and hence (9) and (10); (11) and hence by substitution (12): $P(C/A)$ and $P(C)$ must be equal. But if we take three pairwise incompatible sentences C, D, and E such that $P(C)$, $P(D)$ and $P(E)$ are all positive and if we take A as the disjunction $C \vee D$, then $P(AC)$ and $P(A\bar{C})$ are positive but $P(C/A)$ and $P(C)$ are unequal. So there are no such three sentences. Further, P has at most four different values. Else there would be two different values of P, x and y, strictly intermediate between 0 and 1 and such that $x + y \neq 1$. But then if $P(F) = x$ and $P(G) = y$ it follows that at least three of $P(FG)$, $P(\bar{F}G)$, $P(F\bar{G})$, and $P(\bar{F}\bar{G})$ are positive, which we have seen to be impossible.

If a probability function never assigns positive probability to more than two incompatible alternatives, and hence is at most four-valued, then we may call it a *trivial probability function*. We have proved this theorem: *if a class of probability functions is closed under conditionalizing, then there can be no probability conditional for that class unless the class con-sists entirely of trivial probability functions.* Since some probability func-tions that represent possible systems of belief are not trivial, our indicative conditional is not a probability conditional for the class of all such proba-bility functions. Whatever it may mean, it cannot possibly have a meaning such as to guarantee, for all possible subjective probability functions at once, that the probabilities of conditionals equal the corresponding con-ditional probabilities. These is no such meaning to be had. We shall have to grant that the assertability of indicative conditionals does not go by absolute probability, and seek elsewhere for an explanation of the fact that it goes by conditional probability instead.

THE INDICATIVE CONDITIONAL AS NON-TRUTH-VALUED

Assertability goes in general by probability because probability is prob-ability of truth and the speaker wants to be truthful. If this is not so for indicative conditionals, perhaps the reason is that they have no truth values, no truth conditions, and no probabilities of truth. Perhaps they are governed not by a semantic rule of truth but by a rule of assertability.

We might reasonably take it as the goal of semantics to specify our pre-vailing rules of assertability. Most of the time, to be sure, that can best be done by giving truth conditions plus the general rule that speakers should try to be truthful, or in other words that assertability goes by probability of truth. But sometimes the job might better be done another way: for instance, by giving truth conditions for antecedents and for consequents, but not for whole conditionals, plus the special rule that the assertability of an indicative conditional goes by the conditional subjective probability of the consequent given the antecedent. Why not? We are surely free to institute a new sentence form, without truth conditions, to be used for making it known that certain of one's conditional subjective probabilities are close to 1. But then it should be no surprise if we turn out to have such a device already.

* Adams himself seems to favor this hypothesis about the semantics of indicative conditionals.[6] He advises us, at any rate, to set aside questions about their truth and to concentrate instead on their assertability. There is one complication: Adams does *say* that conditonal probabilities are probabilities of conditionals. Nevertheless he does not mean by this that the indicative conditional is what I have here called a probability conditional; for he does not claim that the so-called "probabilities" of conditionals are probabilities of truth, and neither does he claim that they obey the standard laws of probability. They are probabilities only in name. Adam's position is therefore invulnerable to my triviality results, which were proved by applying standard laws of probability to the probabilities of conditionals.

Would it make sense to suppose that indicative conditionals *do not* have truth values, truth conditions, or probabilities of truth, but that they *do* have probabilities that obey the standard laws? Yes, but only if we first restate those laws to get rid of all mention of truth. We must continue to permit unrestricted compounding of sentences by means of the usual connectives, so that the domain of our probability functions will be a Boolean algebra (as is standardly required); but we can no longer assume that these connectives always have their usual truth-functional interpretations, since truth-functional compounding of non-truth-valued sentences makes no sense. Instead we must choose some deductive system – any standard formalization of sentential logic will do – and characterize the usual connectives by their deductive role in this system. We must replace mention of equivalence, incompatibility, and necessity in laws (3) through (5) by mention to their syntactic substitutes in the chosen system: inter-deducibility, deductive inconsistency, and deducibility. In this way we could describe the probability functions for our language without assuming that all probabilities of sentences, or even any of them, are probabilities of truth. We could still hold that assertability goes in most cases by probability, though we could no longer restate this as a rule that speakers should try to tell the truth.

Merely to deny that probabilities of conditionals are probabilities of truth, while retaining all the standard laws of probability in suitably adapted form, would not yet make it safe to revive the thesis that probabilities of conditionals are conditional probabilities. It was not the connection between truth and probability that led to my triviality results, but only the application of standard probability theory to the probabilities of conditionals. The proofs could just as well have used versions of the laws that mentioned deducibility instead of truth. Whoever still wants to say that probabilities of conditionals are conditional probabilities had better also employ a non-standard calculus of 'probabilities'. He might drop the requirement that the domain of a probability function is a Boolean algebra, in order to exclude conjunctions with conditional conjuncts from the language. Or he might instead limit (4), the law of additivity, refusing to apply it when the disjuncts A and B contain conditional conjuncts. Either maneuver would block my proofs. But if it be granted that the 'probabilities' of conditionals do not obey the standard laws, I do not see what is to be gained by insisting on calling them 'probabilities'. It seems to me that a position like Adams's might best be expressed by saying that indicative conditionals have neither truth values nor probabilities, and by introducing some neutral term such as 'assertability' or 'value' which denotes the probability of truth in the case of nonconditionals and the appropriate conditional probability in the case of indicative conditionals.

I have no conclusive objection to the hypothesis that indicative conditionals are non-truth-valued sentences, governed by a special rule of assertability that does not involve their nonexistent probabilities of truth. I have an inconclusive objection, however: the hypothesis requires too much of a fresh start. It burdens us with too much work still to be done, and wastes too much that has been done already. So far, we have nothing but a rule of assertability for conditionals with truth-valued antecedents and consequents. But what about compound sentences that have such conditionals as constituents? We think we know how the truth conditions for compound sentences of various kinds are determined by the truth conditions of constituent subsentences, but this knowledge would be useless if any of those subsentences lacked truth conditions. Either we need new semantic rules for many familiar connectives and operators when applied to indicative conditionals – perhaps rules of truth, perhaps special rules of assertability like the rule for conditionals themselves – or else we need to explain away all seeming examples of compound sentences with conditional constituents.

THE INDICATIVE CONDITIONAL AS TRUTH-FUNCTIONAL

Fortunately a more conservative hypothesis is at hand. H. P. Grice has given an elegant explanation of some qualitative rules governing the assertability of indicative conditionals.[7] It turns out that a quantitative hypothesis based on Grice's ideas gives us just what we want: the rule that assertability goes by conditional subjective probability.

According to Grice, indicative conditionals *do* have truth values, truth conditions, and probabilities of truth. In fact, the indicative conditional $A \to C$ is simply the truth-functional conditional $A \supset C$. But the assertability of this truth-functional conditional does not go just by $P(A \supset C)$, its subjective probability of truth. It goes by the resultant of that and something else.

It may happen that a speaker believes a truth-functional conditional to be true, yet he ought not to assert it. Its assertability might be diminished for various reasons, but let us consider one in particular. The speaker ought not to assert the conditional if he believes it to be true predominantly because he believes its antecedent to be false, so that its probability of truth consists mostly of its probability of vacuous truth. In this situation, why assert the conditional instead of denying the antecedent? It is pointless to do so. And if it is pointless, then also it is worse than pointless: it is misleading. The hearer, trusting the speaker not to assert pointlessly, will assume that he has not done so. The hearer may then wrongly infer that the speaker has additional reason to believe that the conditional is true, over and above his disbelief in the antecedent.

This consideration detracts from the assertability of $A \supset C$ to the extent that both of two conditions hold: first, that the probability $P(\bar{A})$ of vacuity is high; and second, that the probability $P(\bar{C}A)$ of falsity is a large fraction of the total probability $P(A)$ of non vacuity. The product

$$(13) \quad P(\bar{A}) \cdot (P(\bar{C}A)/P(A))$$

of the degrees to which the two conditions are met is therefore a suitable measure of diminution of assertability. Taking the probability $P(A \supset C)$ of truth, and subtracting the diminution of assertability as measured by (13), we obtain a suitable measure of resultant assertability:

$$(14) \quad P(A \supset C) - P(\bar{A}) \cdot (P(\bar{C}A)/P(A)).$$

But (14) may be simplified, using standard probability theory; and so we find that the resultant assertability, probability of truth minus the diminution given by (13), is equal to the conditional probability $P(C/A)$. That is why assertability goes by conditional probability.

Diminished assertability for such reasons is by no means special to conditionals. It appears also with uncontroversially truth-functional constructions such as negated conjunction. We are gathering mushrooms; I say to you "You won't eat that one and live." A dirty trick: I thought that one was safe and especially delicious, I wanted it myself, so I hoped to dissuade you from taking it without actually lying. I thought it highly probable that my trick would work, that you would not eat the mushroom, and therefore that I would turn out to have told the truth. But though what I said had a high subjective probability of truth, it had a low assertability and it was a misdeed to assert it. Its assertability goes not just by probability but by the resultant of that and a correction term to take account of the pointlessness and misleadingness of denying a conjunction when one believes it false predominantly because of disbelieving one conjunct. Surely few would care to explain the low assertability of what I said by rejecting the usual truth-functional semantics for negation and conjunction, and positing instead a special probabilistic rule of assertability.

There are many considerations that might detract from assertability. Why stop at (14)? Why not add more terms to take account of the diminished assertability of insults, of irrelevancies, of long-winded pomposities, of breaches of confidence, and so forth? Perhaps part of the reason is that, unlike the diminution of assertability when the probability of a conditional is predominantly due to the improbability of the antecedent, these other diminutions depend heavily on miscellaneous features of the conversational context. In logic we are accustomed to consider sentences and inferences in abstraction from context. Therefore it is understandable if, when we philosophize, our judgements of assertability or of assertability-preserving inference are governed by a measure of assertability such as (14), that is $P(C/A)$, in which the more context-dependent dimensions of assertability are left out.

There is a more serious problem, however. What of conditionals that have a high probability predominantly because of the probability of the consequent? If we are on the right track, it seems that there should be a diminution of assertability in this case also, and one that should still show up if we abstract from context: we could argue that in such a case it is pointless, and hence also misleading, to assert the conditional rather than the consequent. This supposed diminution is left out, and I think rightly so, if we measure the assertability of a conditional $A \supset C$ (in abstraction from context) by $P(C/A)$. If A and C are probabilitically independent and each has probability .9, then the probability of the conditional (.91) is predominantly due to the probability of the consequent (.9), yet the conditional probability $P(C/A)$ is high (.9) so we count the conditional as assertable. And it does seem so, at least in some such cases: "I'll probably flunk, and it doesn't matter whether I study; I'll flunk if I do and I'll flunk if I don't."

The best I can do to account for the absence of a marked diminution in the case of the probable consequent is to concede that considerations of conversational pointlessness are not decisive. They create only tendencies toward diminished assertability, tendencies that may or may not be conventionally reinforced. In the case of the improbable antecedent, they are strongly reinforced. In the case of the probable consequent, apparently they are not.

In conceding this, I reduce the distance between my present hypothesis that indicative conditionals are truth-functional and the rival hypothesis that they are non-truth-valued and governed by a special rule of assertability. Truth conditions plus general conversational considerations are not quite the whole story. They go much of the way toward determining the assertability of conditionals, but a separate convention is needed to finish the job. The point of ascribing truth conditions to indicative conditionals is not that we can thereby get rid entirely of special rules of assertability.

Rather, the point of ascribing truth conditions is that we thereby gain at least a prima facie theory of the truth conditions and assertability of compound sentences with conditional constituents. We need not waste whatever general knowledge we have about the way the truth conditions of compounds depend on the truth conditions of their constituents. Admittedly we might go wrong by proceeding in this way. We have found one explicable discrepancy between asssertability and probability in the case of conditionals themselves, and there might be more such discrepancies in the case of various compounds of conditionals. (For instance the assertability of a negated conditional seems not to go by its probability of truth, but rather to vary inversely with the assertability of the conditional.) It is beyond the scope of this paper to survey the evidence, but I think it reasonable to hope that the discrepancies are not so many, or so difficult to explain, that they destroy the explanatory power of the hypothesis that the indicative conditional is truth-functional.

PROBABILITIES OF STALNAKER CONDITIONALS

It is in some of the writings of Robert Stalnaker that we find the fullest elaboration of the thesis that conditional probabilities are probabilities of conditionals.[8] Stalnaker's conditional connective $>$ has truth conditions roughly as follows: a conditional $A > C$ is true iff the least drastic revision of the facts that would make A true would make C true as well. Stalnaker conjectures that this interpretation will make $P(A > C)$ and $P(C/A)$ equal whenever $P(A)$ is positive. He also lays down certain constraints on $P(A > C)$ for the case that $P(A)$ is zero, explaining this by means of an extended concept of conditional probability that need not concern us here.

Stalnaker supports his conjecture by exhibiting a coincidence between two sorts of validity. The sentences that are true no matter what, under Stalnaker's truth conditions, turn out to be exactly those that have positive probability no matter what, under his hypothesis about probabilities of conditionals. Certainly this is weighty evidence, but is it not decisive. Cases are known in modal logic, for instance, in which very different interpretations of a language happen to validate the very same sentences. And indeed our triviality results show that Stalnaker's conjecture cannot be right, unless we confine our attention to trivial probability functions.[9]

But it is almost right, as we shall see. Probabilities of Stalnaker conditionals do not, in general, equal the corresponding conditional probabilities.[10] But they do have some of the characteristic properties of conditional probabilities.

A possible totality of facts corresponds to a possible world; so a revision of facts corresponds to a transition from one world to another. For any given world W and (possible) antecedent A, let W_A be the world we reach by the least drastic revision of the facts of W that makes A true. There is to be no gratuitous revision: W_A may differ from W as much as it must to permit A to hold, but no more. Balancing off respects of similarity and difference against each other according to the importance we attach to them, W_A is to be the closest in overall similarity to W among the worlds where A is true. Then the Stalnaker conditional $A > C$ is true at the world W iff C is true at W_A, the closest A-world to W. (In case the antecedent A is impossible, so that there is no possible A-world to serve as W_A, we take $A > C$ to be vacuously true at all worlds. For simplicity I speak here only of absolute impossibility; Stalnaker works with impossibility relative to worlds.) Let us introduce this notation:

(15) $W(A) = df \begin{cases} 1 \text{ if } A \text{ is true at the world } W \\ 0 \text{ if } A \text{ is false at } W \end{cases}$.

Then we may give the truth conditions for non vacuous Stalnaker conditionals as follows:

(16) $W(A > C) = W_A(C)$, if A is possible.

It will be convenient to pretend, from this point on, that there are only finitely many possible worlds. That will trivialize the mathematics but not distort our conclusions. Then we can think of a probability function P as a distribution of probability over the worlds. Each world W has a probability $P(W)$, and these probabilities of worlds sum to 1. We return from probabilities of worlds to probabilities of sentences by summing the probabilities of the worlds where a sentence is true:

(17) $P(A) = \Sigma_W P(W) \cdot W(A)$.

I shall also assume that the worlds are distinguishable: for any two, some sentence of our language is true at one but not the other. Thus we disregard phenomena that might result if our language were sufficiently lacking in expressive power.

Given any probability function P and any possible A, there is a probability function P' such that, for any world W',

(18) $P'(W') = \Sigma_W P(W) \cdot \begin{cases} 1 \text{ if } W_A \text{ is } W' \\ 0 \text{ otherwise.} \end{cases}$.

Let us say that P' *comes from P by imaging on A*, and call P' the *image of P on A*. Intuitively, the image on A of a probability function is formed by shifting the original probability of each world W over to W_A, the closest A-world to W. Probability is moved around but not created or destroyed, so the probabilities of worlds still sum to 1. Each A-world keeps whatever probability it had originally, since if W is an A-world then W_A is W itself, and it may also gain additional shares of probability that have been shifted away from $\bar{A}$-worlds. The $\bar{A}$-worlds retain none of their original probability, and gain none. All the probability has been concentrated on the A-worlds. And this has been accomplished with no gratuitous movement of probability. Every share stays as close as it can to the world where it was originally located.

Suppose that P' comes from P by imaging on A, and consider any sentence C.

(19) $\quad P'(C) \;=\; \Sigma_{W'} P'(W') \cdot W'(C)$, by (17) applied to P';

$$= \Sigma_{W'} \left(\Sigma_{W} P(W) \cdot \begin{Bmatrix} 1 \text{ if } W_A \text{ is } W' \\ 0 \text{ otherwise} \end{Bmatrix} \right) \cdot W'(C), \text{ by (18)};$$

$$= \Sigma_{W} P(W) \cdot \left(\Sigma_{W'} \begin{Bmatrix} 1 \text{ if } W_A \text{ is } W' \\ 0 \text{ otherwise} \end{Bmatrix} \cdot W'(C) \right), \text{ by algebra};$$

$$= \Sigma_{W} P(W) \cdot W_A(C), \text{ simplifying the inner sum};$$

$$= \Sigma_{W} P(W) \cdot W(A > C), \text{ by (16)};$$

$$= P(A > C), \text{ by (17)}.$$

We have proved this theorem: *the probability of a Stalnaker conditional with a possible antecedent is the probability of the consequent after imaging on the antecedent.*

Conditionalizing is one way of revising a given probability function so as to confer certainty – probability of 1 – on a given sentence. Imaging is another way to do the same thing. The two methods do not in general agree. (Example: let $P(W)$, $P(W')$, and $P(W'')$ each equal $\frac{1}{3}$; let A hold at W and W' but not W''; and let W' be the closest A-world to W''. Then the probability function that comes from P by conditionalizing on A assigns probability $\frac{1}{2}$ to both W and W'; whereas the probability function that comes from P by imaging on A assigns probability $\frac{1}{3}$ to W and $\frac{2}{3}$ to W'.) But though the methods differ, either one can plausibly be held to given minimal revisions: to revise the given probability function as much as must be done to make the given sentence certain, but no more. Imaging P on A gives a minimal revision in this sense: unlike all other revisions of P to make A certain, it involves no gratuitous movement of probability from worlds to dissimilar worlds. Conditionalizing P on A gives a minimal revision in this different sense: unlike all other revisions of P to make A certain, it does not distort the profile of probability ratios, equalities, and inequalities among sentences that imply A.[11]

Stalnaker's conjecture divides into two parts. This part is true: the probability of a nonvacuous Stalnaker conditional is the probability of the consequent, after minimal revision of the original probability function to make the antecedent certain. But it is not true that this minimal revision works by conditionalizing. Rather it must work by imaging. Only when the two methods give the same result does the probability of a Stalnaker conditional equal the corresponding conditional probability.

Stalnaker gives the following instructions for deciding whether or not you believe a conditional.[12]

First, add the antecedent (hypothetically) to your stock of beliefs; second, make whatever adjustments are required to maintain consistency (without modifying the hypothetical belief in the antecedent); finally, consider whether or not the consequent is true.

That is right, for a Stalnaker conditional, if the feigned revision of beliefs works by imaging. However the passage suggests that the thing to do is to feign the sort of revision that would take place if the antecedent really were added to your stock of beliefs. That is wrong. If the antecedent really were added, you should (if possible) revise by conditionalizing. The reasons in favor of responding to new evidence by conditionalizing are equally reasons against responding by imaging instead.

PROBABILITY-REVISION CONDITIONALS

Suppose that the connective → is interpreted in such a way that for any probability function P, and for any sentences A and C,

(20) $P(A \to C) = P_A(C)$, if A is possible,

where P_A is (in some sense) the minimal revision of P that raises the probability of A to 1. Iff so, let us call → a *probability-revision conditional.* Is there such a thing? We have seen that it depends on the method of revision. Conditionalizing yields revisions that are minimal in one sense; and if P_A is obtained (when possible) by conditionalizing, then no probability-revision conditional exists (unless the language is trivial). Imaging yields revisions that are minimal in another sense; and if P_A is obtained by imaging then the Stalnaker conditional is a probability-revision conditional. Doubtless there are still other methods of revision, yielding revisions that are minimal in still other senses than we have yet considered. Are there any other methods which, like imaging and unlike conditionalizing, can give us a probability-revision conditional? There are not, as we shall see. The only way to have a probability-revision conditional is to interpret the conditional in Stalnaker's way and revise by imaging.

Since we have not fixed on a particular method of revising probability functions, our definition of a probability-revision conditional should be understood as tacitly relative to a method. To make this relativity explicit, let us call → a *probability-revision conditional for* a given method iff (20) holds in general when P_A is taken to be the revision obtained by that method.

Our definition of a Stalnaker conditional should likewise be understood as tacitly relative to a method of revising worlds. Stalnaker's truth conditions were deliberately left vague at the point where they mention the minimal revision of a given world to make a given antecedent true. With worlds, as with probability functions, different methods of revision will yield revisions that are minimal in different senses. We can indeed describe any method as selecting the antecedent-world closest in overall similarity to the original world; but different methods will fit this description under different resolutions of the vagueness of similarity, resolutions that stress different respects of comparison. To be explicit, let us call $\rightarrow$ a *Stalnaker conditional for* a given method of revising worlds iff (16) holds in general when W_A is taken to be the revision obtained by that method (and $A \rightarrow C$ is true at all worlds if A is impossible). I spoke loosely of "the" Stalnaker conditional, but henceforth it will be better to speak in the plural of the Stalnaker conditionals for various methods of revising worlds.

We are interested only in those methods of revision, for worlds and for probability functions, that can be regarded as giving revisions that are in some reasonable sense minimal. We have no hope of saying in any precise way just which methods those are, but at least we can list some formal requirements that such a method must satisfy. The requirements were given by Stalnaker for revision of worlds, but they carry over *mutatis mutandis* to revision of probability functions also. First, a minimal revision to reach some goal must be one that does reach it. For worlds, W_A must be a world where A is true; for probability function, P_A must assign to A a probability of 1. Second, there must be no revision when none is needed. For worlds, if A is already true at W then W_A must be W itself; for probability functions, if $P(A)$ is already 1, then P_A must be P. Third, the method must be consistent in its comparisons. For worlds, if B is true at W_A and A is true at W_B then W_A and W_B must be the same; else W_A would be treated as both less and more of a revision of W than is W_B. Likewise for probability functions, if $P_A(B)$ and $P_B(A)$ both are 1, then P_A and P_B must be the same.

Let us call any method of revision of worlds or of probability functions *eligible* iff it satisfies these three requirements. We note that the methods of revising probability functions that we have considered are indeed eligible. Conditionalizing is an eligible method; or, more precisely, conditionalizing can be extended to an eligible method applicable to any probability function P and any possible A. (Choose some fixed arbitrary well-ordering of all probability functions. In case P_A cannot be obtained by conditionalizing because $P(A)$ is zero, let it be the first, according to the arbitrary ordering, of the probability functions that assign to A a probability of 1.) Imaging is also an eligible method. More precisely, imaging on the basis of any eligible method of revising worlds is an eligible method of revising probability functions.

Our theorem of the previous section may be restated as follows. *If →
is a Stalnaker conditional for any eligible method of revising worlds, then
→ is also a probability-revision conditional for an eligible method of revising
probability functions; namely, for the method that works by imaging on
the basis of the given method of revising worlds.* Now we shall prove the
converse: *if → is a probability-revision conditional for an eligible method
of revising probability functions, then → is also a Stalnaker conditional for
an eligible method of revising worlds.* In short, the *probability-revision
conditionals are exactly the Stalnaker conditionals.*

Suppose that we have some eligible method of revising probability functions; and suppose that → is a probability-revision conditional for this
method.

We shall need to find a method of revising worlds; therefore let us consider the revision of certain special probability functions that stand in one-
to-one correspondence with the worlds. For each world W, there is a probability function P that gives all the probability to W and none to any other
world. Accordingly, by (17),

$$(21) \qquad P(A) = \begin{Bmatrix} 1 \text{ if } A \text{ is true at } W \\ 0 \text{ if } A \text{ is false at } W \end{Bmatrix} = W(A)$$

for any sentence A. Call such a probability function *opinionated,* since it
would represent the beliefs of someone who was absolutely certain that the
world W was actual and who therefore held a firm opinion about every
question; and call the world W where P concentrates all the probability
the *belief world of P.*

Our given method of revising probability functions preserved opinion-
ation. Suppose P were opinionated and P_A were not, for some possible A. That
is to say that P_A gives positive probability to two or more worlds. We have
assumed that our language has the means to distinguish the worlds, so there
is some sentence C such that $P_A(C)$ is neither 0 nor 1. But since P is
opinionated, $P(A → C)$ is either 0 or 1, contradicting the hypothesis that
→ is a probability-revision conditional so that $P_A(C)$ and $P(A → C)$ are equal.

Then we have the following method of revising worlds. Given a world
W and possible sentence A, let P be the opinionated probability function
with belief world W, revise P according to our given method of revising
probability functions, and let W_A be the belief world of the resulting
opinionated probability function P_A. Since the given method of revising
probability function is eligible, so is this derived method of revising worlds.

Consider any world W and sentences A and C. Let P be the opinionated probability function with belief world W, and let W_A be as above. Then if A is possible,

$$(22) \quad W(A \to C) \;=\; P(A \to C), \text{by (21)};$$
$$=\; P_A(C), \text{by (20)};$$
$$=\; W_A(C), \text{by (21) applied to } W_A.$$

So $\to$ is a Stalnaker conditional for the derived method of revising worlds. *Quod erat demonstrandum.*[13]

Princeton University

NOTES

[1] Ernest Adams, 'The Logic of Conditionals', *Inquiry* 8 (1965), 166-197; and 'Probability and the Logic of Conditionals', *Aspects of Inductive Logic*, ed. by Jaakko Hintikka and Patrick Suppes, Dordrecht, 1966. I shall not here consider Adams's subsequent work, which differs at least in emphasis.

[2] More precisely, just the ones that satisfy this condition: for any positive ϵ there is a positive δ such that if any probability function gives each premise an assertability within δ of 1 then it also gives the conclusion an assertability within ϵ of 1.

[3] Richard Jeffrey, 'If' (abstract), *Journal of Philosophy* 61 (1964), 702-703; Brian Ellis, 'An Epistemological Concept of Truth', *Contemporary Philosophy in Australia*, ed. by Rober Brown and C. D. Rollins, London, 1969; Robert Stalnaker, 'Probability and Conditionals', *Philosophy of Science*, 37 (1970), 64-80. We shall consider later whether to count Adams as another adherent of the thesis.

[4] I am indebted to Bas van Fraassen for this observation. He has also shown that by judicious selection of the B's we can give $\to$ some further properties that might seem appropriate to a conditional connective. See Bas van Fraassen, 'Probabilities of Conditionals', in *Foundations of Probability Theory, Statistical Inference and Statistical Theories of Science*, Volume I, ed. by W. Harper and C. A. Hooker, D. Reidel, Dordrecht, Holland, 1976, p. 261.

[5] These reasons may be found in Paul Teller, 'Conditionalization and Observation', *Synthese* 26 (1973), 218-258.

[6] 'The Logic of Conditionals'.

[7] H. P. Grice, 'Logic and Conversation', The William James Lectures, given at Harvard University in 1967.

[8] 'Probabilities and Conditionals'. The Stalnaker conditional had been introduced in Robert Stalnaker, 'A Theory of Conditionals', *Studies in Logical Theory*, ed. by Nicholas Rescher, Oxford, 1968. I have discussed the Stalnaker conditional in *Counterfactuals*, Oxford, 1973, pp. 77-83, arguing there that an interpretation quite similar to Stalnaker's is right for counterfactuals but wrong for indicative conditionals.

[9] Once it is recognized that the Stalnaker conditional is not a probability conditional, the coincidence of logics has a new significance. The hypothesis that assertability of

indicative conditionals goes by conditional probabilities, though still sufficiently well supported by direct evidence, is no longer unrivalled as an explanation of our judgements of validity for inferences with indicative conditional premises or conclusions. The same judgements could be explained instead by the hypothesis that the indicative conditional is the Stalnaker conditional and we judge valid those inferences that preserve truth.

[10] Although the probabilities of Stalnaker conditionals and the corresponding conditional probabilities cannot always be equal, they often are. They are equal whenever the conditional (and perhaps some non-conditional state of affairs on which it depends) is probabilistically independent of the antecedent. For example, my present subjective probabilities are such that the conditional probability of finding a penny in my pocket, given that I look for one, equals the probability of the conditional "I look for a penny > I find one." The reason is that both are equal to the absolute probability that there is a penny in my pocket now.

[11] Teller, 'Conditionalization and Observation'.

[12] A Theory of Conditionals', p. 102.

[13] An earlier version of this paper was presented at a Canadian Philosophical Association colloquium on probability semantics for conditional logic at Montreal in June 1972. I am grateful to many friends an colleagues, and especially to Ernest Adams and Robert Stalnaker, for valuable comments.

PART 4

CONDITIONALS FOR DECISION MAKING

(Another Paradigm)

ROBERT C. STALNAKER

LETTER TO DAVID LEWIS
May 21, 1972

It *does* seem to me worth noting that if P is a probability distribution, and if for any A and B, $P_B(A) = P(B > A)$, then P_B is a probability distribution too (excepting the absurd case). What it is good for, I would like to suggest, is deliberation – the calculation of expected utilities.

Let $S_1, \ldots, S_n$ be an exhaustive set of mutually exclusive propositions characterizing the alternative possible outcomes of some contemplated action. Let A be the proposition that I perform the action. My suggestion is that expected utility should be defined as follows:

$$u(A) = P(A > S_1) \times u(S_1) + \cdots + P(A > S_n) \times u(S_n).$$

Why $P(A > S_i)$ rather than $P(S_i/A)$? Because what is relevant to deliberation is a comparison of what will happen if I perform some action with what *would* happen if I instead did something else. A difference between $P(S/A)$ and $P(S)$ represents a belief that A is *evidentially relevant* to the truth of S, but not necessarily a belief that the action has any causal influence on the outcome. That a person performs a certain kind of action can be evidence that makes some state subjectively more probable, even when the action in no way contributes to the state. Suppose that this is true for some action A and desirable state S. Then $P(S/A) > P(S)$, but only an ostrich would count this as any sort of reason inclining one to bring it about that A. To do so would be to act so as to change the evidence, knowing full well that one is in no way changing the facts for which the evidence is evidence.

I am thinking of Nozick's puzzle ("Newcomb's problem", in the Hempel festschrift), which I just discovered, but which I assume you know. My intuitive reaction to this puzzle was the following: there is only one rational choice (assuming there is no backwards causation in the case), and that is to choose the dominating action. But this seems to conflict with the principle of maximizing expected utility. But from my suggested version of the principle, the rational choice follows. The principle of expected utility may be held to be universally applicable.

Since quotient conditionalization is the way to revise your beliefs, it is also rational in the Newcomb problem to bet, after having made the rational choice, that you will fail to get the million dollars. Had you made the other

151

W. L. Harper, R. Stalnaker, and G. Pearce (eds.), *Ifs*, 151–152.

choice, it would have been rational to bet that you would succeed in getting the million dollars. But *this* is no reason to wish that you had chosen differently, since you could have changed only the fair betting odds, not the facts, by acting differently.

The suggested version of the expected utility principle makes it possible for a single principle to account for various mixed cases: the probabilistic dependence may have two components, one causal and one non-causal. The components may reinforce each other, or counteract each other. They might cancel out, leaving the evidence irrelevant, even though there is a believed causal dependence. Also, it may be unknown whether the probabilistic dependence is causal or not. Imagine a man deliberating about whether or not to smoke. There are two, equally likely hypotheses (according to his beliefs) for explaining the statistical correlation between smoking and cancer: (1) a genetic disposition to cancer is correlated with a genetic tendency to the sort of nervous disposition which often inclines one to smoke. (2) Smoking, more or less, causes cancer in some cases. If hypothesis (1) is true, he has no independent way to find out whether or not he has the right sort of nervous disposition. In such a case, it seems clear that the probability of the conditional (if I were to smoke, I would get cancer), and not the conditional probability is what is relevant . . . .

ALLAN GIBBARD AND WILLIAM L. HARPER

COUNTERFACTUALS AND TWO KINDS
OF EXPECTED UTILITY*

1. INTRODUCTION

We begin with a rough theory of rational decision-making. In the first place, rational decision-making involves conditional propositions: when a person weighs a major decision, it is rational for him to ask, for each act he considers, what would happen if he performed that act. It is rational, then, for him to consider propositions of the form 'If I were to do a, then c would happen'. Such a proposition we' shall call a *counterfactual*, and we shall form counterfactuals with a connective '$\square\!\!\rightarrow$' on this pattern: 'If I were to do a, then c would happen' is to be written 'I do a $\square\!\!\rightarrow$ c happens'.

Now ordinarily, of course, a person does not know everything that would happen if he performed a given act. He must resort to probabilities: he must ascribe a probability to each pertinent counterfactual 'I do a $\square\!\!\rightarrow$ c happens'. He can then use these probabilities, along with the desirabilities he ascribes to the various things that might happen if he did a given act, to reckon the expected utility of a. If a has possible outcomes $o_1, \ldots, o_n$, the expected utility of a is the weighted sum

$$\Sigma_i \, prob \, (\text{I do } a \, \square\!\!\rightarrow o_i \text{ obtains}) \mathcal{D} o_i,$$

where $\mathcal{D} o_i$ is the desirability of o_i. On the view we are sketching, then, the probabilities to be used in calculating expected utility are the probabilities of certain counterfactuals.

That is not the story told in familiar Bayesian accounts of rational decision; those accounts make no overt mention of counterfactuals. We shall discuss later how Savage's account (1972) does without counterfactuals; consider first an account given by Jeffrey (1965, pp. 5–6).

A formal Bayesian decision problem is specified by two rectangular arrays (matrices) of numbers which represent probability and desirability assignments to the act-condition pairs. The columns represent a set of incompatible conditions, an unknown one of which actually obtains. Each row of the desirability matrix,

$$d_1 \, d_2 \ldots d_n$$

represents the desirabilities that the agent attributes to the n conditions described by the

W. L. Harper, R. Stalnaker, and G. Pearce (eds.), Ifs, 153–190.
Copyright © 1978 by D. Reidel Publishing Company.

column headings, on the assumption that he is about to perform the act described by the row heading; and the corresponding row of the probability matrix,

$$p_1 \, p_1 \ldots p_n$$

represents the probabilities that the agent attributes to the same n conditions, still on the assumption that he is about to perform the act described by the row heading. To compute the expected desirability of the act, multiply the corresponding probabilities and desirabilities, and add:

$$p_1 d_1 + p_2 d_2 + \ldots + p_n d_n.$$

On the Bayesian model as presented by Jeffrey, then, the probabilities to be used in calculating 'expected desirability' are 'probabilities that the agent attributes' to certain conditions 'on the assumption that he is about to perform' a given act. These, then, are conditional probabilities; they take the form $prob \, (S/A)$, where A is the proposition that the agent is about to perform a given act and S is the proposition that a given condition holds.

On the account Jeffrey gives, then, the probabilities to be used in decision problems are not the unconditional probabilities of certain counterfactuals, but are instead certain conditional probabilities. They take the form $prob \, (S/A)$, whereas on the view we sketched at the outset, they should take the form $prob \, (A \, \square \!\!\rightarrow S)$. Now perhaps, for all we have said so far, the difference between these accounts is merely one of presentation. Perhaps for every appropriate A and S, we have

(1) $prob \, (A \, \square \!\!\rightarrow S) = prob \, (S/A);$

the probability of a counterfactual $A \, \square \!\!\rightarrow S$ always equals the corresponding conditional probability. That would be so if (1) is a logical truth. David Lewis, however, has shown (1976) that on certain very weak and plausible assumptions, (1) is not a logical truth: it does not hold in general for arbitrary propositions A and S.[1] That leaves the possibility that (1) holds at least in all decision contexts: that it holds whenever A is an act an agent can perform and $prob$ gives that agent's probability ascriptions at the time.

In Section 3, we shall state a condition that guarantees the truth of (1) in decision contexts. We shall argue, however, that there are decision contexts in which this condition is violated. The context we shall use as an example is patterned after one given by Stalnaker. We shall follow Stalnaker in arguing that in such contexts, (1) indeed fails, and it is probabilities of counterfactuals rather than conditional probabilities that should be used in calculations of expected utility. The rest of the paper takes up the ramifications for decision theory of the two ways of calculating expected utility. In particular, the two opposing answers to Newcomb's problem (Nozick, 1969) are supported respectively by the two kinds of expected utility maximization we are discussing.

We are working in this paper within the Bayesian tradition in decision theory, in that the probabilities we are using are subjective probabilities, and we suppose an agent to ascribe values to all probabilities needed in calculations of expected utilities. It is not our purpose here to defend this general tradition, but rather to work within it, and to consider two divergent ways of developing it.

2. COUNTERFACTUALS

What we shall be saying requires little in the way of an elaborate theory of counterfactuals. We do suppose that counterfactuals are genuine propositions. For a proposition to be a counterfactual, we do not require that its antecedent be false: on the view we are considering, a rational agent entertains counterfactuals of the form 'I do $a \,\square\!\!\rightarrow S$' both for the act he will turn out to perform and for acts he will turn out not to perform. To say $A \,\square\!\!\rightarrow S$ is not to say that A's holding would bring about S's holding: $A \,\square\!\!\rightarrow S$ is indeed true if A's holding would bring about S's holding, but $A \,\square\!\!\rightarrow S$ is true also if S would hold regardless of whether A held.

These comments by no means constitute a full theory of counterfactuals. In what follows, we shall appeal not to a theory of counterfactuals, but to the reader's intuitions about them — asking the reader to bear clearly in mind that 'I do $a \,\square\!\!\rightarrow S$' is to be read 'If I were to do a, then S would hold'.

It may nevertheless be useful to sketch a theory that would support what we shall be saying; the theory we sketch here is somewhat like that of Stalnaker and Thomason (Stalnaker, 1968; Stalnaker and Thomason, 1970). Let a be an act which I might decide at time t to perform. An a-world will be a possible world which is like the actual world before t, in which I decide to do a at t and do it, and which obeys physical laws from time t on. Let W_a be the a-world which, at t, is most like the actual world at t. Thus W_a is a possible world which unfolds after t in accordance with physical law, and whose initial conditions at time t are minimally different from conditions in the actual world at t in such a way that 'I do a' is true in W_a. The differences in initial conditions should be entirely within the agent's decision-making apparatus. Then 'I do $a \,\square\!\!\rightarrow S$' is true iff S is true in W_a.[2]

Two axioms that hold on this theory will be useful in later arguments. Our first axiom is just a principle of modus ponens for the counterfactual.

AXIOM 1. $(A \,\&\, (A \,\square\!\!\rightarrow S)) \supset S.$

Our second axiom is a Stalnaker-like principle.

AXIOM 2. $(A \mathbin{\square\!\rightarrow} \bar{S}) \equiv \overline{(A \mathbin{\square\!\rightarrow} S)}$.

The rationale for this is that 'I do $a \mathbin{\square\!\rightarrow} S$' is true iff S holds in W_a and 'I do $a \mathbin{\square\!\rightarrow} \bar{S}$' is true iff $\bar{S}$ holds in W_a. We shall also appeal to a consequence of these axioms.

CONSEQUENCE 1. $A \supset [(A \mathbin{\square\!\rightarrow} S) \equiv S]$.

We do not regard Axiom 2 and Consequence 1 as self-evident. Our reason for casting the rough theory in a form which gives these principles is that circumstances where these can fail involve complications which it would be best to ignore in preliminary work.[3] Our appeals to these Axioms will be rare and explicit. For the most part in treating counterfactuals we shall simply depend on a normal understanding of the way counterfactuals apply to the situations we discuss.

3. TWO KINDS OF EXPECTED UTILITY

We have spoken on the one hand of expected utility calculated from the probabilities of counterfactuals, and on the other hand of expected utility calculated from conditional probabilities. In what follows, we shall not distinguish between an act an agent can perform and the proposition that says that he is about to perform it; acts will be expressed by capital letters early in the alphabet. An act will ordinarily have a number of alternative outcomes, where an *outcome* of an act is a single proposition which, for all the agent knows, expresses *all* the consequences of that act which he cares about. An outcome, then, is a specification of what might eventuate which is complete in the sense that any further specification of detail is irrelevant to the agent's concerns, and it specifies something that, for all the agent knows, might really happen if he performed the act. The agent, we shall assume, ascribes a magnitude $\mathscr{D}O$ to each outcome O. He knows that if he performed the act, one and only one of its outcomes would obtain, although he does not ordinarily know which of its outcomes that would be.

Let $O_1, \ldots, O_m$ be the outcomes of act A. The *expected utility* of A *calculated from probabilities of counterfactuals* we shall call $\mathscr{U}(A)$; it is given by the formula

$$\mathscr{U}(A) = \Sigma_j \, prob \, (A \mathbin{\square\!\rightarrow} O_j) \, \mathscr{D}O_j.$$

The *expected utility* of A *calculated from conditional probabilities* we shall call $\mathscr{V}(A)$; it is given by the formula

$$\mathscr{V}(A) = \Sigma_j \, prob \, (O_j/A) \mathscr{D}O_j.$$

Perhaps the best mnemonic for distinguishing $\mathscr{U}$ from $\mathscr{V}$ is this: we shall be advocating the use of counterfactuals in calculating expected utility, and we shall claim that $\mathscr{U}(A)$ is the genuine expected utility of A. $\mathscr{V}(A)$, we shall claim, measures instead the welcomeness of the news that one is about to perform A. Remember $\mathscr{V}(A)$, then, as the *value* of A *as news*, and remember $\mathscr{U}(A)$ as what the authors regard as the genuine expected utility of A.

Now clearly $\mathscr{U}(A)$ and $\mathscr{V}(A)$ will be the same if

(2) $prob \, (A \,\square\!\!\rightarrow O_j) = prob \, (O_j/A)$

for each outcome O_j. Unless (2) holds for every O_j such that $\mathscr{D}O_j \neq 0$, $\mathscr{U}(A)$ and $\mathscr{V}(A)$ will be the same only by coincidence. We know from Lewis's work (1976) that (2) does not hold for all propositions A and O_j; can we expect that (2) will hold for the appropriate propositions?

One assumption, together with the logical truth of Consequence 1, will guarantee that (2) holds for an act and its outcomes. Here and throughout, we suppose that the function *prob* gives the probability ascriptions of an agent who can immediately perform the act in question, and that $prob \, \phi = 1$ for any logical truth ϕ.

CONDITION 1 *on act A and outcome O_i.* The counterfactual $A \,\square\!\!\rightarrow O_i$ is stochastically independent of the act A. That is to say,

$$prob(A \,\square\!\!\rightarrow O_i/A) = prob \, (A \,\square\!\!\rightarrow O_i).$$

(Read *prob* $(A \,\square\!\!\rightarrow O_i/A)$ as the conditional probability of $A \,\square\!\!\rightarrow O_i$ on A.)

ASSERTION 1. Suppose Consequence 1 is a logical truth. If A and O_i satisfy Condition 1, and $prob \, (A) > 0$, then

$$prob \, (A \,\square\!\!\rightarrow O_i) = prob \, (O_i/A).[4]$$

Proof. Since Consequence 1 is a logical truth, for any propositions P and Q,

$$prob \, (P \supset [(P \,\square\!\!\rightarrow Q) \equiv Q]) = 1.$$

Hence if *prob* $P > 0$, then

$$prob \, ([(P \,\square\!\!\rightarrow Q) \equiv Q]/P) = 1;$$

$$\therefore prob\ (P \,\square\!\!\rightarrow Q/P) = prob\ (Q/P).$$

From this general truth we have

$$prob\ (A \,\square\!\!\rightarrow O_i/A) = prob\ (O_i/A),$$

and from this and Condition 1, it follows that

$$prob\ (A \,\square\!\!\rightarrow O_i) = prob\ (O_i/A).$$

That proves the Assertion.

Condition 1 is that the counterfactuals relevant to decision be stochastically independent of the acts contemplated. Stochastic independence is the same as epistemic independence. For $prob\ (A \,\square\!\!\rightarrow O_i/A)$ is the probability it would be rational for the agent to ascribe to the counterfactual $A \,\square\!\!\rightarrow O_i$ on learning A and nothing else — on learning that he was about to perform that act. Thus to say that $prob\ (A \,\square\!\!\rightarrow O_i/A) = prob\ (A \,\square\!\!\rightarrow O_i)$ is to say that learning that one was about to perform the act would not change the probability one ascribes to the proposition that if one were to perform the act, outcome O_i would obtain. We shall use the terms 'stochastic independence' and 'epistemic independence' interchangeably.

The two kinds of expected utility $\mathscr{U}$ and $\mathscr{V}$ can also be characterized in a way suggested by Jeffrey's account of the Bayesian model. Let acts $A_1, \ldots,$ A_m be open to the agent. Let states $S_1, \ldots, S_n$ partition the possibilities in the following sense. For any propositions $S_1, \ldots, S_n$, the truth-function $aut(S_1, \ldots, S_n)$ will be their exclusive disjunction: $aut(S_1, \ldots, S_n)$ holds in and only in circumstances where exactly one of $S_1, \ldots, S_n$ is true. Let the agent know $aut\ (S_1, \ldots, S_n)$. For each act A_i and state S_j, let him know that if he did A_i and S_j obtained, the outcome would be O_{ij}. Let him ascribe each outcome O_{ij} a desirability $\mathscr{D}O_{ij}$. This will be a *matrix formulation* of a decision problem; its defining features are that the agent knows that $S_1, \ldots, S_n$ partition the possibilities, and in each of these states $S_1, \ldots, S_n$, each act open to the agent has a unique outcome. A set $\{S_1, \ldots, S_n\}$ of states which satisfy these conditions will be called *the states of a matrix formulation* of the decision problem in question.

Both $\mathscr{U}$ and $\mathscr{V}$ can be characterized in terms of a matrix formulation:

$$\mathscr{U}(A_i) = \Sigma_j\ prob\ (A_i \,\square\!\!\rightarrow S_j)\ \mathscr{D}O_{ij};$$
$$\mathscr{V}(A_i) = \Sigma_j\ prob\ (S_j/A_i)\ \mathscr{D}O_{ij}.$$

If $\mathscr{D}O_{ij}$ can be regarded as the desirability the agent attributes to S_j 'on the assumption that' he will do A_i, then $\mathscr{V}(A_i)$ is the desirability of A_i as characterized in the account we quoted from Jeffrey.

On the basis of these matrix characterizations of $\mathcal{U}$ and $\mathcal{V}$, we can state another sufficient condition for the $\mathcal{U}$-utility and $\mathcal{V}$-utility of an act to be the same.

CONDITION 2 *on act* A_i, *states* $S_1, \ldots, S_n$, *and the function prob*. For each A_i and S_j,

$$prob\ (A_i \,\Box\!\!\rightarrow S_j/A_i) = prob\ (A_i \,\Box\!\!\rightarrow S_j).$$

ASSERTION 2. Suppose Consequence 1 is a logical truth. If a decision problem satisfies Condition 2 for act A_i, then $\mathcal{U}(A_i) = \mathcal{V}(A_i)$. The proof is like that of Assertion 1.

4. ACT-DEPENDENT STATES IN THE SAVAGE FRAMEWORK

Savage's representation of decision problems (1954) is roughly the matrix formulation just discussed. Ignorance is represented as ignorance about which of a number of states of the world obtains. These states are mutually exclusive, and as specific as the problem requires (p. 15). The agent ascribes desirability to 'consequences', or what we are calling *outcomes*. For each act open to the agent, he knows what outcome obtains for each state of the world; if he does not, the problem must be reformulated so that he does. Savage indeed defines an act as a function from states to outcomes (Savage, 1954, p. 14).

It is a consequence of the axioms Savage gives that a rational agent is disposed to choose as if he ascribed a numerical desirability to each outcome and a numerical probability to each state, and then acted to maximize expected utility, where the expected utility of an act A is

(3) $\Sigma_S\ prob\ (S)\mathcal{D}O(A, S).$

(Here $O(A, S)$ is the outcome of act A in state S.) Another consequence of Savage's axioms is the principle of dominance: If for every state S, the outcome of act A in S is more desirable than the outcome of B in S, then A is preferable to B.

Consider this misuse of the Savage apparatus; it is of a kind discussed by Jeffrey (1965, pp. 8–10).

CASE 1. David wants Bathsheba, but since she is the wife of Uriah, he fears that summoning her to him would provoke a revolt. He reasons to himself as follows: 'There are two possibilities: R, that there will be a revolt, and $\bar{R}$, that there won't be. The outcomes and their desirabilities are given in Matrix 1, where B is that I take Bathsheba and A is that I abstain from her. Whether or not there is a revolt, I prefer having Bathsheba to not having her, and so taking Bathsheba dominates over abstaining from her.

	R	$\bar{R}$
A	$R\bar{B}(0)$	$\bar{R}\bar{B}(9)$
B	$RB(1)$	$\bar{R}B(10)$

Matrix 1

This argument is of course fallacious: dominance requires that the states in question be independent of the acts contemplated, whereas taking Bathsheba may provoke revolt. To apply the Savage framework to a decision problem, one must find states of the world which are in some sense act-independent.

We now pursue a suggestion by Jeffrey on how to deal with states that are act-dependent. Construct four new conditionalized[5] states:

S_{00} : There would be no revolt whatever I did.
S_{01} : A would not elicit revolt, whereas B would.
S_{10} : A would elicit revolt, whereas B would not.
S_{11} : There would be a revolt whatever I did.

If these states hold independently of A and B, we can now work from Matrix 2 without fallacy. Since in Matrix 2 neither row dominates, the decision must be made on the basis of probabilities ascribed to the states $S_{00}, \ldots, S_{11}$.

	S_{00}	S_{01}	S_{10}	S_{11}
A	$\bar{R}\bar{B}(9)$	$\bar{R}\bar{B}(9)$	$R\bar{B}(0)$	$R\bar{B}(0)$
B	$\bar{R}B(10)$	$RB(1)$	$\bar{R}B(10)$	$RB(1)$

Matrix 2

What should the probabilities of these states be? One possible answer would be this: Each of the four states $S_{00}, \ldots, S_{11}$ can be expressed as a conjunction of counterfactuals. S_{01}, for instance, is the proposition $(A \;\square\!\!\!\rightarrow \bar{R})$ & $(B \;\square\!\!\!\rightarrow R)$. The probability of S_{01}, then, is simply the probability of this proposition, $prob\,([A \;\square\!\!\!\rightarrow \bar{R}] \;\&\; [B \;\square\!\!\!\rightarrow R])$.

The expected utility of an act can now be calculated in the standard way given by (3). The expected utility of A, for instance, will be

(4) $\Sigma_S\, prob\,(S)\,\mathscr{D}O(A,S),$

where the summation is over the new states S, and $O(A, S)$ is the outcome of A in state S.

Does this procedure give the correct expected utility for the act? What it gives as the expected utility of A, we can show, is $\mathcal{U}(A)$ — at least that is what it gives if Axiom 2 is part of the logic of counterfactuals. For (4) expands to

$$prob\ (S_{00})\mathcal{D}O(A, S_{00}) + prob\ (S_{01})\mathcal{D}O(A, S_{01}) +$$
$$prob\ (S_{10})\mathcal{D}O(A, S_{10}) + prob\ (S_{11})\mathcal{D}O(A, S_{11}).$$

We have

$$O(A, S_{00}) = O(A, S_{01}) = \bar{R}\bar{B};$$
$$O(A, S_{10}) = O(A, S_{11}) = R\bar{B}.$$

Thus since S_{00} and S_{01} are mutually exclusive, (4) becomes

$$prob\ (S_{00} \vee S_{01})\mathcal{D}\bar{R}\bar{B} + prob\ (S_{10} \vee S_{11})\mathcal{D}R\bar{B}.$$

Now $S_{00} \vee S_{01}$ is $([A\ \Box\!\!\rightarrow \bar{R}]\ \&\ [B\ \Box\!\!\rightarrow \bar{R}]) \vee [A\ \Box\!\!\rightarrow \bar{R}]\ \&\ [B\ \Box\!\!\rightarrow R])$, and in virtue of the logical truth of Axiom 2, this is $A\ \Box\!\!\rightarrow \bar{R}$. Similarly, $S_{10} \vee S_{11}$ is $A\ \Box\!\!\rightarrow R$. Thus (4) becomes

$$prob\ (A\ \Box\!\!\rightarrow \bar{R})\mathcal{D}\bar{R}\bar{B} + prob\ (A\ \Box\!\!\rightarrow R)\mathcal{D}R\bar{B},$$

which is $\mathcal{U}(A)$. This proof can of course be generalized.

We have considered one way to construct conditionalized states from act-dependent states; it is a way that makes use of counterfactuals. Suppose, though, we want to avoid the use of counterfactuals and rely instead on conditional probabilities. Jeffrey, as we understand him, suggests the following: ascribe to each new, conditionalized state the product of the pertinent conditional probabilities. We shall call this probability $prob^*$; thus, for instance,

$$prob^*(S_{01}) = prob\ (\bar{R}/A)\ prob\ (R/B),$$

and corresponding formulas hold for the other new states S_{00}, S_{10}, and S_{11}.

Using $prob^*$, we can again calculate expected utility in the standard way given by (3). The expected utility of A, for instance, will be

(5) $\Sigma_S\ prob^*(S)\mathcal{D}O(A, S),$

where again the summation is over the new states S_{00}, S_{01}, S_{10}, and S_{11}. Now (5), it can be shown, has the value $\mathcal{V}(A)$. For (5) is the sum of terms

$$prob^*\ (S_{00})\mathcal{D}O(A, S_{00}) = prob\ (\bar{R}/A)\ prob\ (\bar{R}/B)\mathcal{D}\bar{R}\bar{B},$$
$$prob^*\ (S_{01})\mathcal{D}O(A, S_{01}) = prob\ (\bar{R}/A)\ prob\ (R/B)\mathcal{D}R\bar{B},$$

$$prob^* (S_{10})\mathscr{D}O(A, S_{10}) = prob\ (R/A)\,prob\ (\bar{R}/B)\mathscr{D}R\bar{B},$$
$$prob^* (S_{11})\mathscr{D}O(A, S_{11}) = prob\ (R/A)\,prob\ (R/B)\mathscr{D}R\bar{B}.$$

Thus (5) equals

$$[prob\ (\bar{R}/B) + prob\ (R/B)]\ prob\ (\bar{R}/A)\mathscr{D}\bar{R}\bar{B}$$
$$+ [prob\ (\bar{R}/B) + prob\ (R/B)]\ prob\ (R/A)\mathscr{D}R\bar{B}$$
$$= prob\ (\bar{R}/A)\mathscr{D}\bar{R}\bar{B} + prob\ (R/A)\mathscr{D}R\bar{B},$$

and this is $\mathscr{V}(A)$.

Where, then, a decision problem is misformulated in the Savage framework with act-dependent states, we now have two ways of reformulating the problem with conditionalized states. The first way is to express each conditionalized state as a conjunction of counterfactuals. If the expected utility of an act A is then calculated in the standard manner and Axiom 2 holds, the result is $\mathscr{U}(A)$. The second way to reformulate the problem is to ascribe to each new conditionalized state the product of the pertinent conditional probabilities. If the expected utility of an act A is then calculated in the standard manner, the result is $\mathscr{V}(A)$. If Axiom 2 holds, then, the two reformulations yield respectively the two kinds of expected utility we have been discussing.

So far we have given the two reformulations only for an example. Here is the way the two methods of reformulation work in general. Let acts A_1, $\ldots, A_m$ be open to the agent, let states $S_1, \ldots, S_n$ not all be act-independent, and for each A_i and S_j, let the outcome of act A_i in S_j be O_{ij}. For each possible sequence $T_1, \ldots, T_m$ consisting of states in $\{S_1, \ldots, S_n\}$, there will be a new, conditionalized state $S(T_1, \ldots, T_m)$. The outcome of an act A_i in the new state $S(T_1, \ldots, T_m)$ will simply be the outcome of A_i in the old state T_i. What has been said so far applies to both methods. Now, according to the first method of reformulation, this new state $S(T_1, \ldots, T_m)$ will be

$$(A_1 \ \Box\!\!\rightarrow T_1)\ \&\ \ldots\ \&\ (A_m \ \Box\!\!\rightarrow T_m),$$

and hence, of course, its probability will be the probability of this proposition. According to the second method of reformulation, the probability of new state $S(T_1, \ldots, T_m)$ will be

$$prob\ (T_1/A_1) \times \ldots \times prob\ (T_m/A_m).$$

Once the problem is reformulated, expected utility is to be calculated in the standard way by formula (3).

Are these two ways of reformulating a decision problem equivalent or distinct? They are, of course, equivalent if Axiom 2 hold and $\mathscr{U}(A_i) = \mathscr{V}(A_i)$

for each act A_i, since the first method yields $\mathcal{U}(A_i)$ if Axiom 2 holds and the second method yields $\mathcal{V}(A_i)$. We already know that Condition 2 and the logical truth of Consequence 1 guarantee that $\mathcal{U}(A_i) = \mathcal{V}(A_i)$. Therefore, we may conclude that if Condition 2 holds and Axioms 1 and 2 are logical truths, the two reformulations are equivalent. Condition 2, recall, is that the counterfactuals $A_i \,\square\!\!\rightarrow S_j$ are epistemically act-independent: that for each of the old, act-dependent states in terms of which the problem is formulated, learning that one is about to perform a given act will not change the probability one ascribes to the proposition that if one were to perform that act, that state would obtain.

The upshot of the discussion is this. For the Savage apparatus to apply to a decision problem, the states of the decision matrix must be independent of the acts. We have considered two ways of dealing with a problem stated in terms of act-dependent states; both ways involve reformulating the problem in terms of new states which are act-independent. Given the logical truth of Axioms 1 and 2, a sufficient condition for the equivalence of the two reformulations is that the counterfactuals $A_i \,\square\!\!\rightarrow S_j$ be epistemically act-independent.

5. ACT-DEPENDENT COUNTERFACTUALS

Should we expect Condition 2 to hold? In the case of David, it seems that we should. Suppose David somehow learned that he was about to send for Bathsheba; that would give him no reason to change the probability he ascribes to the proposition 'If I were to send for Bathsheba, there would be a revolt'. Similarly, if David learned that he was about to abstain from Bathsheba, that would give him no reason to change the probability he ascribes to the proposition 'If I were to abstain from Bathsheba, there would be a revolt'. In the case of David, it seems, the pertinent counterfactuals are epistemically act-independent, and hence for each act he can perform, the $\mathcal{U}$-utility and the $\mathcal{V}$-utility are the same.

When, however, a common factor is believed to affect both behaviour and outcome, Condition 2 may fail, and $\mathcal{U}$-utility may diverge from $\mathcal{V}$-utility. The following case is patterned after an example used by Stalnaker to make the same point.[6]

CASE 2. Solomon faces a situation like David's, but he, unlike David, has studied works on psychology and political science which teach him the following: Kings have two basic personality types, charismatic and uncharismatic.

A king's degree of charisma depends on his genetic make-up and early child-hood experiences, and cannot be changed in adulthood. Now charistmatic kings tend to act justly and uncharismatic kings unjustly. Successful revolts against charismatic kings are rare, whereas successful revolts against un-charismatic kings are frequent. Unjust acts themselves, though, do not cause successful revolts; the reason that uncharismatic kings are prone to successful revolts is that they have a sneaky, ignoble bearing. Solomon does not know whether or not he is charismatic; he does know that it is unjust to send for another man's wife.

Now in this case, Condition 2 fails for states R and $\bar{R}$. The counterfactual $B \;\square\!\!\rightarrow R$ is not epistemically independent of B: we have

$$prob\,(B \;\square\!\!\rightarrow R/B) > prob\,(B \;\square\!\!\rightarrow R).$$

For the conditional probability of anything on B is the probability Solomon would rationally ascribe to it if he learned that B. Since he knows that B's holding would in no way tend to bring about R's holding, he always ascribes the same probability to $B \;\square\!\!\rightarrow R$ as to R. Hence both $prob\,(B \;\square\!\!\rightarrow R) = prob\,(R)$ and $prob\,(B \;\square\!\!\rightarrow R/B) = prob\,(R/B)$. Now if Solomon learned that B, he would have reason to think that he was uncharismatic, and thus revolt-prone. Hence $prob\,(R/B) > prob\,(R)$, and therefore

(6) $prob\,(B \;\square\!\!\rightarrow R/B) = prob\,(R/B) > prob\,(R) = prob\,(B \;\square\!\!\rightarrow R).$

Here, then, the counterfactual is not epistemically act-independent.

(6) states also that $prob\,(B \;\square\!\!\rightarrow R) < prob\,(R/B)$, so that in this case, the probability of the counterfactual does not equal the corresponding con-ditional probability. By similar argument we could show that $prob\,(A \;\square\!\!\rightarrow R) > prob\,(R/A)$. Indeed in this case a $\mathscr{U}$-maximizer will choose to send for his neighbour's wife whereas a $\mathscr{V}$-maximizer will choose to abstain from her – although we shall need to stipulate the case in more detail to prove the latter.

Consider first $\mathscr{U}$-maximization. We have that

$$\mathscr{U}(B) = prob\,(B \;\square\!\!\rightarrow \bar{R})\mathscr{D}\bar{R}B + prob\,(B \;\square\!\!\rightarrow R)\mathscr{D}RB;$$
$$\mathscr{U}(A) = prob\,(A \;\square\!\!\rightarrow \bar{R})\mathscr{D}\bar{R}\bar{B} + prob\,(A \;\square\!\!\rightarrow R)\mathscr{D}R\bar{B}.$$

We have argued that $prob\,(B \;\square\!\!\rightarrow R) = prob\,(R)$. Similarly, $prob\,(A \;\square\!\!\rightarrow R) = prob\,(R)$, and so $prob\,(A \;\square\!\!\rightarrow R) = prob\,(B \;\square\!\!\rightarrow R)$. Likewise $prob\,(A \;\square\!\!\rightarrow \bar{R}) = prob\,(B \;\square\!\!\rightarrow \bar{R})$. We know that $\mathscr{D}\bar{R}B > \mathscr{D}\bar{R}\bar{B}$ and $\mathscr{D}RB > \mathscr{D}R\bar{B}$. There-fore $\mathscr{U}(B) > \mathscr{U}(A)$. This is in effect an argument from dominance, as we shall discuss in Section 8.

Now consider $\mathcal{V}$-maximization. Learning that A would give Solomon reason to think he was charismatic and thus not-revolt prone, whereas learning that B would give him reason to think that he was uncharismatic and revolt-prone. Thus $prob\ (R/B) > prob\ (R/A)$. Suppose the difference between these probabilities is greater than $1/9$, so that where $prob\ (R/A) = \alpha$ and prob $(R/B) = \alpha + \epsilon$, we have $\epsilon > 1/9$. From Matrix 1, we have

$$\mathcal{V}(A) = prob\ (\bar{R}/A)\mathcal{D}\bar{R}\bar{B} + prob\ (R/A)\mathcal{D}R\bar{B} = 9(1-\alpha) + 0.$$
$$\mathcal{V}(B) = prob\ (\bar{R}/B)\mathcal{D}\bar{R}B + prob\ (R/B)\mathcal{D}RB$$
$$= 10(1-\alpha-\epsilon) + 1(\alpha + \epsilon).$$

Therefore $\mathcal{V}(A)-\mathcal{V}(B) = 9\epsilon - 1$, and since $\epsilon > 1/9$, this is positive. We have shown that if $\epsilon > 1/9$, then although $\mathcal{U}(B) > \mathcal{U}(A)$, we have $\mathcal{V}(A) > \mathcal{V}(B)$. Thus $\mathcal{U}$-maximaization and $\mathcal{V}$-maximization in this case yield conflicting prescriptions.

Which of these prescriptions is the rational one? It seems clear that in this case it is rational to perform the $\mathcal{U}$-maximizing act: unjustly to send for the wife of his neighbor. For Solomon cares only about getting the woman and avoiding revolt. He knows that sending for the woman would not cause a revolt. To be sure, sending for her would be an indication that Solomon lacked charisma, and hence an indication that he will face a revolt. To abstain from the woman for this reason, though, would be knowingly to bring about an indication of a desired outcome without in any way bringing about the desired outcome itself. That seems clearly irrational.

For those who find Solomon too distant in time and place or who mistrust charisma, we offer the case of Robert Jones, rising young executive of International Energy Conglomerate Incorporated. Jones and several other young executives have been competing for a very lucrative promotion. The company brass found the candidates so evenly matched that they employed a psychologist to break the tie by testing for personality qualities that lead to long run successful performance in the corporate world. The test was administered to the candidates on Thursday. The promotion decision is made on the basis of the test and will be announced on Monday. It is now Friday. Jones learns, through a reliable company grapevine, that all the candidates have scored equally well on all factors except ruthlessness and that the promotion will go to whichever of them has scored highest on this factor, but he cannot find out which of them this is.

On Friday afternoon Jones is faced with a new problem. He must decide whether or not to fire poor old John Smith, who failed to meet his sales quota this month because of the death of his wife. Jones believes that Smith will come up to snuff after he gets over his loss provided that he is treated leniently, and that he can convince the brass that leniency to Smith will benefit the company. Moreover, he believes that this would favorably impress the brass with his astuteness. Unfortunately, Jones has no way to get in touch with them until after they announce the promotion on Monday.

Jones knows that the ruthlessness factor of the personality test he has taken accurately predicts his behaviour in just the sort of decision he now faces. Firing Smith is good evidence that he has passed the test and will get the promotion, while leniency is good evidence that he has failed the test and will not get the promotion. We suppose that the utilities and probabilities correspond to those facing Solomon. $\mathcal{V}$-maximizing recommends firing Smith, while $\mathcal{U}$-maximizing recommends leniency. Firing Smith would produce evidence that Jones will get his desired promotion. It seems clear, however, that to fire Smith for this reason despite the fact that to do so would in no way help to bring about the promotion and would itself be harmful, is irrational.

6. THE SIGNIFICANCE OF $\mathcal{U}$ AND $\mathcal{V}$

From the Solomon example, it should be apparent that the $\mathcal{V}$-utility of an act is a measure of the welcomeness of the news that one is about to perform that act. Such news may tend to be welcome because the act is likely to have desirable consequences, or tend to be unwelcome because the act is likely to have disagreeable consequences. Those, however, are not the only reasons an act may be welcome or unwelcome: an act may be welcome because its being performed is an indication that the world is in a desired state. Solomon, for instance, would welcome the news that he was about to abstain from his neighbor's wife, but he would welcome it not because he thought just acts any more likely to have desirable consequences than unjust acts, but because he takes just acts to be a sign of charisma, and he thinks that charisma may bring about a desired outcome.

$\mathcal{U}$-utility, in contrast, is a measure of the expected efficacy of an act in bringing about states of affairs the agent desires; it measures the expected value of the consequences of an act. That can be seen in the case of Solomon. The $\mathcal{U}$-utility of sending for his neighbor's wife is greater than that of abstaining, and that is because he knows that sending for her will bring about a consequence he desires − having the woman − and he knows that it will not bring about any consequences he wishes to avoid: in particular, he knows that it will not bring about a revolt.

What is it for an act to bring about a consequence? Here are two possible answers, both formulated in terms of counterfactuals.

In the first place, roughly following Sobel (1970, p. 400) we may say that act A brings about state S if $A \mathbin{\Box\!\!\rightarrow} S$ holds, and for some alternative A^* to A, $A^* \mathbin{\Box\!\!\rightarrow} S$ does not hold.[7] (An *alternative* to A is another act open to the agent on the same occasion). Now on this analysis, the $\mathcal{U}$-utility of an act as we have defined it is the sum of the expected value of its consequences plus a term which is the same for all acts open to the agent on the occasion in question; this latter term is the expected value of unavoidable outcomes. A state S is *unavoidable* iff for every act A^* open to the agent, $A^* \mathbin{\Box\!\!\rightarrow} S$ holds. Thus $A \mathbin{\Box\!\!\rightarrow} S$ holds iff S is a consequence of A or S is unavoidable. Hence in particular, for any outcome O,

$$prob\ (A \mathbin{\Box\!\!\rightarrow} O) = prob\ (O \text{ is a consequence of } A) \\ + prob\ (O \text{ is unavoidable}),$$

and so we have

$$\mathcal{U}(A) = \Sigma_O\, prob\ (A \mathbin{\Box\!\!\rightarrow} O)\mathcal{D}O \\ = \Sigma_O\, prob\ (O \text{ is a consequence of } A)\mathcal{D}O \\ + \Sigma_O\, prob\ (O \text{ is unavoidable})\mathcal{D}O.$$

The first term is the expected value of the consequences of A, and the second term is the same for all acts open to the agent. Therefore on this analysis of the term 'consequence', $\mathcal{U}$-utility is maximal for the act or acts whose consequences have maximal expected value.

Here is a second possible way of analyzing what it is to be a consequence. When an agent chooses between two acts A and B, what he really needs to know is not what the consequences of A are and what the consequences of B are, but rather what the consequences are of A as opposed to B and *vice versa*. Thus for purposes of decision-making, we can do without an analysis of the clause 'S is a consequence of A', and analyze instead the clause 'S is a consequence of A as opposed to B'. This we can analyze as

$$(A \mathbin{\Box\!\!\rightarrow} S) \,\&\, \sim (B \mathbin{\Box\!\!\rightarrow} S).$$

Now on this analysis, $\mathcal{U}(A) > \mathcal{U}(B)$ iff the expected value of the consequences of A as opposed to B exceeds the expected value of the consequences of B as opposed to A. For any state S, $A \mathbin{\Box\!\!\rightarrow} S$ holds iff either S is a consequence of A as opposed to B or $(A \mathbin{\Box\!\!\rightarrow} S) \,\&\, (B \mathbin{\Box\!\!\rightarrow} S)$ holds.

Thus $\qquad \mathcal{U}(A) = \Sigma_O \, prob \, (A \, \square\!\!\rightarrow O) \mathcal{D}O$
$\qquad\qquad\quad = \Sigma_O \, prob \, (O$ is a consequence of A as opposed to $B) \mathcal{D}O$
$\qquad\qquad\qquad + \Sigma_O \, prob \, ([A \, \square\!\!\rightarrow O] \, \& \, [B \, \square\!\!\rightarrow O]) \mathcal{D}O$
$\qquad\;\; \mathcal{U}(\mathrm{B}) = \Sigma_O \, prob \, (O$ is a consequence of B as opposed to $A) \mathcal{D}O$
$\qquad\qquad\qquad + \Sigma_O \, prob \, ([A \, \square\!\!\rightarrow O] \, \& \, [B \, \square\!\!\rightarrow O]) \mathcal{D}O.$

The second term is the same in both cases, and so $\mathcal{U}(A) > \mathcal{U}(B)$ iff

$\qquad \Sigma_O \, prob \, (O$ is a consequence of A as opposed to $B) \mathcal{D}O >$
$\qquad \Sigma_O \, prob \, (O$ is a consequence of B as opposed to $A) \mathcal{D}O.$

The left side is the expected value of the consequences of A as opposed to B; the right side is the expected value of the consequences of B as opposed to A. Thus for any pair of alternatives, to prefer the one with the higher $\mathcal{U}$-utility is to prefer the one the consequences of which as opposed to the other have the greater expected value.

We can now ask whether $\mathcal{U}$ or $\mathcal{V}$ is more properly called the 'utility' of an act. The answer seems clearly to be $\mathcal{U}$. The 'utility' of an act should be its expected genuine efficacy in bringing about states of affairs the agent wants, not the degree to which news of the act ought to cheer the agent. Since $\mathcal{U}$-utility is a matter of what the act can be expected to bring about whereas $\mathcal{V}$-utility is a matter of the welcomeness of news, $\mathcal{U}$-utility seems best to capture the notion of utility.

Jeffrey (1965, pp. 73–4) writes, 'If the agent is deliberating about performing act A or act B, and if AB is impossible, there is no effective difference between asking whether he prefers A to B as a news item or as an act, for he makes the news'. It should now be clear why it may sometimes be rational for an agent to choose an act B instead of an act A, even though he would welcome the news of A more than that of B. The news of an act may furnish evidence of a state of the world which the act itself is known not to produce. In that case, though the agent indeed makes the news of his act, he does not make all the news his act bespeaks.

7. TWO SURE THING PRINCIPLES

CASE 3. Upon his accession to the throne, Reoboam wonders whether to announce that he will reign severely or to announce that he will reign leniently. He will be bound by what he announces. He slightly prefers a short severe reign to a short lenient reign, and he slightly prefers a long severe reign to a long lenient reign. He strongly prefers a long reign of any kind to a short reign of any kind. Where L is that he is lenient and D, that he is deposed early, his utilities are as in the Matrix 3.

$$
\begin{array}{c|cc}
 & D & \bar{D} \\
\hline
L & 0 & 80 \\
\bar{L} & 10 & 100 \\
\end{array}
$$

Matrix 3

The wise men of the kingdom give him these findings of behavioural science: There is no correlation between a king's severity and the length of his reign. Severity, nevertheless, often causes early deposition. The reason for the lack of correlation between severity and early deposition is that on the one hand, charismatic kings tend to be severe, and on the other hand, lack of charisma tends to elicit revolts. A king's degree of charisma cannot be changed in adulthood. There is at present no indication of whether Reoboam is charismatic or not.

These findings were based on a sample of 100 kings, 48 of whom had their reigns cut short by revolt. On post mortem examination of the pineal gland, 50 were found to have been charismatic and 50 uncharismatic. 80% of the charismatic kings had been severe and 80% of the uncharismatic kings had been lenient. Of the charismatic kings, 40% of those who were severe were deposed whereas only 20% of those who were lenient were deposed. Of the uncharismatic kings, 80% of those who were severe were deposed whereas only 55% of those who were lenient were deposed. The totals were as in Table 1. This is Reoboam's total evidence on the subject.[8]

TABLE 1

	Charismatic	Uncharismatic	Total
Severe	16 deposed (40%) 24 long-reigned	8 deposed (80%) 2 long-reigned	24 deposed (48%) 26 long reigned
Lenient	2 deposed (20%) 8 long-reigned	22 deposed (55%) 18 long-reigned	24 deposed (48%) 26 long reigned

Reoboam's older advisors argue from a sure thing principle. There are two possibilities, they say: that Reoboam is charismatic and that he is uncharismatic; what he does now will not affect his degree of charisma. On the assumption that he is charismatic, it is rational to prefer lenience. For since

40% of severe charismatic kings are deposed, the expected utility of severity in that case would be

$$0.4\mathcal{D}SD + 0.6\mathcal{D}S\bar{D} = 0.4 \times 10 + 0.6 \times 80 = 52,$$

whereas since only 20% of lenient charismatic kings are deposed, the expected utility of lenience in that case would be

$$0.2\mathcal{D}LD + 0.8\mathcal{D}L\bar{D} = 0.2 \times 0 + 0.8 \times 10 = 64.$$

On the assumption that he is uncharismatic, it is again rational to prefer lenience. For since 80% of severe uncharismatic kings are deposed, the expected utility of severity in this case would be

$$0.8\mathcal{D}SD + 0.2\mathcal{D}S\bar{D} = 0.8 \times 10 + 0.2 \times 100 = 28,$$

whereas since only 55% of lenient uncharismatic kings are deposed, the expected utility of lenience in this case would be

$$0.55\mathcal{D}LD + 0.45\mathcal{D}L\bar{D} = 0.55 \times 0 + 0.45 \times 80 = 36.$$

Thus in either case, lenience is to be preferred, and so by a sure thing principle, it is rational to prefer lenience in the actual case.

Reoboam's youthful friends argue that on the contrary, sure thing considerations prescribe severity. Severity is indeed the dominant strategy. There are two possibilities: D, that Reoboam will be deposed, and $\bar{D}$, that he will not be. These two states are stochastically independent of the acts contemplated: both *prob* (D/S) and *prob* (D/L) are 0.48. Therefore, his youthful friends urge, one can without fallacy use the states D and $\bar{D}$ in an argument from dominance. On the assumption that he will be deposed, he prefers to be severe, and likewise on the assumption that he will not be deposed, he prefers to be severe. Thus by dominance, it is rational for him to prefer severity.

Here, then, are two sure thing arguments which lead to contrary prescriptions. One argument appeals to the finding that charisma is causally independent of the acts contemplated; the other appeals to the finding that being deposed is stochastically independent of the acts. The old advisors and youthful companions are in effect appealing to different versions of a sure thing principle, one of which requires causal independence and the other of which requires stochastic independence. The two versions lead to incompatible conclusions.

The sure thing principle is this: if a rational agent knows *aut* $(S_1, \ldots, S_n)$ and prefers A to B in each case, then he prefers A to B. If the propositions

$S_1, \ldots, S_n$ are required to be states in a matrix formulation of the decision problem, so that each pair of state and act determine a unique outcome, the sure thing principle becomes the principle of dominance to be discussed in Section 8; the principle of dominance is thus a special case of the sure thing principle. Now the principle of dominance, we have said, requires a proviso that the states in question be act-independent. The sure thing principle should presumably include the same proviso. The sure thing principle, then, should be this: If a rational agent knows that precisely one of the propositions S_1, $\ldots, S_n$ holds and prefers act A to act B in each case, and if in addition the propositions $S_1, \ldots, S_n$ are independent of the acts A and B, then he prefers A to B.

The problem in the case of Reoboam is that his two groups of advisors appeal to different kinds of independence to reach opposing conclusions. The older advisors appeal to causal independence; they cite the finding that a king's degree of charisma is unaffected by his adult actions. His youthful companions appeal to stochastic independence; they cite the finding that there is no correlation between severity in kings and revolt. The two appeals yield opposite conclusions.

It seems, then, that the sure thing principle comes in two different versions, one of which requires that the propositions in question be causally independent of the acts, and the other of which requires the propositions to be stochastically independent of the acts.

The principle to which the youthful companions appeal can be put as follows.

DEFINITION. Act A *is sure against* act B *with stochastic independence of* $S_1, \ldots, S_n$ iff the following hold. The agent knows that independently of the choice between A and B, propositions $S_1, \ldots, S_n$ partition the possibilities; that is to say, $prob\,(aut\,(S_1, \ldots, S_n)/A) = 1$ and $prob\,(aut\,(S_1, \ldots, S_n)/B) = 1$. The propositions $S_1, \ldots, S_n$ are epistemically independent of the choice between A and B, in the sense that for each, $prob\,(S_i/A) = prob\,(S_i/B)$. Finally, for each of these propositions S_i it would be rational to prefer A to B if it were known that S_i held.

Sure-thing with Stochastic Independence. If act A is sure against act B with stochastic independence, then it is rational to prefer A to B.

The principle to which the older advisors appeal will take longer to formulate. The proviso for this version will be that the propositions S_1, $\ldots, S_n$ be causally independent of the choice between A and B; this can be formulated in terms of counterfactuals. To say that a state S_i is causally independent of the choice between A and B is to say that S_i would hold if A were performed iff S_i would hold if B were performed: $(A \mathbin{\square\!\!\rightarrow} S_i) \equiv (B \mathbin{\square\!\!\rightarrow} S_i)$. We now want to suppose that for each state S_i, A would be preferred to B given, in some sense, knowledge of S_i. This knowledge of S_i should not simply be knowledge that S_i holds, but knowledge that S_i holds independently of the choice between A and B: that $(A \mathbin{\square\!\!\rightarrow} S_i) \mathbin{\&} (B \mathbin{\square\!\!\rightarrow} S_i)$. We can now state the principle.

For each S_i let $S_i{}^*$ be $(A \mathbin{\square\!\!\rightarrow} S_i) \mathbin{\&} (B \mathbin{\square\!\!\rightarrow} S_i)$.

DEFINITION. *A is sure against B with causal independence of* $S_1, \ldots, S_n$ iff the following hold. The agent knows $aut(S_1{}^*, \ldots, S_n{}^*)$, and for each S_i it would be rational to prefer A to B if $S_i{}^*$ were known to hold.[9]

(Note that since for each S_i, $(A \mathbin{\square\!\!\rightarrow} S_i) \equiv (B \mathbin{\square\!\!\rightarrow} S_i)$ follows from aut $(S_i{}^*, \ldots, S_n{}^*)$, this guarantees that our agent knows that each S_i is causally independent of the choice between A and B.) We can now state the principle to which the older advisors appeal.

Sure-thing with Causal Independence. If A is sure against B with causal independence, then it is rational to prefer A to B.

In the case of Reoboam, we have seen, Sure-thing with Stochastic Independence prescribes severity and Sure-thing with Causal Independence prescribes lenience. Now to us it seems clear that the only rational action in this case is that prescribed by Sure-thing with Causal Independence. It is rational for Reoboam to prefer lenience because severity tends to bring about deposition and he wants not to be deposed much more strongly than he wants to be severe. To be guided by Sure-thing with Stochastic Independence in this case is to ignore the finding that severity tends to bring about revolt – to ignore that finding simply because severity is not on balance a *sign* that revolt will occur. To choose to be severe is to act in a way that tends to bring about a dreaded consequence, simply because the act is not a sign of the consequence. That seems to us to be irrational.

The two versions of the sure thing principle we have discussed correspond to the two kinds of utility discussed earlier. Sure Thing with Stochastic Independence follows from the principle that an act is rationally preferred to another iff it has greater $\mathcal{V}$-utility, whereas Sure Thing with Causal Independence follows from the principle that an act is rationally preferred to another iff it maximizes $\mathcal{U}$-utility.

ASSERTION. Suppose that in any possible situation, it is rational to prefer an act A to an act B iff the $\mathcal{U}$-utility of A is greater than that of B. Then Sure Thing with Causal Independence holds.

Proof. Suppose A is sure against B with causal independence of $S_1, \ldots, S_n$, and that in any possible circumstance, it would be rational to prefer A to B iff A's $\mathcal{U}$-utility were greater than B's. The Assertion will be proved if we show from these assumptions that $\mathcal{U}(A) > \mathcal{U}(B)$.

Since A is sure against B with causal independence of $S_1, \ldots, S_n$, for each S_i it would be rational to prefer A to B if S_i^* were known to hold. Therefore if S_i^* were known to hold, the $\mathcal{U}$-utility of A would be greater than that of B. Now the $\mathcal{U}$-utility that A would have if S_i^* were known is

$$\Sigma_O \, prob \, (A \;\square\!\!\rightarrow O/S_i^*) \mathcal{D}O.$$

Call this $\mathcal{U}_i^*(A)$, and define $\mathcal{U}_i^*(B)$ in a like manner. We have supposed that for each $S_i, \mathcal{U}_i^*(A) > \mathcal{U}_i^*(B)$.

Now by definition of the function $\mathcal{U}$,

$$\mathcal{U}(A) = \Sigma_O \, prob \, (A \;\square\!\!\rightarrow O) \mathcal{D}O;$$

Since A is sure against B with causal independence of $S_1, \ldots, S_n$, it is known that $aut \, (S_1^*, \ldots, S_i^*)$ holds. By the probability calculus, then, for each outcome O

$$prob \, (A \;\square\!\!\rightarrow O) = \Sigma_i \, prob \, (A \;\square\!\!\rightarrow O/S_i^*) \, prob/S_i^*.$$

Therefore

$$\begin{aligned}\mathcal{U}(A) &= \Sigma_O[\Sigma_i \, prob \, (A \;\square\!\!\rightarrow O/S_i^*) \, prob \, S_i^*] \mathcal{D}O \\ &= \Sigma_i \, prob \, S_i^* [\Sigma_O prob \, (A \;\square\!\!\rightarrow O/S_i^*) \mathcal{D}O], \\ &= \Sigma_i \mathcal{U}_i^*(A) \, prob \, S_i^*.\end{aligned}$$

By a like argument,

$$\mathcal{U}(B) = \Sigma_i \mathcal{U}_i^*(B) \, prob \, S_i^*.$$

Since for each $S_i, \mathcal{U}_i^*(A) > \mathcal{U}_i^*(B)$, it follows that $\mathcal{U}(A) > \mathcal{U}(B)$, and the Assertion is proved.

ASSERTION. Suppose that in any possible circumstance, it is rational to prefer an act A to an act B iff the $\mathcal{V}$-utility of A is greater than that of B. Then Sure Thing with Stochastic Independence holds.

Proof. Suppose A is sure against B with causal independence of $S_1, \ldots, S_n$, and that in any possible circumstances, it would be rational to prefer A

to B iff A's $\mathcal{V}$-utility is greater than B's. The Assertion will be proved if we show from these assumptions that $\mathcal{V}(A) > \mathcal{V}(B)$.

Now since A is sure against B with stochastic independence of $S_1, \ldots, S_n$, for each S_i it would be rational to prefer A to B if S_i were known to hold. Therefore, if S_i were known to hold, then the $\mathcal{V}$-utility of A would be greater than that of B. Now the $\mathcal{V}$-utility that A would have if S_i were known to hold is

$$\Sigma_O \, prob \, (O/AS_i)\mathcal{D}O.$$

Call this $\mathcal{V}_i^*(A)$, and define $\mathcal{V}_i^*(B)$ correspondingly. We have that for each S_i, $\mathcal{V}_i^*(A) > \mathcal{V}_i^*(B)$. Now by definition of the function $\mathcal{V}$,

$$\mathcal{V}(A) = \Sigma_O \, prob \, (O/A)\mathcal{D}O.$$

Since A is sure against B with stochastic independence of $S_1, \ldots, S_n$, we have $prob \, (aut \, (S_1, \ldots, S_n)/A) = 1$, and so by the probability calculus, for each O,

$$\mathcal{V} prob \, (O/A) = \Sigma_i \, prob \, (O/AS_i) \, prob \, (S_i/A).$$

Hence

$$\begin{aligned}
\mathcal{V}(A) &= \Sigma_O[\Sigma_i \, prob \, (O/AS_i) \, prob \, (S_i/A)]\mathcal{D}O \\
&= \Sigma_i \, prob \, (S_i/A)[\Sigma_O \, prob \, (O/AS_i)\mathcal{D}O] \\
&= \Sigma_i \, prob \, (S_i/A)\mathcal{V}_i^*(A).
\end{aligned}$$

By a like argument,

$$\mathcal{V}(B) = \Sigma_i \, prob \, (S_i/B)\mathcal{V}_i^*(B).$$

Since for each S_i, $prob \, (S_i/A) = prob \, (S_i/B)$ and $\mathcal{V}_i^*(A) > \mathcal{V}_i^*(B)$ it follows that $\mathcal{V}(A) > \mathcal{V}(B)$, and the Assertion is proved.

8. TWO KINDS OF DOMINANCE

We have said that the principle of dominance is the sure thing principle restricted to a special case, and that the sure thing principle has two versions, one of which holds for $\mathcal{U}$-maximization and the other for $\mathcal{V}$-maximization. There should, then, be two versions of the principle of dominance, one for each kind of utility maximization. The principles can be formulated as follows.

DEFINITION. Let $S_1, \ldots, S_n$ be the states of a standard decision matrix, and let A and B be acts. Then A *strongly dominates* B *with respect to*

$S_1, \ldots, S_n$ if for each S_i, the outcome of A in S_i is more desirable than the outcome of B in S_i.

Principle of Dominance with Causal Independence. Suppose act A strongly dominates act B with respect to states $S_1, \ldots, S_n$. If for each state S_i, the agent knows that $(A \,\square\!\!\rightarrow S_i) \equiv S_i$ and $(B \,\square\!\!\rightarrow S_i) \equiv S_i$, then it is rational for him to prefer A to B.

Principle of Dominance with Stochastic Independence. Suppose act A strongly dominates act B with respect to states $S_1, \ldots, S_n$. If for each state S_i, $prob\,(S_i/A) = prob\,(S_i) = prob\,(S_i/B)$, then it is rational for him to prefer A to B.

The Principle of Dominance with Causal Independence holds if rationality requires maximization of $\mathcal{U}$, and the Principle of Dominance with Stochastic Independence holds if rationality requires maximization of $\mathcal{V}$.[10]

Although these two principles are respective consequences of two principles of expected utility maximization which may conflict, they cannot themselves conflict. For suppose A strongly dominates B with respect to some set of states $S_1, \ldots, S_n$. Then the worst outcome of A is more desirable than some outcome of B. For the worst outcome of A is the outcome of A in some state S_i, and since A strongly dominates B with respect to $S_1, \ldots, S_n$, the outcome of A in S_i is more desirable than the outcome of B in S_i. Thus the worst outcome of A is more desirable than the worst outcome of B. It cannot be the case, then, that B strongly dominates A with respect to some other set of states $T_1, \ldots, T_n$. For if that indeed were the case, then, we have seen, the worst outcome of B would be more desirable than the worst outcome of A. We have seen that if A strongly dominates B with respect to a set of states, then there is no set of states with respect to which B strongly dominates A. For that reason, the two principles of dominance we have stated will never yield conflicting prescriptions for a simple decision problem.

In a weaker form, however, dominance indeed can be exploited to yield conflicting prescriptions.

DEFINITION. Let $S_1, \ldots, S_n$ be the states of a standard decision matrix, and let A and B be acts. *A weakly dominates B with respect to $S_1, \ldots, S_n$* iff for each state S_i, the outcome of A in S_i is at least as desirable as the outcome of B in S_i, and for some state S_i with $prob\,(S_i) > O$, the outcome of A in S_i is more desirable than the outcome of B in S_i.

We now get two Principles of Weak Dominance by substituting 'weakly dominates' for 'strongly dominates' in the two Principles of Dominance stated above.

CASE 4. A subject is presented with two boxes, one to the left and one to the right. He must choose between two acts:

A_L Take the box on the left.
A_R Take the box on the right.

The experimenter has already done one of the following.

M_{11} Place a million dollars in each box.
M_{01} Place a million dollars in the box on the right and nothing in the box on the left.
M_{00} Place nothing in either box.

He has definitely not placed money in the left box without placing money in the right box. Now the experimenter has predicted the behavior of the subject, and before making his prediction, he has used a random device to select one of the following three strategies.

(i)   Reward choice of left box: M_{11} if A_L is predicted; M_{00} if A_R is predicted.
(ii)   Ensure payment: M_{11} if A_L is predicted; M_{01} if A_R is predicted.
(iii)   Ensure non-payment: M_{01} if A_L is predicted; M_{00} if A_R is predicted.

The subject knows all this, and believes in the accuracy of the experimenter's predictions with complete certainty.

The Principle of Weak Dominance with Causal Independence prescribes taking the box on the right. The three states M_{11}, M_{01}, and M_{00} are causally independent of the act the subject performs. The possible outcomes are shown in the table, where 1 is getting the million dollars and 0 is not getting it.

	M_{11}	M_{01}	M_{00}
A_L	1	0	0
A_R	1	1	0

M_{01} has non-zero probability, since if A_L was predicted it would result from the experimenter's using strategy (iii) and if A_R was predicted, it would result from the experimenter's using strategy (ii). Thus A_R weakly dominates

A_L with respect to M_{11}, M_{01}, M_{00}, and the Principle of Weak Dominance with Causal Independence prescribes taking the box on the right.

The Principle of Weak Dominance with Stochastic Independence, in contrast, prescribes taking the box on the left.

The possibilities can be partitioned as follows:

S_1 the experimenter predicts correctly and follows strategy (i).
S_2 S_1 does not hold and the subject wins a million dollars.
S_3 S_1 does not hold and the subject wins nothing.

The payoffs are given in the table.

	S_1	S_2	S_3
A_L	1	1	0
A_R	0	1	0

Now $prob\,(S_1) \neq 0$, and hence A_L weakly dominates A_R with respect to S_1, S_2, S_3. Moreover, the states S_1, S_2, and S_3 are stochastically independent of A_L and A_R. For the subject knows that the experimenter has selected his strategy independently of his prediction, by means of a random device; hence learning that he was about to perform A_L, say, would not affect the probability he ascribes to the experimenter's having had any given strategy. By the subject's probability function, then, which strategy the experimenter has used is stochastically independent of the subject's act. Now the subject believes that the experimenter has predicted correctly and used strategy (i), (ii), or (iii). Hence he thinks that S_1 holds iff the experimenter has used strategy (i), that S_2 holds iff the experimenter has used strategy (ii), and that S_3 holds if the experimenter has used strategy (iii). Hence under his probability function, states S_1, S_2, and S_3 are stochastically independent of A_L and A_R. Thus the Principle of Weak Dominance with Stochastic Independence applies, and it prescribes taking the box on the left.

Some readers may object in Case 4 to the subject's complete certainty that the experimenter has predicted correctly. It is possible to construct a conflict between the two principles of weak dominance without requiring such certainty, but the example becomes more complicated.

CASE 5. Same as Case 4, except for the following.

The subject ascribes a probability of 0.8 to the experimenter's having predicted correctly, and this probability is independent of the subject's choice of A_L or A_R. Thus where C is 'the experimenter has predicted correctly',

$prob\ (C/A_L) = 0.8$ and $prob\ (C/A_R) = 0.8$.

The experimenter has chosen among the following three strategies by means of a random device.

(i) M_{11} if A_L is predicted; M_{00} if A_R is predicted.

(ii*) M_{11} if A_L is predicted; M_{01} or M_{00}, with equal probability, if A_R is predicted.

(iii*) M_{11} or M_{01}, with equal probability, if A_L is predicted; M_{00} if A_R is predicted.

He has followed (i) with a probability 0.5, (ii*) with a probability 0.25, and (iii*) with a probability 0.25.

In Case 5, as in Case 4, the states M_{11}, M_{01}, and M_{00}, are causally independent of the acts A_R and A_L, and from the Principle of Weak Dominance with Causal Independence and the facts of the case, it follows that it is rational to prefer A_R to A_L.

Now let states S_1, S_2, and S_3 be as before: S_1 is that the experimenter predicts correctly and follows strategy (i); S_2 is that S_1 does not hold and the subject receives a million dollars; S_3 is that S_1 does not hold nd the subject receives nothing. As in Case 4, if S_1, S_2, and S_3 are stochastically independent of A_L and A_R, then from the Principle of Weak Dominance with Stochastic Independence and the facts of the case, it follows that it is rational to prefer A_L to A_R. It is clear that S_1 is stochastically independent of the acts A_L and A_R; we now show that S_2 and S_3 are as well: that $prob\ (S_2/A_L) = prob\ (S_2/A_R)$ and $prob\ (S_3/A_L) = prob\ (S_3/A_R)$.

There are two possible acts, two possible experimenter's predictions, and three possible experimenter's strategies, some of which may involve the flip of a coin. Call a combination of act, prediction, experimenter's strategy, and result of coin flip if it matters, a *case*. For each case, the Table 2 shows.

(1) The state M_{11}, M_{01}, or M_{00} which would hold in that case.

(2) The conditional probability of the case given the act.

(3) The outcome in that case: 1 for getting the million dollars, 0 for not.

(4) The state S_1, S_2, or S_3 which holds in that case.

The conditional probability $prob\ (S_2/A_L)$ is then obtained by adding up the conditional probabilities given A_L of cases in which S_2 holds; a like procedure gives $prob\ (S_3/A_L)$, $prob\ (S_2/A_R)$, and $prob\ (S_3/A_R)$.

The conclusion of Table 2 is that the states S_1, S_2, and S_3 are indeed epistemically independent of the acts A_L and A_R. Since A_L weakly dominates A_R with respect to states S_1, S_2, and S_3, it follows that A_L weakly dominates A_R with respect to stochastically independent states. We already know that A_R weakly dominates A_L with respect to causally independent states M_{11}, M_{01}, and M_{00}. In Case 5, then, the two principles of weak dominance are in conflict.

TABLE 2

	A_L Performed		A_R Performed	
	A_L Predicted 0.8	A_R Predicted 0.2	A_L Predicted 0.2	A_R Predicted 0.8
Strategy (i) 0.5	M_{11} 0.4 1 S_1	M_{00} 0.1 0 S_3	M_{11} 0.1 1 S_2	M_{00} 0.4 0 S_1
Strategy (ii*) 0.25	M_{11} 0.2 1 S_2	M_{01} 0.025 0 S_3	M_{11} 0.05 1 S_2	M_{01} 0.1 1 S_2
		M_{00} 0.025 0 S_3		M_{00} 0.1 0 S_3
Strategy (iii*) 0.25	M_{11} 0.1 1 S_2	M_{00} 0.05 0 S_3	M_{11} 0.025 1 S_2	M_{00} 0.2 0 S_3
	M_{01} 0.1 0 S_3		M_{01} 0.025 1 S_2	
Totals	prob $(S_2/A_L) = 0.3$ prob $(S_3/A_L) = 0.3$		prob $(S_2/A_R) = 0.3$ prob $(S_3/A_R) = 0.3$	

9. ACT-INDEPENDENCE IN THE SAVAGE FORMULATION

In Section 4, we said that to apply the Savage framework to a decision problem, one must find states of the world that are in some sense act-independent. In the last section, we distinguished two kinds of independence, causal and epistemic. Which kind is needed in the Savage formulation of decision problems?

The answer is that the Savage formulation has both a $\mathcal{U}$-maximizing interpretation and a $\mathcal{V}$-maximizing interpretation. On the $\mathcal{U}$-maximizing interpretation, the states must be causally independent of the acts, whereas on the $\mathcal{V}$-maximizing interpretation, the states must be epistemically independent of the acts. That is to say, if the states are causally act-independent, then utility as calculated by the Savage method is $\mathcal{U}$-utility, whereas if the states are epistemically act-independent, then utility as calculated by the Savage method is $\mathcal{V}$-utility. If the states are both causally and epistemically act-independent, then the $\mathcal{U}$-utility of each act equals its $\mathcal{V}$-utility. Thus the Savage formulation itself is not committed to either kind of utility: the kind of utility it yields depends on the way it is applied to decision problems.

The expected utility of an act A in the Savage theory is

$$(3) \qquad \Sigma_S \, prob \, (S)\mathcal{D}O(A, S).$$

If the states S are all known to be causally independent of A, so that for each state S, the agent knows that $(A \,\square\!\!\rightarrow S) \equiv S$, then for each S, we have $prob \, (S) = prob \, (A \,\square\!\!\rightarrow S)$. (3) thus becomes

$$\Sigma_S \, prob \, (A \,\square\!\!\rightarrow S)\mathcal{D}O(A, S),$$

and this, we said in Section 3, is $\mathcal{U}(A)$. If, on the other hand, the states S are stochastically independent of A, so that for each S, $prob \, (S) = prob \, (S/A)$, then (3) becomes

$$\Sigma_S \, prob \, (S/A)\mathcal{D}O(A, S),$$

which is $\mathcal{V}(A)$.

10. NEWCOMB'S PROBLEM

The Newcomb paradox discussed by Nozick (1969) has the same structure as the case of Solomon discussed in Section 3. Nozick treats it as a conflict between the principle of expected utility maximization and the principle of dominance. On the views we have propounded in this paper, the problem is rather a conflict between two kinds of expected utility maximization. The problem is this. There are two boxes, transparent and opaque; the transparent box contains a thousand dollars. The agent can perform A_1, taking just the contents of the opaque box, or A_2, taking the contents of both boxes. A predictor has already placed a million dollars in the opaque box if he predicted A_1 and nothing if he predicted A_2. The agent knows all this, and he knows the predictor to be highly reliable in that both *prob* (he has predicted A_1/A_1) and *prob* (he has predicted A_2/A_2) are close to one.

To show how the expected utility calculations work, we must add detail to the specification of the situation. Suppose, somewhat unrealistically, that getting no money has a utility of zero, getting \$1000 a utility of 10, that getting \$1,000,000 has a utility of 100, and that getting \$1,001,000 has a utility of 101. Let M be 'there are a million dollars in the opaque box', and suppose *prob* $(M/A_1) = 0.9$ and *prob* $(M/A_2) = 0.1$. The calculation of $\mathscr{V}(A_1)$ and $\mathscr{V}(A_2)$ is familiar.

$$\mathscr{V}(A_1) = prob\ (M/A_1)\mathscr{D}\$1{,}000{,}000 + prob\ (\bar{M}/A_1)\mathscr{D}\$0$$
$$= 0.9\ (100) + 0.1(0) = 90.$$
$$\mathscr{V}(A_2) = prob\ (M/A_2)\mathscr{D}\$1{,}001{,}000 + prob\ (\bar{M}/A_2)\mathscr{D}\$1000$$
$$= 0.1(101) + 0.9(10) = 19.1$$

Maximization of $\mathscr{V}$, as is well known, prescribes taking only the contents of the opaque box.[11]

$\mathscr{U}(A_1)$ and $\mathscr{U}(A_2)$ depend on the probability of M, which in turn depends on the probabilities of A_1 and A_2. For any probability of M, though, we have $\mathscr{U}(A_2) > \mathscr{U}(A_1)$. For let the probability of M be μ; then since M is causally act-independent, *prob* $(A_1 \;\square\!\!\rightarrow M) = \mu$ and *prob* $(A_2 \;\square\!\!\rightarrow M) = \mu$. Therefore

$$\mathscr{U}(A_1) = prob\ (A_1 \;\square\!\!\rightarrow M)\mathscr{D}\$1{,}000{,}000 + prob\ (A_1 \;\square\!\!\rightarrow\bar{M})\mathscr{D}\$0$$
$$= 100\mu + 0(1-\mu) = 100\mu.$$
$$\mathscr{U}(A_2) = prob\ (A_2 \;\square\!\!\rightarrow M)\mathscr{D}\$1{,}001{,}000 + prob\ (A_2 \;\square\!\!\rightarrow\bar{M})\mathscr{D}\$1000$$
$$= 101\mu + 10(1-\mu) = 91\mu + 10.$$

Thus $\mathscr{U}(A_2) - \mathscr{U}(A_1) = 10 - 9\mu$, and since $\mu \leqslant 1$, this is always positive. Therefore whatever probability M may have, $\mathscr{U}(A_2) > \mathscr{U}(A_1)$, and $\mathscr{U}$-maximization prescribes taking both boxes.

To some people, this prescription seems irrational.[12] One possible argument against it takes roughly the form 'If you're so smart, why ain't you rich?' $\mathscr{V}$-maximizers tend to leave the experiment millionaires whereas $\mathscr{U}$-maximizers do not. Both very much want to be millionaires, and the $\mathscr{V}$-maximizers usually succeed; hence it must be the $\mathscr{V}$-maximizers who are making the rational choice. We take the moral of the paradox to be something else: If someone is very good at predicting behavior and rewards predicted irrationality richly; then irrationality will be richly rewarded.

To see this, consider a variation on Newcomb's story: the subject of the experiment is to take the contents of the opaque box first and learn what it is; he then may choose either to take the thousand dollars in the second box or not to take it. The predictor has an excellent record, and a thoroughly accepted theory to back it up. Most people find nothing in the first box and then take the contents of the second box. Of the million subjects tested, 1% have found a million dollars in the first box, and strangely enough only 1% of these − 100 in 10,000 − have gone on to take the thousand dollars they could each see in the second box. When those who leave the thousand dollars are later asked why they do so, they say things like 'If I were the sort of person who would take the thousand dollars in that situation, I wouldn't be a millionaire'.

On both grounds of $\mathcal{U}$-maximization and of $\mathcal{V}$-maximization, these new millionaires have acted irrationally in failing to take the extra thousand dollars. They know for certain that they have the million dollars; therefore the $\mathcal{V}$-utility of taking the thousand as well is 101, whereas the $\mathcal{V}$-utility of not taking it is 100. Even on the view of $\mathcal{V}$-maximizers, then, this experiment will almost always make irrational people and only irrational people millionaires. Everyone knows so at the outset.

Return now to the unmodified Newcomb situation, where the subject must take or pass up the thousand dollars before he sees whether the opaque box is full or empty. What happens if the subject knows not merely that the predictor is highly reliable, but that he is infallible? The argument that the $\mathcal{U}$-utility of taking both boxes exceeds that of taking only one box goes through unchanged. To some people, however, it seems especially apparent in this case that it is rational to take only the opaque box and irrational to take both. For in this case the subject is certain that he will be a millionaire if and only if he takes only the opaque box. If in the case where the predictor is known to be infallible it is irrational to take both boxes, then, $\mathcal{U}$-maximization is not always the rational policy.

We maintain that $\mathcal{U}$-maximization is rational even in the case where the predictor is known to be infallible. True, where R is 'I become a millionaire', the agent knows in this case that R holds if A_1 holds: he knows the truth-functional proposition $R \equiv A_1$. From this proposition, however, it does not follow that he *would* be a millionaire if he did A_1, or that he *would* be a non-millionaire if he did A_2.

If the subject knows for sure that he will take just the opaque box, then he knows for sure that the million dollars is in the opaque box, and so he knows for sure that he will be a millionaire. But since he knows for sure that the million dollars is already in the opaque box, he knows for sure that even if he were to take both boxes, he would be a millionaire. If, on the other hand, the subject knows for sure that he will take both boxes, then he knows for sure that the opaque box is empty, and so he knows for sure that he will be a non-millionaire. But since in this case he knows for sure that the opaque box is empty, he knows for sure that even if he were to take just the opaque box, he would be a non-millionaire.

If the subject does not know what he will do, then what he knows is this: either he will take just the opaque box and be a millionaire, or he will take both boxes and be a non-millionaire. From this, however, it follows neither that (i) if he took just the opaque box, he would be a millionaire, nor that (ii) if he took both boxes he would be a non-millionaire. For (i), the subject knows, is true iff the opaque box is filled with a million dollars, and (ii), the subject knows, is true iff the opaque box is empty. Thus, if (i) followed from what the agent knows, he could conclude for certain that the opaque box contains a million dollars, and if (ii) followed from what the agent knows, he could conclude that the opaque box is empty. Since the subject, we have supposed, does not know what he will do, he can conclude neither that the opaque box contains a million dollars nor that it is empty. Therefore neither (i) nor (ii) follows from what the subject knows.

Rational choice in Newcomb's situation, we maintain, depends on a comparison of what would happen if one took both boxes with what would happen if one took only the opaque box. What the agent knows for sure is this: if he took both boxes, he would get a thousand dollars more than he would if he took only the opaque box. That, on our view, makes it rational for someone who wants as much much as he can get to take both boxes, and irrational to take only one box.

Why, then, does it seem obvious to many people that if the predictor is known to be infallible, it is rational to take only the opaque box and irrational to take both boxes? We have three possible explanations. The first is that a person may have a tendency to want to bring about an indication of a desired state of the world, even if it is known that the act that brings about the indication in no way brings about the desired state itself. Taking just the opaque box would be a sure indication that it contained a million dollars, even though taking just the opaque box in no way brings it about that the box contains a million dollars.

The second possible explanation lies in the force of the argument 'If you're so smart, why ain't you rich?' That argument, though, if it holds good, should apply equally well to the modified Newcomb situation, with a predictor who is known to be highly accurate but fallible. There the conclusion of the argument seems absurd: according to the argument, having already received the million dollars, one should pass up the additional thousand dollars one is free to take, on the grounds that those who are disposed to pass it up tend to become millionaires. Since the argument leads to an absurd conclusion in one case, there must be something wrong with it.

The third possible explanation is the fallacious inference we have just discussed, from

> Either I shall take one box and be a
> millionaire, or I shall take both boxes
> and be a non-millionaire

to the conclusion

> If I were to take one box, I would be a
> millionaire, and if I were to take both
> boxes, I would be a non-millionaire.

If, to someone who is free of fallacies, it is still intuitively apparent that the subject should take only the opaque box, we have no further arguments to give him. If in addition he thinks the subject should take only the opaque box even in the case where the predictor is known to be somewhat fallible, if he also thinks that in the modified Newcomb situation the subject, on receiving the extra million dollars, should take the extra thousand, if he also thinks that it is rational for Reoboam to be severe, and if he also thinks it is rational for Solomon to abstain from his neighbor's wife, then he may genuinely have the intuitions of a $\mathcal{V}$-maximizer: $\mathcal{V}$-maximization then provides a systematic account of his intuitions. If he thinks some of these things but not all of them, then we leave it to him to provide a systematic account of his views. Our own views are systematically accounted for by $\mathcal{U}$-maximization.

11. STABILITY OF DECISION

When a person decides what to do, he has in effect learned what he will do, and so he has new information. He will adjust his probability ascriptions accordingly. These adjustments may affect the $\mathcal{U}$-utility of the various acts open to him.

Indeed, once the person decides to perform an act A, the $\mathcal{U}$-utility of A will be equal to its $\mathcal{V}$-utility.[13] Or at least this holds if Consequence 1 in Section 2, that $A \supset [(A \mathrel{\Box\!\!\rightarrow} C) \equiv C]$, is a logical truth. For we saw in the proof of Assertion 1 that if Consequence 1 is a logical truth, then for any pair of propositions P and Q, $prob\,(P \mathrel{\Box\!\!\rightarrow} Q/P) = prob\,(Q/P)$. Now let $\mathcal{U}_A$ (A) be the $\mathcal{U}$-utility of act A as reckoned by the agent after he has decided for sure to do A, let *prob* give the agent's probability ascriptions before he has decided what to do. Let $prob_A$ give the agent's probability ascriptions after he has decided for sure to do A. Then for any proposition P, $prob_A$ $(P) = prob\ (P/A)$. Thus

$$\mathcal{U}_A(A) = \Sigma_O\, prob_A\ (A\ \square\!\!\rightarrow O)\mathcal{D}O$$
$$= \Sigma_O\, prob\ (A\ \square\!\!\rightarrow O/A)\mathcal{D}O$$
$$= \Sigma_O\, prob\ (O/A)\mathcal{D}O$$
$$= \mathcal{V}(A).$$

The $\mathcal{V}$-utility of an act, then, is what its $\mathcal{U}$-utility would be if the agent knew he were going to perform it.

It does not follow that once a person knows what he will do, $\mathcal{V}$-maximization and $\mathcal{U}$-maximization give the same prescriptions. For although for any act A, $\mathcal{U}_A(A) = \mathcal{V}(A)$, it is not in general true that for alternatives B to A, $\mathcal{U}_A(B) = \mathcal{V}(B)$. Thus in cases where $\mathcal{U}(A) < \mathcal{U}(B)$ but $\mathcal{V}(A) > \mathcal{V}(B)$, it is consistent with what we have said to suppose that $\mathcal{U}_A(A) < \mathcal{U}_A(B)$. In such a case, $\mathcal{V}$-maximization prescribes A regardless of what the agent believes he will do, but even if he believes he will do A, $\mathcal{U}$-maximization prescribes B. The situation is this:

$$\mathcal{U}_A(B) > \mathcal{U}_A(A) = \mathcal{V}(A) > \mathcal{V}(B).$$

Even though, once an agent knows what he will do, the distinction between the $\mathcal{U}$-utility of that act and its $\mathcal{V}$-utility disappears, the distinction between $\mathcal{U}$-maximization and $\mathcal{V}$-maximization remains.

That deciding what to do can affect the $\mathcal{U}$-utilities of the acts open to an agent raises a problem of stability of decision for $\mathcal{U}$-maximizers. Consider the story of the man who met death in Damascus.[14] Death looked surprised, but then recovered his ghastly composure and said, 'I am coming for you tomorrow'. The terrified man that night bought a camel and rode to Aleppo. The next day, death knocked on the door of the room where he was hiding and said 'I have come for you'.

'But I thought you would be looking for me in Damascus', said the man.

'Not at all', said death 'that is why I was surprised to see you yesterday. I knew that today I was to find you in Aleppo'.

Now suppose the man knows the following. Death works from an appointment book which states time and place; a person dies if and only if the book correctly states in what city he will be at the stated time. The book is made up weeks in advance on the basis of highly reliable predictions. An appointment on the next day has been inscribed for him. Suppose, on this basis, the man would take his being in Damascus the next day as strong evidence that his appointment with death is in Damascus, and would take his being in Aleppo the next day as strong evidence that his appointment is in Aleppo.

Two acts are open to him: A, go to Aleppo, and D, stay in Damascus. There are two possibilities: S_A, death will seek him in Aleppo, and S_D, death will seek him in Damascus. He knows that death will find him if and only if death looks for him in the right city, so that, where L is that he lives, he knows $(D \,\square\!\!\rightarrow L) \equiv S_A$ and $(A \,\square\!\!\rightarrow L) \equiv S_D$. He ascribes conditional probabilities $prob\,(S_A/A) \approx 1$ and $prob\,(S_D/D) \approx 1$; suppose these are both 0.99 and that $prob\,(S_D/A) = 0.01$ and $prob\,(S_A/D) = 0.01$. Suppose $\mathscr{D}(L) = -100$ and $\mathscr{D}(\bar{L}) = 0$. Then where α is $prob\,(A)$, his probability of going to Aleppo, and $1 - \alpha$ is his probability of going to Damascus,

$$prob\,(A \,\square\!\!\rightarrow L) = prob\,(S_D) = \alpha\,prob\,(S_D/A) + (1-\alpha)\,prob\,(S_D/D)$$
$$= 0.01\alpha + 0.99(1 - \alpha) = 0.99 - 0.98\alpha$$
$$prob\,(A \,\square\!\!\rightarrow \bar{L}) = prob\,(S_A) = 1 - prob\,(S_D) = 0.01 + 0.98\alpha.$$

Thus

$$\mathscr{U}(A) = prob\,(A \,\square\!\!\rightarrow L)\mathscr{D}(L) + prob\,(A \,\square\!\!\rightarrow \bar{L})\mathscr{D}(\bar{L})$$
$$= (0.01 + 0.98\alpha)\,(-100) = -1 - 98\alpha.$$

By a like calculation, $\mathscr{U}(D) = -99 + 98\alpha$. Thus if $\alpha = 1$, then $\mathscr{U}(D) = -1$ and $\mathscr{U}(A) = -99$, and thus $\mathscr{U}(D) > \mathscr{U}(A)$. If $\alpha = 0$, then $\mathscr{U}(D) = -99$ and $\mathscr{U}(A) = -1$, so that $\mathscr{U}(A) > \mathscr{U}(D)$. Indeed we have $\mathscr{U}(D) > \mathscr{U}(A)$ whenever $prob\,(A) > 1/2$, and $\mathscr{U}(A) > \mathscr{U}(D)$ whenever $prob\,(D) > 1/2$.

What are we to make of this? If the man ascribes himself equal probabilities of going to Aleppo and staying in Damascus, he has equal grounds for thinking that death intends to seek him in Damascus and that death intends to seek him in Aleppo. If, however, he decides to go to Aleppo, he then has strong grounds for expecting that Aleppo is where death already expects him to be, and hence it is rational for him to prefer staying in Damascus. Similarly, deciding to stay in Damascus would give him strong grounds for thinking that he ought to go to Aleppo: once he knows he will stay in Damascus, he can be almost sure that death already expects him in Damascus, and hence that if he had gone to Aleppo, death would have sought him in vain.

$\mathscr{V}$-maximization does not lead to such instability. What happens to $\mathscr{V}$-utility when an agent knows for sure what he will do is somewhat unclear. Standard probability theory offers no interpreation of $prob_A(O/B)$ where $prob\,(B/A) = O$, and so on the standard theory, once an agent knows for sure what he will do, the $\mathscr{V}$-utility of the alternatives ceases to be well-defined.

What we can say about $\mathscr{V}$-utility is this: as long as an act's being performed has non-zero probability, its $\overline{\mathscr{V}}$-utility is independent of its probability and the probabilities of alternatives to it. For the $\mathscr{V}$-utility of an act A depends on conditional probabilities of the form *prob* (O/A). This is just the probability the agent would ascribe to O on learning A for sure, and that is independent of how likely he now regards A. Whereas, then, the $\mathscr{U}$-utility of an act may vary with its probability of being performed, its $\mathscr{V}$-utility does not. $\mathscr{U}$-maximization, then, may give rise to a kind of instability which $\mathscr{V}$-maximization precludes: in certain cases, an act will be $\mathscr{U}$-maximal if and only if the probability of its performance is low.

Is this a reason for preferring $\mathscr{V}$-maximization? We think not. In the case of death in Damascus, rational decision does seem to be unstable. Any reason the doomed man has for thinking he will go to Aleppo is a reason for thinking he would live longer if he stayed in Damascus, and any reason he has for thinking he will stay in Damascus is reason for thinking he would live longer if he went to Aleppo. Thinking he will do one is reason for doing the other. That there can be cases of unstable $\mathscr{U}$-maximization seems strange, but the strangeness lies in the cases, not in $\mathscr{U}$-maximization: instability of rational decision seems to be a genuine feature of such cases.

12. APPLICATIONS TO GAME THEORY

Game theory provides many cases where $\mathscr{U}$-maximizing and $\mathscr{V}$-maximizing diverge; perhaps the most striking of these is the prisoners' dilemma, for which a desirability matrix is shown.

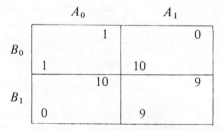

Here A_0 and B_0 are respectively A's and B's options of confessing, while A_1 and B_1 are the options of not confessing. The desirabilities reflect these facts: (1) if both confess, they both get long prison terms; (2) if one confesses and the other doesn't, then the confessor gets off while the other gets an even longer prison term; (3) if neither confesses, both get off with very light sentences.

Suppose each prisoner knows that the other thinks in much the same way he does. Then his own choice gives him evidence for what the other will do. Thus, the conditional probability of a long prison term on his confessing is greater than the conditional probability of a long prison term on his not confessing. If the difference between these two conditional probabilities is sufficiently great, then $\mathscr{V}$-maximizing will prescribe not confessing.

The $\mathscr{V}$-utilities of the acts open to B will be as follows.

$$\mathscr{V}(B_0) = prob\ (A_0/B_0) \times 1 + prob\ (A_1/B_0) \times 10$$
$$\mathscr{V}(B_1) = prob\ (A_0/B_1) \times 0 + prob\ (A_1/B_1) \times 9.$$

If $prob\ (A_1/B_1) - prob\ (A_1/B_0)$ is sufficiently great (in this case 1/9 or more), then $\mathscr{V}$-maximizing recommends that B take option B_1 and not confess. If the probabilities for A are similar, then $\mathscr{V}$-maximizing also recommends not confessing for A. The outcome if both $\mathscr{V}$-maximize is $A_1 B_1$, the optimal one of mutual co-operation.[15]

For a $\mathscr{U}$-maximizer, dominance applies because his companion's choice is causally independent of his own. Therefore, $\mathscr{U}$-maximizing yields the classical outcome of the prisoners' dilemma. This suggests that $\mathscr{U}$-maximizing and not $\mathscr{V}$-maximizing corresponds to the kind of utility maximizing commonly assumed in game theory.

University of Michigan and University of Western Ontario.

NOTES

* An earlier draft of this paper was circulated in January 1976. A much shorter version was presented to the 5th International Congress of Logic, Methodology, and Philosophy of Science, London, Ontario, August 1975. There, and at the earlier University of Western Ontario research colloquium on Foundations and Applications of Decision Theory we benefited from discussions with many people; in particular we should mention Richard Jeffrey, Isaac Levi, Barry O'Neill and Howard Sobel.

[1] Lewis first presented this result at the June 1972 meeting of the Canadian Philosophical Association.

[2] Although the rough treatment of counterfactuals we propose is similar is many respects to the theories developed by Stalnaker and Lewis, it differs from them in some important respects. Stalnaker and Lewis each base their accounts on comparisons of overall similarity of worlds. On our account, what matters is comparative similarity of worlds at the instant of decision. Whether a given a-world is selected as W_a depends not at all on how similar the future in that world is to the actual future; whatever similarities the future in W_a may have to the actual future will be a semantical consequence of laws of nature, conditions in W_a at the instant of decision, and actual conditions at that

instant. (Roughly, then, they will be consequences of laws of nature and the similarity of W_a to the actual world at the instant of decision.) We consider only worlds in which the past is exactly like the actual past, for since the agent cannot now alter the past, those are the only worlds relevant to his decision. Lewis (1973, p. 566 and in conversation) suggests that a proper treatment of overall similarity will yield as a deep consequence of general facts about the world the conditions we are imposing by fiat.

[3] In characterizing our conditional we have imposed the Stalnaker-like constraint that there is a unique world W_a which would eventuate from performing a at t. Our rationale for Axiom 2 depends on this assumption and on the assumption that if a is actually performed then W_a is the actual world itself. Consequence 1 is weaker than Axiom 2, and only depends on the second part of this assumption. In circumstances where these assumptions break down, it would seem to us that using conditionals to compute expected utility is inappropriate. A more general approach is needed to handle such cases.

[4] This is stated by Lewis (1975, note 10).

[5] This is our understanding of a proposal made by Jeffrey at the colloquium on Foundations and Applications of Decision Theory, University of Western Ontario, 1975. J. H. Sobel shows (in an unpublished manuscript) that, for all we have said, these new, conditionalized states may not themselves be act-independent. This section is slightly changed in light of Sobel's result.

[6] Meeting of the Canadian Philosophical Association, 1972. Nozick gives a similar example (1969, p. 125).

[7] Sobel actually uses '$A* \:\square\!\!\rightarrow S$ does hold' where we use '$A* \:\square\!\!\rightarrow S$ does not hold'. With Axiom 2, these are equivalent.

[8] We realize that a Bayesian king presented with these data would not ordinarily take on degrees of belief that exactly match the frequencies given in the table; nevertheless, with appropriate prior beliefs and evidence, he would come to have those degrees of belief. Assume that he does.

[9] Under these conditions, if A and B are the only alternatives, then S_i* holds if and only if S_i holds. If there are other alternatives, it may be that neither A nor B is performed and S_i holds without either $A \:\square\!\!\rightarrow S_i$ or $B \:\square\!\!\rightarrow S_i$. In that case, what matters is not whether it would be rational to prefer A to B knowing that S_i holds, but whether it would be rational to prefer A to B knowing $(A \:\square\!\!\rightarrow S_i)$ & $(B \:\square\!\!\rightarrow S_i)$.

[10] Nozick (1969) in effect endorses the Principle of Dominance with Stochastic Independence (p. 127), but not $\mathscr{V}$-maximization: in cases of the kind we have been considering, he considers the recommendations of $\mathscr{V}$-maximization 'perfectly wild' (p. 126). Nozick also states and endorses the principle of dominance with causal independence (p. 132).

[11] For $\mathscr{V}$-maximizing treatments of Newcomb's problem, see Bar Hillel and Margalit (1972) and Levi (1975).

[12] Levi (1975) reconstructs Nozick's argument for taking both boxes in a way which uses $prob(M)$ rather than $prob(M/A_1)$ and $prob(M/A_2)$ as the appropriate probabilities for computing expected utility in Newcomb's problem. This agrees with $\mathscr{U}$-maximizing in that the same probabilities are used for computing expected utility for A_1 as for A_2, and results in the same recommendation to take both boxes. Levi is one of the people to whom this recommendation seems irrational.

[13] We owe this point to Barry O'Neill.

[14] A version of this story quoted from Somerset Maugham's play *Sheppey* (New York, Doubleday 1934) appears on the facing page of John O'Hara's novel *Appointment in Samarra*. (New York, Random House 1934). The story is undoubtedly much older.

[15] Nozick (1969), Brams (1975), Grofman (1975) and Rapoport (1975), have all suggested a link between Newcomb's problem and the Prisoners Dilemma. Brams, Grofman and Rapoport all endorse co-operative solutions, Rapoport (1975, p. 619) appears to endorse ✔-maximizing.

BIBLIOGRAPHY

Bar Hillel, M. and Margalit, A., 'Newcomb's Paradox Revisited', *British Journal for the Philosophy of Science* **23** (1972), 295–304.

Brams, S.J., 'Newcomb's Problem and Prisoners' Dilemma', *Journal of Conflict Resolution* **19**, 4, December 1975.

Grofman, B., 'A Comment on "Newcomb's Problem and the Prisoners' Dilemma" ', Manuscript, September 1975.

Jeffrey, R. C., *The Logic of Decision*, McGraw Hill, New York, 1965.

Levi, I., 'Newcomb's Many Problems', *Theory and Decision* **6** (1975), 161–75.

Lewis, D. K., 'Probabilities of Conditionals and Conditional Probabilities', *Philosophical Review* **85** (1976), 297–315.

Lewis, D. K., *Counterfactuals*, Harvard University Press, Cambridge, Massachusetts, 1973a.

Lewis, D. K., 'The Counterfactual Analysis of Causation', *Journal of Philosophy* Volume LXX, Number 17, (1973b), 556–567.

Nozick, R., 'Newcomb's Problem and Two Principles of Choice', in Nicholas Rescher (ed.), *Essays in Honor of Carl G. Hempel*, Reidel, Dordrecht-Holland, 1969.

Rapoport A., 'Comment on Brams's Discussion of Newcomb's Paradox', *Journal of Conflict Resolution* **19**, 4, December 1975.

Savage, L. J., *The Foundations of Statistics*, Dover, New York, 1972 (original edition 1954).

Sobel, J. H., 'Utilitarianisms: Simple and General', *Inquiry* **13** (1970), 394–449.

Stalnaker, R., 'A Theory of Conditionals', in *Studies in Logical Theory*, American Philosophical Quarterly Monograph Series, No. 2, 1968.

Stalnaker, R. and Thomason, R., 'A Semantic Analysis of Conditional Logic', *Theoria* **36** (1970), 23–42.

PART 5

INDICATIVE VS. SUBJUNCTIVE
CONDITIONALS

ROBERT C. STALNAKER

INDICATIVE CONDITIONALS

"Either the butler or the gardener did it. Therefore, if the butler didn't do it, the gardener did." This piece of reasoning – call it the *direct argument* – may seem tedious, but it is surely compelling. Yet if it is a valid inference, then the indicative conditional conclusion must be logically equivalent to the truth-functional material conditional,[1] and *this* conclusion has consequences that are notoriously paradoxical. The problem is that if one accepts the validity of the intuitively reasonable direct argument from the material conditional to the ordinary indicative conditional, then one must accept as well the validity of many arguments that are intuitively absurd. Consider, for example, "the butler did it; therefore, if he didn't, the gardener did." The premiss of this argument entails the premiss of the direct argument, and their conclusions are the same. Therefore, if the direct argument is valid, so is this one. But this argument has no trace of intuitive plausibility. Or consider what may be inferred from the *denial* of a conditional. Surely I may deny that if the butler didn't do it, the gardener did without affirming the butler's guilt. Yet if the conditional is material, its negation entails the truth of its antecedent. It is easy to multiply paradoxes of the material conditional in this way – paradoxes that must be explained away by anyone who wants to defend the thesis that the direct argument is valid. Yet anyone who denies the validity of that argument must explain how an invalid argument can be as compelling as this one seems to be.

There are thus two strategies that one may adopt to respond to this puzzle: defend the material conditional analysis and explain away the paradoxes of material implication, or reject the material conditional analysis and explain away the force of the direct argument.[2] H. P. Grice, in his William James lectures,[3] pursued the first of these strategies, using principles of conversation to explain facts about the use of conditionals that seem to conflict with the truth-functional analysis of the ordinary indicative conditional. I will follow the second strategy, defending an alternative semantic analysis of conditionals according to which the

193

W. L. Harper, R. Stalnaker, and G. Pearce (eds.), Ifs, 193–210.

conditional entails, but is not entailed by, the corresponding material conditional. I will argue that, although the premiss of the direct argument does not *semantically entail* its conclusion, the inference is nevertheless a *reasonable inference*. My main task will be to define and explain a concept of reasonable inference which diverges from semantic entailment, and which justifies this claim.

Grice's strategy and mine have this in common: both locate the source of the problem in the mistaken attempt to explain the facts about assertion and inference solely in terms of the semantic content, or truth conditions, of the propositions asserted and inferred. Both attempt to explain the facts partly in terms of the semantic analysis of the relevant notions, but partly in terms of pragmatic principles governing discourse. Both recognize that since assertion aims at more than truth, and inference at more than preserving truth, it is a mistake to reason too quickly from facts about assertion and inference to conclusions about semantic content and semantic entailment.

My plan will be this: first, I will try to explain, in general terms, the concept of reasonable inference and to show intuitively how there can be reasonable inferences which are not entailments. Second, I will describe a formal framework in which semantic concepts like content and entailment as well as pragmatic concepts like assertion and inference can be made precise. Third, within this framework, I will sketch the specific semantic analysis of conditionals, and state and defend some principles relating conditional sentences to the contexts in which they are used. Fourth, I will show that, according to these analyses, the direct argument is a reasonable inference. Finally, I will look at another puzzling argument involving reasoning with conditionals – an argument for fatalism – from the point of view of this framework.

I

Reasonable inference, as I shall define it, is a pragmatic relation: it relates speech acts rather than the propositions which are the contents of speech acts. Thus it contrasts with entailment which is a purely semantic relation. Here are rough informal definitions of the two notions: first, reasonable inference: an inference from a sequence of assertions or suppositions (the premisses) to an assertion or hypothetical assertion (the conclusion) is *reasonable* just in case, in every context in which the premisses could ap-

propriately be asserted or supposed, it is impossible for anyone to accept
the premisses without committing himself to the conclusion; second, entailment: a set of propositions (the premisses) *entails* a proposition (the
conclusion) just in case it is impossible for the premisses to be true without
the conclusion being true as well. The two relations are obviously different since they relate different things, but one might expect them to be
equivalent in the sense that an inference would be reasonable if and only
if the set of propositions expressed in the premisses entailed the proposition expressed in the conclusion. If this equivalence held, then the pragmatic concept of inference would of course have no interest. I shall argue
that, and try to show why, the equivalence does not hold. Before discussing the specific framework in which this will be shown, let me try to
explain in general terms how it is possible for an inference to be reasonable,
in the sense defined, even when the premisses do not entail the conclusion.

The basic idea is this: many sentences are context dependent; that is,
their semantic content depends not just on the meanings of the words in
them, but also on the situations in which they are uttered. Examples are
familiar: quantified sentences are interpreted in terms of a domain of
discourse, and the domain of discourse depends on the context; the refer-
ents of first and second person pronouns depend on who is speaking, and
to whom; the content of a tensed sentence depends on when it is uttered.
Thus context constrains content in systematic ways. But also, the fact
that a certain sentence is uttered, and a certain proposition expressed, may
in turn constrain or alter the context. There are two ways this may happen:
first, since particular utterances are appropriate only in certain contexts,
one can infer something about a context from the fact that a particular
utterance is made (together with the assumption that the utterance is
appropriate); second, the expression of a proposition alters the context, at
the very least by changing it into a context in which that proposition has
just been expressed. At any given time in a conversation, the context will
depend in part on what utterances have been made, and what propositions expressed, previously in the conversation. There is thus a two way
interaction between contexts of utterance and the contents of utterances.
If there are general rules governing this interaction, these rules may give
rise to systematic relations between propositions expressed at different
points in a conversation, relations which are mediated by the context.
Such relations may become lost if one ignores the context and considers

propositions in abstraction from their place in a discourse. It is because entailment relates propositions independently of their being asserted, supposed or accepted, while reasonable inference concerns propositions which are expressed and accepted, that the two relations may diverge.

These general remarks are not an attempt to show that the notions of entailment and reasonable inference do in fact diverge, but only an attempt to point to the source of the divergence that will be shown. To show the divergence, I must say what contexts are, or how they are to be represented formally. I must say, for some specific construction (here, conditionals) how semantic content is a function of context. And I must state and defend some rules which relate contexts to the propositions expressed in them.

II

The framework I will use begins with, and takes for granted, the concept of a possible world. While model theory based on possible worlds is generally agreed to be a powerful and mathematically elegant tool, its intuitive content and explanatory power are disputed. It is argued that a theory committed to the existence of such implausible entities as possible worlds must be false. Or at least the theory cannot do any philosophical work unless it can provide some kind of substantive answer to the question, what is a possible world? Possible worlds are certainly in need of philosophical explanation and defence, but for the present I will make just a brief remark which will perhaps indicate how I understand this basic notion.[4]

It is a common and essential feature of such activities as inquiring, deliberating, exchanging information, predicting the future, giving advice, debating, negotiating, explaining and justifying behavior, that the participants in the activities seek to distinguish, in one way or another, among alternative situations that may arise, or might have arisen. Possible worlds theory, as an explanatory theory of rational activity, begins with the notion of an alternative way that things may be or might have been (which is all that a possible world is) not because it takes this notion to be unproblematic, but because it takes it to be fundamental to the different activities that a theory of rationality seeks to characterize and relate to each other. The notion will get its content, not from any direct answer to the question, what is a possible world? or from any reduction of that notion to something more basic or familiar, but from its role in the explanations of such a theory. Thus it may be that the best philosophical defense that one can give for possible worlds is to use them in the development of substantive theory.

Taking possible worlds for granted, we can define a *proposition* as a function from possible worlds into truth values.[5] Since there are two truth values, this means that a proposition is any way of dividing a set of possible worlds into two parts – those for which the function yields the value true, and those for which it yields the value false. The motivation for this representation of propositions is that, as mentioned above, it is an essential part of various rational activities to distinguish among alternative possible situations, and it is by expressing and adopting attitudes toward propositions that such distinctions are made.

How should a context be defined? This depends on what elements of the situations in which discourse takes place are relevant to determining what propositions are expressed by context dependent sentences and to explaining the effects of various kinds of speech acts. The most important element of a context, I suggest, is the common knowledge, or presumed common knowledge and common assumption of the participants in the discourse.[6] A speaker inevitably takes certain information for granted when he speaks as the common ground of the participants in the conversation. It is this information which he can use as a resource for the communication of further information, and against which he will expect his speech acts to be understood. The presumed common ground in the sense intended – the *presuppositions* of the speaker – need not be the beliefs which are really common to the speaker and his audience; in fact, they need not be beliefs at all. The presuppositions will include whatever the speaker finds it convenient to take for granted, or to pretend to take for granted, to facilitate his communication. What is essential is not that the propositions presupposed in this sense be believed by the speaker, but rather that the speaker believe that the presuppositions are common to himself and his audience. This is essential since they provide the context in which the speaker intends his statements to be received.

In the possible worlds framework, we can represent this background information by a set of possible worlds – the possible worlds not ruled out by the presupposed background information. I will call this set of possible worlds the *context set*.[7] Possible worlds within the set are situations among which the speaker intends his speech acts to distinguish. I will sometimes talk of propositions being *compatible with* or *entailed by* a context. This means, in the first case, that the proposition is true in some of the worlds in the context set, and in the second case that the proposition is true in all of the worlds in the context set. Intuitively, it means, in the first case, that it is at least an open question in the context whether or not the proposition is true, and in the second case, that the proposition is presupposed, or accepted, in the context.

Propositions, then, are ways of distinguishing among any set of possible worlds, while context sets are the sets of possible worlds among which a speaker means to distinguish when he expresses a proposition.

III

The semantic analysis of conditionals that I will summarize here is developed and defended more fully elsewhere.[8] The analysis was constructed primarily to account for counterfactual conditionals – conditionals whose antecedents are assumed by the speaker to be false – but the analysis was intended to fit conditional sentences generally, without regard to the attitudes taken by the speaker to antecedent or consequent or his purpose in uttering them, and without regard to the grammatical mood in which the conditional is expressed.

The idea of the analysis is this: a conditional statement, *if A, then B*, is an assertion that the consequent is true, not necessarily in the world as it is, but in the world as it would be if the antecedent were true. To express this idea formally in a semantic rule for the conditional, we need a function which takes a proposition (the antecedent) and a possible world (the world as it is) into a possible world (the world as it would be if the antecedent were true). Intuitively, the *value* of the function should be that world in which the antecedent is true which is most similar, in relevant respects, to the actual world (the world which is one of the *arguments* of the function). In terms of such a function – call it 'f' – the semantic rule for the conditional may be stated as follows: a conditional *if A, then B*, is true in a possible world i just in case B is true in possible world $f(A, i)$.[9]

It may seem that little has been accomplished by this analysis, since it just exchanges the problem of analyzing the conditional for the problem of analyzing a semantic function which is equally problematic, if not more so. In one sense this is correct: the analysis is not intended as a reduction of the conditional to something more familiar or less problematic, and it should not satisfy one who comes to the problem of analyzing conditionals with the epistemological scruples of a Hume or a Goodman. The aim of the analysis is to give a perspicuous representation of the formal structure of conditionals – to give the *form* of their truth conditions. Even if nothing substantive is said about how antecedents select counter factual possible worlds, the analysis still has non-trivial, and in some cases surprising, consequences for the logic of conditionals.

But what more can be said about this selection function? If it is to be based on *similarity* in some respect or other, then it must have certain formal properties. It must be a function that determines a coherent ordering of the possible worlds that are selected. And, since whatever the respects of similarity are that are relevant, it will always be true that something is more similar to itself than to anything else, the selection function must be one that selects the actual world whenever possible, which means whenever the antecedent is true in the actual world. Can anything more substantive be said about the relevant respects of similarity on which the selection is based? Not, I think, in the *semantic* theory of conditionals. Relevant respects of similarity are determined by the context, and the semantics abstracts away from the context by taking it as an unexplained given. But we can, I think, say something in a pragmatic theory of conditional statements about how the context constrains the truth conditions for conditionals, at least for indicative conditionals.

I cannot *define* the selection function in terms of the context set, but the following constraint imposed by the context on the selection function seems plausible: if the conditional is being evaluated at a world in the context set, then the world selected must, if possible, be within the context set as well (where C is the context set, if $i \in C$, then $f(A, i) \in C$). In other words, all worlds within the context set are closer to each other than any worlds outside it. The idea is that when a speaker says "If A," then everything he is presupposing to hold in the actual situation is presupposed to hold in the hypothetical situation in which A is true. Suppose it is an open question whether the butler did it or not, but it is established and accepted that whoever did it, he or she did it with an ice pick. Then it may be taken as accepted and established that if the butler did it, he did it with an ice pick.

The motivation of the principle is this: normally a speaker is concerned only with possible worlds within the context set, since this set is defined as the set of possible worlds among which the speaker wishes to distinguish. So it is at least a normal expectation that the selection function should turn first to these worlds before considering *counterfactual* worlds – those presupposed to be non-actual. Conditional statements can be directly relevant to their primary uses – deliberation, contingency planning, making hedged predictions – only if they conform to this principle.

Nevertheless, this principle is only a defeasible presumption and not a universal generalization. For some special purposes a speaker may want to make use of a selection function which reaches outside of the context set, which is to say he may want to suspend temporarily some of the presuppositions made in that context. He may do so provided that he indicates in some way that his selection function is an exception to the presumption. Semantic determinants like domains and selection functions are a function of the speaker's intentions; that is why we must allow for exceptions to such pragmatic generalizations. But they are a function of the speaker's intention. *to communicate* something, and that is why it is essential that it be conveyed to the audience that an exception is being made.

I take it that the subjunctive mood in English and some other languages is a conventional device for indicating that presuppositions are being suspended, which means in the case of subjunctive *conditional* statements, that the selection function is one that may reach outside of the context set. Given this conventional device, I would expect that the pragmatic principle stated above should hold without exception for indicative conditionals.

In what kinds of cases would a speaker want to use a selection function that might reach outside of the context set? The most obvious case would be one where the antecedent of the conditional statement was counterfactual, or incompatible with the presuppositions of the context. In that case one is forced to go outside the context set, since there are no possible worlds in it which are eligible to be selected. But there are non-counterfactual cases as well.[10] Consider the argument, "The murderer used an ice pick. But if the butler had done it, he wouldn't have used an ice pick. So the murderer must have been someone else."[11] The subjunctive conditional premiss in this *modus tollens* argument cannot be counterfactual since if it were the speaker would be blatantly begging the question by presupposing, in giving his argument, that his conclusion was true. But that premiss does not conform to the constraint on selection functions, since the consequent denies the first premiss of the argument, which presumably is accepted when the second premiss is given.

Notice that if the argument is restated with the conditional premiss in the indicative mood, it is anomalous.

My second example of a subjunctive non-counterfactual conditional which violates the constraint is adapted from an example given by Alan Anderson many years ago.[12] "If the butler had done it, we would have found just the clues which we in fact found." Here a conditional is presented as evidence for the truth of its antecedent. The conditional cannot be counterfactual, since it would be self-defeating to presuppose false what one is trying to show true. And it cannot conform to the constraint on selection functions since if it did, it would be trivially true, and so no evidence for the truth of the antecedent. Notice, again that when recast into the indicative mood, the conditional seems trivial, and does not look like evidence for anything.

The generalization that all indicative conditionals conform to the pragmatic constraint on selection functions has the following consequence about appropriateness conditions for indicative conditionals: *It is appropriate to make an indicative conditional statement or supposition only in a context which is compatible with the antecedent*. In effect, this says that *counterfactual* conditionals must be expressed in the subjunctive. This follows since indicative conditionals are those which must conform to the constraint, while counterfactuals are, by definition, those which cannot.

I need just one more assumption in order to show that the direct argument is a reasonable inference – an assumption about conditions of appropriateness for making assertions. The generalization that I will state is a quite specific one concerning disjunctive statements. I am sure it is derivable from more general conversational principles of the kind that Grice has discussed, but since I am not sure exactly what form such general principles should take, I will confine myself here to a generalization which has narrow application, but which can be clearly stated and easily defended. The generalization is this: *a disjunctive statement is appropriately made only in a context which allows either disjunct to be true without the other*. That is, one may say *A or B* only in a situation in which both *A and not-B* and *B and not-A* are open possibilities. The point is that each disjunct must be making some contribution to determining what is said. If the context did not satisfy this condition, then the assertion of the disjunction would be equivalent to the assertion of one of the disjuncts alone. So the disjunctive assertion would be pointless, hence misleading, and therefore inappropriate.[13]

IV

All of the ingredients of the solution to the puzzle are now assembled and ready to put together. It may seem that this is a rather elaborate apparatus for such a simple puzzle, but each of the elements – propositions and contexts, the semantic analysis of conditionals, the pragmatic constraint on conditionals, and the generalization about appropriateness – is independently motivated. It is not that this apparatus has been assembled just to solve the little puzzle; it is rather that the puzzle is being used to illustrate, in a small way, the explanatory capacity of the apparatus.

The argument we began with has the form *A or B, therefore, if not-A, then B*. This inference form is a reasonable inference form just in case every context in which a premiss of that form could appropriately be asserted or explicitly supposed, and in which it is accepted, is a context which entails the proposition expressed by the corresponding conclusion. Now suppose the premiss, *A or B*, is assertable and accepted. By the constraint on the appropriateness of disjunctive statements, it follows that the context is compatible with the conjunction of *not-A* with *B*. Hence the antecedent of the conditional conclusion, *not-A*, is compatible with the context. Now it follows from the pragmatic constraint on selection functions that if a proposition *P* is *compatible* with the context, and another proposition *Q* is *accepted* in it, or *entailed* by it, then the conditional, *if P, then Q*, is entailed by it as well. So, since *not-A* is compatible with the context, and the premiss *A or B* is accepted, the conditional, *if not-A, then A or B*, must be accepted as well. But this conditional proposition entails the conclusion of the argument, *if not-A, then B*. So the inference is a reasonable one.

Since the argument works the other way as well, it follows that the indicative conditional and the material conditional are equivalent in the following sense: in any context where either might appropriately be asserted, the one is accepted, or entailed by the context, if and only if the other is accepted, or entailed by the context. This equivalence explains the plausibility of the truth-functional analysis of indicative conditionals, but it does not justify that analysis since the two propositions coincide only in their assertion and acceptance conditions, and not in their truth conditions. The difference between the truth conditions of the two propositions will show itself if one looks at acts and attitudes other than assertion and acceptance. To take the simplest case, it may be reasonable to deny a conditional, even when not denying the corresponding material conditional. For example, I know *I* didn't do it, so I know that it is false that if the butler didn't do it, I did. But since I don't know whether the butler did it or not, I am in no position to deny the material conditional, which is equivalent to the disjunction, either the butler did it or I did. I may even think that that disjunction is very probably true.

There are two other familiar inference forms involving conditionals which are judged to be reasonable, although invalid, by this analysis: contraposition and the hypothetical syllogism. It was one of the surprising consequences of the *semantic* analysis sketched above that these inferences are, in general, invalid. Nevertheless, these consequences count in favor of the semantic analysis rather than against it since there are clear counterexamples to both inference forms. But all the counterexamples involve subjunctive conditionals which are counterfactual-conditionals whose antecedents are presupposed to be false. Now we can explain why there are no purely indicative counterexamples, and also why the arguments have the appearance of validity which they have. Both argument forms can be shown to be reasonable inferences, given that all conditionals involved are indicative, and given the assumption that indicative conditionals always conform to the pragmatic constraint on selection functions.[14]

V

I want to conclude by looking at a notorious argument involving indicative conditionals. The argument for fatalism is, I will argue, unreasonable as well as invalid. But it gains its appearance of force from the fact that it is an artful sequence of steps, each one of which has the form of a reasonable or of a valid inference. The trick of the argument, according to the diagnosis I will give, is that it exploits the changing context in an illegitimate way. Subordinate conclusions, legitimately drawn within their own subordinate contexts, are illegitimately detached from those contexts and combined outside of them. To make clear what I mean, let me sketch the argument. The specific form it takes, and the example used to present it, are taken from Michael Dummett's discussion of fatalism in his paper, 'Bringing about the Past.'[15] The setting of the example is wartime Britain during an air raid. I reason as follows: "Either I will be killed in this raid or I will not be killed. Suppose that I will. Then even if I take precautions I will be killed, so any precautions I take will be ineffective. But suppose I am not going to be killed. Then I won't be killed even if I neglect all precautions; so, on this assumption, no precautions are necessary to avoid being killed. Either way, any precautions I take will be either ineffective or unnecessary, and so pointless"

To give an abstract representation of the argument, I will let K mean "I will be killed," P mean "I take precautions," Q mean "precautions are ineffective," and R mean "precautions are unnecessary." The argument, reduced to essentials, is this:

1. K or not-K
2. $\quad K$
3. $\quad$ If P, K
4. $\quad Q$
5. $\quad$ not-K
6. $\quad$ if not-P, not-K
7. $\quad R$
8. Q or R

Now I take it that the main problem posed by this argument is not to say what is wrong with it, but rather to explain its illusion of force. That is, it is not enough to say that step x is invalid and leave it at that, even if that claim is correct. One must explain why anyone should have thought that it was valid. Judged by this criterion, Dummett's analysis of the argument does not solve the problem, even though, I think, what he says about the argument is roughly correct. Dummett argues that any sense of the conditional which will validate the inference from 2 to 3 (and 5 to 6) must be too weak to validate the inference from 3 to 4 (and 6 to 7). Hence, however the conditional is analyzed, the argument as a whole cannot be valid. Dummett's argument to this conclusion is convincing, but it would be a full solution to the problem only if he supplemented it by showing that there *are* in our language distinct senses of the conditional that validate each of those steps. This I do not think he can do, since I do not think the force of the argument rests on an equivocation between two senses of the conditional.

According to the semantic and pragmatic analyses sketched above, there is *one* sense of the conditional according to which the inference from 2 to 3 is a *reasonable inference*,[16] and which is also strong enough to justify the inference from 3 to 4. The fallacy, according to the diagnosis, is thus in neither of the steps that Dummett questions. Both of the sub-arguments are good arguments in the sense that anyone who was in a position to accept the premiss, while it remained an open question whether or not the antecedent of the conditional was true, would be in a position to accept the conclusion. That is, if I were in a position to accept that I were going to be killed even though I hadn't yet decided whether or not to take precautions, then I would surely be reasonable to conclude that taking precautions would be pointless. Likewise if I knew or had reason to accept that I would not be killed.

The problem with the argument is in the final step, an inference which seems to be an instance of an unproblematically valid form – constructive dilemma which has nothing essential to do with conditionals. The argument form that justifies step 8 is this: *A or B*; *C* follows from *A*; *D* follows from *B*; therefore, *C or D*. It is correct that the conclusion follows *validly* from the premiss provided that the subarguments are *valid*. But it is not correct that the conclusion is a *reasonable inference* from the premiss, provided that the subarguments are *reasonable inferences*. In the fatalism argument, the subarguments are reasonable, but not valid, and this is why the argument fails. So it is a confusion of validity with reasonable inference on which the force of the argument rests.

VI

One final remark: my specific motivation for developing this account of indicative conditionals is of course to solve a puzzle, and to defend a particular semantic analysis of conditionals. But I have a broader motivation which is perhaps more important. That is to defend, by example, the claim that the concepts of pragmatics (the study of linguistic contexts) can be made as mathematically precise as any of the concepts of syntax and formal semantics; to show that one can recognize and incorporate into abstract theory the extreme context dependence which is obviously present in natural language without any sacrifice of standards of rigor.[17] I am anxious to put this claim across because it is my impression that semantic theorists have tended to ignore or abstract away from context dependence at the cost of some distortion of the phenomena, and that this practice is motivated not by ignorance or misperception of the phenomenon of context dependence, but rather by the belief that the phenomenon is not appropriately treated in a formal theory. I hope that the analysis of indicative conditionals that I have given, even if not correct in its details, will help to show that this belief is not true.

Cornell University

NOTES

* The ideas in this paper were developed over a number of years. During part of this time my research was supported by the National Science Foundation, grant number GS-2574; more recently it was supported by the John Simon Guggenheim Memorial Foundation.
[1] The argument in the opposite direction – from the indicative conditional to the material conditional – is uncontroversially valid.
[2] This does not exhaust the options. Three other possible strategies might be mentioned. (1) Defend the direct argument, not by accepting the truth-functional analysis of the conditional, but by rejecting the truth-functional analysis of the disjunction. (2) Give a three-valued interpretation of the indicative conditional, assigning the neutral value when the antecedent is false. (3) Interpret the indicative conditional as a conditional assertion rather than the assertion of a conditional proposition. Alternative (1) might disarm this particular puzzle, but it seems *ad hoc* and would not help with other persuasive arguments for the material conditional analysis. Alternative (2) would conflict with some basic and otherwise plausible pragmatic generalizations such as that one should not make an assertion unless one has good reason to think that it is true. Alternative (3) seems to me the most promising and plausible alternative to the account I will develop, but to make it precise, I think one needs much of the framework of a pragmatic theory that I shall use in my account.
[3] Photo-copies have been widely circulated; part of it has been recently published in: D. Davidson and G. Harman (eds.), *The Logic of Grammar*, Dickenson, Encino, Cal., 1975, pp. 64–75.
[4] See David Lewis, *Counterfactuals*, Harvard University Press, Cambridge, 1973, pp. 84–91 for a defense of realism about possible worlds.
[5] See M. J. Cresswell, *Logics and Languages*, Methuen, London, 1973, pp. 23–24, and Stalnaker, 'Pragmatics,' in G. Harman and D. Davidson (eds.), *Semantics of Natural Languages*, Reidel, Dordrecht, 1972, pp. 381–82 for brief discussions of the intuitive motivation of this definition of proposition.

[6] For a fuller discussion and defense of this concept, see Stalnaker, 'Presuppositions', *Journal of Philosophical Logic*, (1973), 447–457.

[7] Elsewhere, I have called this set the *presupposition set*, but this terminology proved misleading since it suggested a set of presuppositions – propositions presupposed – rather than a set of possible worlds. The terminology adopted here was suggested by Lauri Karttunen.

[8] Stalnaker, 'A Theory of Conditionals', in N. Rescher (ed.), *Studies in Logical Theory*, Blackwell, Oxford, 1968, pp. 98–112, and Stalnaker and R. H. Thomason, 'A Semantic Analysis of Conditional Logic', *Theoria*, **36** (1970), 23–42. See also Lewis, *Counterfactuals*. The formal differences between Lewis's theory and mine are irrelevant to the present issue.

[9] If A is the impossible proposition – the one true in *no* possible world – then there will be no possible world which can be the value of the function, $f(A, i)$, and so the function is left undefined for this case. To take care of this special case, the theory stipulates that all conditionals with impossible antecedents are true.

[10] I was slow to see this despite the existence of clear examples in the literature. Comments by John Watling in a discussion of an earlier version of this paper helped me to see the point.

[11] This is Watling's example.

[12] 'A Note on Subjunctive and Counterfactual Conditionals', *Analysis* **12** (1951), 35–38.

[13] As with the pragmatic constraint on selection functions, there may be exceptions to this generalization. One exception is a statement of the form A *or* B *or both*. (I assume that the meaning of 'or' is given by the truth table for inclusive disjunction.) But statements which conflict with the principle must satisfy two conditions if they are to be appropriate. First, the statement must wear on its face that it is an exception so that it cannot be misleading. Second, there must be some explanation available of the purpose of violating the generalization, so that it will not be pointless. In the case of the statement A *or* B *or both*, it is clear from the logical relation between the last disjunct and the others that it must be an exception, so it satisfies the first condition. The explanation of the point of adding the redundant third disjunct is this: the disjunctive statement, A *or* B, requires that A *and not-B* and B *and not-A* be compatible with the context, but leaves open whether A *and* B is compatible with the context. The addition of the third disjunct, while adding nothing to the *assertive content* of the statement, does change the appropriateness conditions of the statement, and thus serves to indicate something about the context, or about the presuppositions of the speaker.

[14] Strictly, the inference to the contrapositive is reasonable only relative to the further assumption that the indicative *conclusion* is not inappropriate.

[15] *Philosophical Review* **73** (1964), 338–359.

[16] As with contraposition, the inference from 2 to 3 is reasonable only relative to the further assumption that the conclusion of the inference is appropriate, which means in this case, only relative to the assumption that P, the antecedent of the conditional, is compatible with the context. This assumption is obviously satisfied since the setting of the argument is a deliberation about whether or not to make P true.

[17] I recognize, of course, that the definitions and generalizations presented here are nothing like a rigorous formal theory. But some parts of the apparatus (in particular, the semantics for conditionals) have been more carefully developed elsewhere, and I believe it is a relatively routine matter to state most of the definitions and generalizations which are new in precise model theoretic terms. Just to show how it might go, I will give in an appendix a very abstract definition of a logical concept of reasonable inference.

APPENDIX

Entailment and reasonable inference relate propositions and speech acts, respectively, but in both cases, given an appropriate language, one can define corresponding logical notions – notions of entailment and reasonable inference which relate formulas, or sentences independently of their specific interpretations.

Let **L** be a language which contains sentences. A *semantic interpretation* of the language will consist of a set of possible worlds and a function which assigns *propositions* (functions from possible worlds into truth-values) to the sentences, relative to *contexts*. The formal semantics for the language will define the class of legitimate interpretations by saying, in the usual way, how the interpretation of complex expressions relates to the interpretation of their parts. A *context* is an n-tuple, the first term of which is a *context set* (a set of possible worlds). The other terms are whatever else, if anything, is necessary to determine the propositions expressed by the sentences.

Notation: I will use P, P_1, P_2, etc. as meta-variables for sentences, ϕ, ϕ_1, ϕ_2, etc. as meta-variables for propositions (for convenience, I will identify a proposition with the set of possible worlds for which it takes the value true); k, k_1, k_2, etc. will be variables ranging over contexts. $S(k)$ will denote the context set of the context k. $\|P\|_k$ will denote the proposition expressed by P in context k under the interpretation in question. (Reference to the interpretation is supressed in the notation.)

Entailment. One may define several notions of entailment. The basic notion is a language independent relation between propositions: ϕ_1 entails ϕ_2 if and only if ϕ_2 includes ϕ_1. The *logical* concept of entailment, entailment-in-**L**, is a relation between *sentences* of **L**: P_1 entails P_2 if and only if for all interpretations and all contexts k, $\|P_1\|_k$ entails $\|P_2\|_k$. Logical entailment is entailment in virtue of the logical structure of the sentences. Similarly, the logical concept of reasonable inference will identify the inferences which are reasonable in virtue of the logical structure of the sentences.

Pragmatic interpretations. To define the logical notion of reasonable inference, we need to expand the concept of an interpretation. A *pragmatic interpretation* of **L** will consist of a semantic interpretation, an *appropriateness relation*, and a *change function*. The appropriateness relation A is a two place relation whose arguments are a sentence of **L** and a context. $A(P, k)$ says that the assertive utterance of P in context k is appropriate. The change function g is a two place function taking a sentence of **L** and a context into a context. Intuitively, $g(P, k)$ denotes the context that results from the assertive utterance of P in context k.

Since **L** is unspecified here, I leave these notions almost completely unconstrained, but it is easy to see how the generalizations about disjunctive and conditional statements would be stated as postulates which give some substance to these notions as applied to a language containing these kinds of statements. Just as the semantics for a specific language will include semantic rules specifying the elements of the context and placing constraints on the allowable semantic interpretations, so the pragmatic theory for a specific language will include rules constraining the two distinctively pragmatic elements of a pragmatic interpretation, as well as the relations among the elements of the context.

I will give here just two constraints which will apply to any language intended to model a practice of assertion.

1. $A(P, k)$ only if $\| P \|_k \cap S(k) \neq 0$.

One cannot appropriately assert a proposition in a context incompatible with it.

2. $S(g(P, k)) = S(k) \cap \| P \|_k$.

Any assertion changes the context by becoming an additional presupposition of subsequent conversation. (In a more careful formulation the second of these would be qualified, since assertions can be rejected or contradicted. But in the absence of rejection, I think it is reasonable to impose this constraint.)

Both the appropriateness relation and the change function can be generalized to apply to finite sequences of sentences in the following way: Let σ be a finite sequence of sentences of **L**, $P_1, P_2, \ldots P_n$. Let $k_1, k_2, \ldots k_n$ be a sequence of contexts defined in terms of σ and a context k as follows: $k_1 = k$; $k_{i+1} = g(k_i, P_i)$. Then $A(\sigma, k)$ if and only if, for all i from 1 to n, $A(P_i, k_i)$. $g(\sigma, k) =_{df} k_n$.

Reasonable inference. The inference from a sequence of sentences of **L**, σ, to a sentence of **L**, *P* is *reasonable-in-***L** if and only if for all interpretations and all contexts k such that $A(\sigma, k)$, $S(g(\sigma, k))$ entails $\|P\|_{g(\sigma, k)}$.

Note that there is no language independent concept of reasonable inference analogous to the language independent notion of entailment. The reason is that, while we have in the theory a notion of proposition that can be characterized independently of any language in which propositions are expressed, we have no corresponding non-linguistic concept of statement, or assertion. One could perhaps be defined, but it would not be a simple matter to do so, since the identity conditions for assertion types will be finer than those for propositions. The reason for this is that different sentences may have different appropriateness conditions even when they express the same proposition.

ALLAN GIBBARD

TWO RECENT THEORIES OF CONDITIONALS

In recent years, two new and fundamentally different accounts of conditionals and their logic have been put forth, one based on nearness of possible
worlds (Stalnaker, 'A Theory of Conditionals', 1968, this volume, pp. 41–
55; Lewis, *Counterfactuals,* 1973) and the other based on subjective conditional probabilities (Adams, *The Logic of Conditionals,* 1975). The two
accounts, I shall claim, have almost nothing in common, They do have a
common logic within the domain on which they both pronounce, but that,
as far as I can discover, is little more than a coincidence. Each of these disparate accounts, though, has an important application to natural language,
or so I shall argue. Roughly, Adams' probabilistic account is true of indicative conditionals, and a nearness of possible worlds account is true of
subjunctive conditionals. If that is so, the apparent similarity of these two
'if' constructions hides a profound semantical difference.

1. THE TWO ACCOUNTS

I begin with a rough and simplified sketch of the two accounts and relationships between them. First, some terminology: I shall use 'proposition' as
a theory-laden word, the theory being a representation of subjective probability, or *credence.* On this representation, we start with a set t of all
epistemically possible worlds (or *worlds*). Any proposition is identified with
a set of worlds, the worlds *in* which the proposition is *true.* Not every subset of t need be a 'proposition'; rather the 'propositions' comprise a fixed
set $\mathscr{F}$ of subsets of t. $\mathscr{F}$ is required to be a 'field of sets' whose 'universal
set' is t. $\mathscr{F}$ is a *field of sets* iff $\mathscr{F}$ is a set of sets and, where $t = \bigcup \mathscr{F}$,
$t \in \mathscr{F}$ and $\mathscr{F}$ is closed under the operations of union and t-complemen-
tation. t is called the *universal set* of $\mathscr{F}$. Members of $\mathscr{F}$ are called *propositions* of $\mathscr{F}$, and members of t are called *worlds* of $\mathscr{F}$. A person's
credences, or degrees of belief, are represented by real numbers from zero
to one, and the function ρ which gives them is a probability measure on
$\mathscr{F}$: a non-negative real-valued function whose domain is $\mathscr{F}$, such that $\rho(t) = 1$
and where propositions a and b are disjoint, $\rho(a \cup b) = \rho(a) + \rho(b)$. (See
Kyburg, 1980, pp. 14–18). When $\rho(a) \neq 0$, $\rho(b/a)$ is defined as the quotient

211

W. L. Harper, R. Stalnaker, and G. Pearce (eds.), Ifs, 211–247.

$\rho(a \cap b)/\rho(a)$; when $\rho(a) = 0$, $\rho(b/a)$ need not be defined for purposes here. Where ρ gives a subject's credences, $\rho(b/a)$ is his conditional credence in b given a; it is the credence he would give b were he to learn a and nothing else. Logical notation will be used: '&' or juxtaposition for intersection, 'v' for union, '$-a$' or '$\bar{a}$' for the t-complement of a, and '$a \supset b$' for $\bar{a} \vee b$. We let f be the empty set, and represent entailment set-theoretically: a *entails* b iff $a \subseteq b$.

On the Adams account (1975), an indicative conditional need not express a proposition in this sense. Indicative conditionals are rather to be understood through their conditions of acceptance or assertability, and where a and b are propositions, one accepts the indicative conditional 'If a, then b' iff one's conditional credence in b given a is sufficiently high. On this basis, a logic for conditionals can be constructed.

Consider first the logic of propositions. For finite sets of propositions, the notions of consistency and consequence can be formulated in terms of probabilities, and the notions so formulated turn out to be equivalent to the notions in their standard formulation. Where $\mathscr{F}$ is a field of sets and $\mathscr{A}$ is a finite set of propositions of $\mathscr{F}$, we can define

> $\mathscr{A}$ is *p-consistent* iff for every $\delta > 0$, there is a probability measure ρ on $\mathscr{F}$ such that for every $A \in \mathscr{A}$, $\rho(A) > 1 - \delta$ (Adams, 1975, p. 51).

Where $\mathscr{A}$ is a finite set of propositions of $\mathscr{F}$ and B is a proposition of $\mathscr{F}$, we can define

> B is a *p-consequence* of $\mathscr{A}$ iff for every $\epsilon > 0$ there is a $\delta > 0$ such that for any probability measure on ρ on $\mathscr{F}$ with $\rho(A) > 1 - \delta$ for each $A \in \mathscr{A}$, we have $\rho(B) > 1 - \epsilon$.

(If there is such a δ, it turns out, then ϵ/n, where n is the number of propositions in $\mathscr{A}$, is such a δ.) It can then be shown that $\mathscr{A}$ is p-consistent iff it is consistent, and B is a p-consequence of $\mathscr{A}$ iff B is a consequence of $\mathscr{A}$. (Adams, 1975, pp. 57–58).

Turn now to conditionals, and consider $\rho(a \rightarrow b)$ just to be the conditional probability $\rho(b/a)$. The definitions of 'p-consistent' and 'p-consequence' can then be applied without change to conditionals and sets of conditionals, or to mixed sets of conditionals and propositions. (Indeed in these definitions, we can represent any proposition a by the conditional $t \rightarrow a$, since for any probability measure ρ, $\rho(a) = \rho(a/t) = \rho(t \rightarrow a)$.)

Note that this account applies only to conditionals constructed from propositions, with $\rightarrow$ the main connective. Where a, b, and c are propositions, the account deals with $a \rightarrow b$, but not with $(a \rightarrow b) \rightarrow c$, $a \rightarrow (b \rightarrow c)$, $a \& (b \rightarrow c)$, $a \vee (b \rightarrow c)$, $-(a \rightarrow b)$, and the like. The account is not one of conditionals embedded in longer sentences. Formally, we might consider a conditional simply to be an ordered pair of propositions; in any case, for all the Adams account tells us, a conditional is not itself a proposition and cannot be treated as one.[1]

On the Stalnaker nearness account (1968), in contrast, a conditional is a proposition. What proposition it is is determined by a *selection function* (or *s-function*), which we may think of as picking out, for each non-empty proposition a and world w, the world in a (or *a-world*) nearest to w. The conditional 'If a then b' says that the nearest a-world to the actual world is a b-world. In other words, for a given s-function σ, the conditional $a \mathbin{\Box\!\!\rightarrow}_\sigma b$ determined by σ is the set $\{w \mid \sigma(a, w) \in b\}$. Formally, σ is an *s-function* for $\mathscr{F}$ iff to every non-empty proposition a and world w, σ assigns a world, and these conditions are satisfied for every world w, proposition $a \neq f$, and proposition b.

(S1) $\sigma(a, w) \in a$.

(S2) If $w \in a$, then $\sigma(a, w) = w$.

(S3) If $a \subseteq b$ and $\sigma(b, w) \in a$, then $\sigma(a, w) = \sigma(b, w)$.

A Stalnaker conditional as I have defined it is a set of worlds but may not be a proposition. That is to say, let $\mathscr{F}$ be a field of sets, a and b be propositions of $\mathscr{F}$ with $a \neq f$, and σ be an s-function for $\mathscr{F}$; then $a \mathbin{\Box\!\!\rightarrow}_\sigma b$ is a subset of t but may not be a proposition of $\mathscr{F}$. σ will be called an *internal s-function* for $\mathscr{F}$ if σ is an s-function for $\mathscr{F}$ and for every a, $b \in \mathscr{F}$ with $a \neq f$,

(S4) $a \mathbin{\Box\!\!\rightarrow}_\sigma b$ is a proposition.

Given propositions a and b, different s-functions σ will yield different sets of worlds as the value of $a \mathbin{\Box\!\!\rightarrow}_\sigma b$. The choice of an s-function, Stalnaker says, is a pragmatic matter, which is determined by context (1968), pp. 109–111; this volume, pp. 51–52, 1975)

Stalnaker's account allows conditionals to be embedded, so that $(a \rightarrow b) \rightarrow c$, $a \rightarrow (b \rightarrow c)$, $a \vee (b \rightarrow c)$, and the like are allowed. Let us confine ourselves for the moment, though, to expressions of the kind Adams countenances, and return to treating conditionals as ordered pairs of propositions. That way, we can compare what the Stalnaker theory has to say with what the Adams theory has to say within the more limited domain of the Adams theory. Where $\mathscr{F}$ is a field of sets, then, a *conditional* $a \to b$ of $\mathscr{F}$ will be an ordered pair $\langle a, b \rangle$ of propositions of $\mathscr{F}$, with $a \neq f$.

Note first that, as with the Adams theory, we may identify a proposition a with a conditional $t \to a$. For any *s*-function σ, $a = (t \;\Box\!\!\!\to_\sigma b)$; that follows from (S2). Thus instead of talking about logical relations among propositions and conditionals here, we may speak simply of conditionals.

Now using the Stalnaker machinery, we can give new characterizations of consistency and consequence of sets of conditionals. Let $A_1, \ldots, A_n$, C be conditionals of $\mathscr{F}$, and let $\mathscr{A} = \{A_1, \ldots, A_n\}$. Henceforth, where A is a conditional $a \to b$, write $a \;\Box\!\!\!\to_\sigma b$ as A^σ. $\mathscr{A}$ is *s-consistent* iff for some *s*-function σ for $\mathscr{A}$, the set $\mathscr{A}^\sigma = \{A_1^\sigma, \ldots, A_n^\sigma\}$ is consistent. Here for any given σ, $A_1^\sigma, \ldots, A_n^\sigma$ are sets of epistemically possible worlds, and so to say that $\mathscr{A}^\sigma$ is consistent is just to say that it has a non-empty intersection. C is an *s-consequence* of $\mathscr{A}$ iff for every *s*-function σ for $\mathscr{F}$, C^σ is a consequence of $\mathscr{A}^\sigma$, or in other words, iff for every such σ,

$$A_1^\sigma \; \& \; \ldots \; \& \; A_n^\sigma \subseteq C^\sigma.$$

Modified versions of these definitions restrict consideration to *s*-functions which are internal: $\mathscr{A}$ is *s-consistent in the strong sense* iff for some internal *s*-function σ for $\mathscr{F}$, $\mathscr{A}^\sigma$ is consistent, and C is an *s-consequence of $\mathscr{A}$ in the weak sense* iff for every internal *s*-function σ for $\mathscr{F}$, C^σ is a consequence of $\mathscr{A}^\sigma$. Theorem 2 of the Appendix shows that these definitions are equivalent to the ones they modify.

Now at least for finite sets of conditionals, the relations of *p*-consequence and *s*-consequence coincide, as do the properties of *p*-consistency and *s*-consistency. That I shall prove in Section 3 and the Appendix. Thus even if the two theories are incompatible with each other, if one explains the logic of conditionals on their common domain, the other will appear to do so equally well.

2. NEARNESS AND PROBABILITIES: A PRIMER

Why do the two accounts yield the same logic? It might seem that they do so because both employ the same fundamental idea, that of a minimal change or minimal revision. They do so, though, in different ways. Here are the maxims that guide the two treatments. Adams' maxim: to decide whether to believe a conditional $a \to b$, hypothetically revise your beliefs in a minimal way so as to believe a, and then see if you believe b. Stalnaker's maxim:

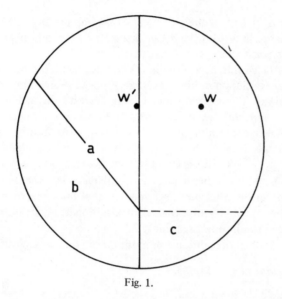

Fig. 1.

to decide whether a conditional $a \to b$ is true in a world w, change w in a minimal way so that a is true, and then see whether b is true in that changed world. Adams' system, then, involves changes in states of belief, whereas Stalnaker's involves changes in a world – that is, in an epistemically possible world. Now a world, as the term is used here, is not a state of belief. A world is rather a maximally specific way things might be. A state of belief, in contrast, is no more specific than one's beliefs are opinionated; it can be represented by a probability measure. Take an illustrative contrast: any world is either one in which Richard III had the two young princes killed or one in which he did not, but I give some credence to both possibilities. Thus my credence measure – the probability measure that represents my state of belief – assigns positive credence both to the set of worlds in which he had them killed and to the set of worlds in which he did not.[2]

The logic of states of belief can be illustrated by Venn diagrams.[3] In a Venn diagram, worlds are represented by points and propositions by regions, with the area of each region proportional to its probability. The entire region of the diagram is a proposition k of probability one. In Figure 1, the entire circle represents k, the left half of the circle represent a proposition a, and the lower left part of that half-circle represents a proposition b which entails a. A minimal revision of k to accomodate a simply involves erasing the right half of the circle and expanding the left half uniformly. A minimal change in w to make a true is a shift from world w to that a-world w' which, in some sense, is most like w.

Likeness of worlds need not, in a Venn diagram, be represented by geometric nearness. Suppose, however, that σ indeed is an s-function for which $\sigma(a, w)$ is always the a-world geometrically nearest to w in Figure 1. Then c is the set of $\bar{a}$-worlds the nearest a-world to which is a b-world. Thus $b \vee c$ is the Stalnaker conditional $a \mathbin{\square\!\!\rightarrow}_\sigma b$ for this σ. Interesting facts can now be read off the diagram. In Figure 1, the conditional probability $\rho(b/a)$ is approximately $\frac{1}{2}$, since ab is roughly half the area of a. $\rho(b \vee c)$, which is $\rho(a \mathbin{\square\!\!\rightarrow}_\sigma b)$, on the other hand, is considerably less than $\frac{1}{2}$. We thus see that nothing in the Stalnaker logic requires that $\rho(a \mathbin{\square\!\!\rightarrow}_\sigma b) = \rho(b/a)$ when $\rho(a) \neq 0$; the probability of a Stalnaker conditional may be distinct from the corresponding conditional probability.[4]

Moreover, for any conditional proposition $a \Rightarrow b$ to be such that

(1) $\rho(a \Rightarrow b) = \rho(b/a)$,

$a \Rightarrow b$ must divide the $\bar{a}$-worlds in exactly the proportion in which b divides the a-worlds. In other words, if in Figure 2, c is the $\bar{a}$-part of $a \Rightarrow b$, we must have

(2) $\rho(c)/\rho(\bar{a}) = \rho(b)/\rho(a)$.

That should be obvious from Figure 2. The proof assumed only that $a \Rightarrow b$ is a genuine proposition, and that it is true in every ab-world and false in every $a\bar{b}$-world, so that

(3) $a \mathbin{\&} (a \Rightarrow b) = ab$.

The question is then how $a \Rightarrow b$ should divide the $\bar{a}$-worlds. For to treat a conditional $a \rightarrow b$ as a proposition $a \Rightarrow b$ is to suppose that in every possible world in which $\bar{a}$ holds, there is a fact of the matter whether $a \Rightarrow b$ holds in that world: the conditional is either true or false in that world. Now we have $(a \Rightarrow b) \doteq (b \vee c)$; thus since b and c are disjoint, $\rho(a \Rightarrow b) = \rho(b) + \rho(c)$. Since $b \subseteq a$, we have $\rho(b/a) = \rho(b)/\rho(a)$. Thus since $\rho(a) \neq 0$, (1) becomes

$$\rho(b) + \rho(c) = \rho(b)/\rho(a).$$

Given that $1 - \rho(a) \neq 0$, this is algebraically equivalent to

$$\frac{\rho(c)}{1 - \rho(a)} = \frac{\rho(b)}{\rho(a)},$$

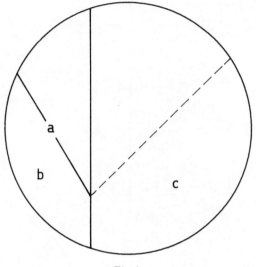

Fig. 2.

and with $\rho(\bar{a})$ substituted for $1 - \rho(a)$, this is (2), which was to be proved. c, then, is carved out of $\bar{a}$ in such a way as to make th ratio $\rho(c)/\rho(\bar{a})$ equal to $\rho(b)/\rho(a)$. I shall call this the *Fundamental Consequence* of requirements (1) and (3). A like argument shows that (2) and (3) entail (1); thus given (3), we have that (2) and (1) are equivalent.

It is hard to imagine a natural way of choosing c that would satisfy the Fundamental Consequence. We have already seen that not all Stalnaker conditionals do: it is not the case that for every ρ, a, b, and σ,

(4) $\rho(a \mathbin{\Box\!\!\rightarrow}_\sigma b) = \rho(b/a)$ if $\rho(a) \neq 0$.

David Lewis (1976, pp. 300–303; this volume, pp. 131–134) has proved something stronger: that except in utterly trivial cases, there is no σ such that for every ρ, a, and b, (4) holds. Lewis's result, indeed, is even stronger than this. Let $\Rightarrow$ be any two-place propositional function: function which, to any two propositions $a \neq f$ and b assigns a proposition $a \Rightarrow b$. We do not require that $\Rightarrow$ be a Stalnaker conditional. We do suppose, for some fixed ρ, that $\rho(a \Rightarrow b) = \rho(b/a)$ whenever $\rho(a) \neq 0$, and that the same holds for the probability measure obtained from ρ by conditionalizing on any proposition c of non-zero probability. In other words, we suppose the following.

(PC)   For any $a, b, c \in \mathscr{F}$, if $\rho(ac) \neq 0$, then $\rho(a \Rightarrow b/c) = \rho(b/ac)$.

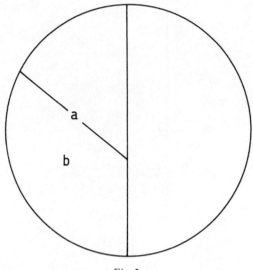

Fig. 3.

A probability measure will be called *non-trivial* iff there are at least three mutually disjoint propositions to which it assigns non-zero probability. Lewis showed[5] that (PC) holds for no non-trivial probability measure ρ.

The proof is best seen by means of Figure 3. There $\bar{a}$ and b are two of the three mutually disjoint propositions whose existence is assured by the non-triviality of ρ; thus $a\bar{b}$, b, and $\bar{a}$ partition the space, all have non-zero probability, and $ab = b$. Now from (PC),

$$\rho(a \Rightarrow b/\bar{b}) = \rho(b/a\bar{b}) = 0.$$

Thus $a \Rightarrow b$ and $\bar{b}$ intersect at most in a set of measure zero, so that $\rho(a \Rightarrow b) \leqslant \rho(b)$. But this is absured, since from (PC) and $ab = b$,

$$\rho(a \Rightarrow b) = \rho(b/a) = \rho(b)/\rho(a),$$

and since $\rho(a) < 1$, it follows that $\rho(a \Rightarrow b) > \rho(b)$.

We began with the question of why two fundamentally different theories of conditionals, Adams' and Stalnaker's, yield the same logic on their common domains. From Lewis' proof, we know that the answer cannot be this: that for some s-function σ, we have $\rho(a \,\square\!\!\rightarrow\, \sigma\, b) = \rho(b/a)$ for every probability measure ρ and pair of propositions a and b with $\rho(a) \neq 0$. The answer cannot even be this: that for some s-function σ and probability measure ρ,

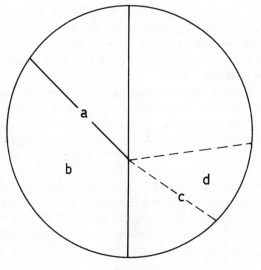

Fig. 4.

we have $\rho_c(a \,\Box\!\!\to_\sigma b) = \rho_c(b/a)$ for all propositions a, b, c with $\rho(ac) \neq 0$, where ρ_c is the probability measure obtained from ρ by conditionalizing on c. We might, however, look for a weaker ground for the sameness of the two logics. Might it be the case, that for at least some internal s-function σ, that there is a fixed probability measure ρ such that, for any two propositions a and b with $\rho(a) \neq 0$, we have $\rho(a \,\Box\!\!\to_\sigma b) = \rho(b/a)$?

Stalnaker (1976, pp. 303–304) has shown that even to this question the answer is no.[6] For no non-trivial probability measure ρ on field of sets $\mathcal{F}$ is there an internal s-function σ such that for every conditional $a \to b$ of $\mathcal{F}$, the following hold.

(i) $a \,\&\, (a \,\Box\!\!\to_\sigma b) = ab$

(ii) $\rho(a \,\Box\!\!\to_\sigma b) = \rho(b/a)$ if $\rho(a) \neq 0$.

For suppose, as in Figure 4, there are three mutually disjoint propositions of non-zero probability, two of which are $\bar{a}$ and b. Again let c be $\bar{a} \,\&\, (a \,\Box\!\!\to_\sigma b)$. From (2), the Fundamental Consequence of (i) and (ii), we know that c must have non-zero probability. Now consider the conditional $\bar{c} \,\Box\!\!\to_\sigma a\bar{b}$. For this, the Fundamental Consequence is that where $d = c \,\&\, (\bar{c} \,\Box\!\!\to_\sigma a\bar{b})$, we must have $\rho(d)/\rho(c) = \rho(a\bar{b})/\rho(\bar{c})$. Since $\rho(a\bar{b}) \neq 0$ and $\rho(c) \neq 0$, we have $\rho(d) \neq 0$, and there is a d-world – that is, a world

w in which both c and $\bar{c} \,\Box\!\!\rightarrow_\sigma a\bar{b}$ obtain. c, though, is precisely the set of $\bar{a}$-worlds whose nearest a-world is a b-world, and since $a \subseteq \bar{c}$, c cannot contain any worlds whose nearest $\bar{c}$-world is an $a\bar{b}$-world. For if the nearest $\bar{c}$-world to a c-world is an a-world, it is the nearest a-world, and hence not a b-world. That completes Stalnaker's proof.

Note that the proof involves treating conditionals, and truth-functions involving conditionals, as themselves components of conditionals to which the requirement that the probability of a conditional equal the corresponding conditional probability applies. Van Fraassen succeeds in giving the requirement a more narrow scope. Start off with a finite field $\mathscr{F}$ of sets and a fixed probability measure ρ over $\mathscr{F}$. The resulting probability space, Van Fraassen shows, can be embedded in a larger probability space with an internal s-function σ such that the probability of any conditional $a \,\Box\!\!\rightarrow_\sigma b$ formed from propositions in the original set equals the corresponding conditional probability.[7] This result provides the key to answering the question of why the Adams and Stalnaker logics are the same in their common domain.

The rough idea of the Van Fraassen construction is this: Let there be many possible worlds, and apart from the requirement that each world be nearest to itself, decide nearness by chance. That is to say, where w is a world, whenever $w \notin a$ and $b \subseteq a$, let the chance that the nearest a-world to w is a b-world simply be $\rho(b/a)$. I shall call such an s-function a ρ-*random* s-function. Then indeed where σ is such an s-function and $c = \bar{a} \,\&\, (a \,\Box\!\!\rightarrow_\sigma b)$, we have $\rho(c)/\rho(\bar{a}) = \rho(b)/\rho(a)$: the Fundamental Consequence is satisfied. For $\rho(c)/\rho(\bar{a})$ is just the proportion of $\bar{a}$ worlds whose nearest a-world is a b-world, and by selecting the nearest a-world to any $\bar{a}$-world randomly, we guarantee that this proportion is just $\rho(b)/\rho(a)$.

3. THE EQUIVALENCE OF THE TWO LOGICS

The possibility of a ρ-random s-function helps to explain why the logic of p-consequence and s-consequence are the same: using such s-functions, we can show that any s-consequence is a p-consequence. Let $A_1, \ldots, A_n, C$ be conditionals, let $\mathscr{A} = \{A_1, \ldots, A_n\}$, and let C be an s-consequence of $\mathscr{A}$. We are to prove that C is a p-consequence of $\mathscr{A}$. First we need to prove that embedding a field of sets in a larger one does not affect the relation of s-consequence; in the Appendix, that is Theorem 1 and the main result is Theorem 3. Now given $\epsilon > 0$, let $\delta = \epsilon/n$, and suppose $\rho(A) > 1 - \delta$ for every $A \in \mathscr{A}$. What we are to show is that $\rho(C) > 1 - \epsilon$.

Here is a sketch of the proof, with questions of embedding ignored. Let σ be a ρ-random s-function. Since C is an s-consequence of $\mathscr{A}$, C^{σ} is a consequence of $\{A_1^{\sigma}, \ldots, A_n^{\sigma}\}$, and so since $\rho(A^{\sigma}) > 1 - \delta$ for each $A \in \mathscr{A}$, we have $\rho(C^{\sigma}) > 1 - \epsilon$, and hence $\rho(C) > 1 - \epsilon$. Supposing that C is an s-consequence of $\mathscr{A}$, we have shown that C is a p-consequence of $\mathscr{A}$.

One way to prove the converse is this. Adams (1975, pp. 60–61) gives a set of inference rules that are complete for p-consequence: any p-consequence is derivable by those rules. Inspection of those rules reveals that each is sound for s-consequence: any conclusion derived by those rules is an s-consequence of its premises. Therefore any p-consequence is an s-consequence.[8]

I now sketch an independent proof that any p-consequence is an s-consequence. In the first place, for any finite set $\mathscr{A}$ of conditionals and conditional $c \to d$, $c \to d$ is a p-consequence of $\mathscr{A}$ iff $\mathscr{A} \cup \{c \to \bar{d}\}$ is not p-consistent, and $c \to d$ is an s-consequence of $\mathscr{A}$ iff $\mathscr{A} \cup \{c \to \bar{d}\}$ is not s-consistent. The first is noted by Adams, and both are straightforward consequences of the definitions. Therefore to show that any p-consequence is an s-consequence, it will suffice to show that any s-consistent set of conditionals is p-consistent. For suppose we have shown this, and suppose $c \to d$ is a p-consequence of $\mathscr{A}$. Then $\mathscr{A} \cup \{c \to \bar{d}\}$ is not p-consistent. Hence by supposition, $\mathscr{A} \cup \{c \to \bar{d}\}$ is not s-consistent, and so $c \to d$ is an s-consequence of $\mathscr{A}$. Hence to complete the proof that p- and s-consequence coincide, we need only prove that any s-consistent set of conditionals is p-consistent, which is Theorem 5 in the Appendix.

The idea of its proof is to show how, given a $\delta > 0$ and an s-function σ such that the set $\mathscr{A}^{\sigma}$ is consistent, to construct a probability measure ρ such that all the antecedants have non-zero probability and all the conditional probabilities $\rho(b_i/a_i)$ are greater than $1 - \delta$. Here is a sketch of the procedure. Since $\mathscr{A}^{\sigma}$ is consistent, we can let w^* be a world in its intersection; thus all of $a_1 \mathbin{\square\!\!\rightarrow}_{\sigma} b_1, \ldots, a_n \mathbin{\square\!\!\rightarrow}_{\sigma} b_n$ hold at w^*. We now order the antecedants $a_1, \ldots, a_n$ by their distance from w^*, and consider the sequence $w_1, \ldots, w_m$ of worlds nearest to w^* in the sequence of antecedants so ordered. We can let our probability measure ρ give non-zero probability only to worlds $w_1, \ldots, w_m$ in such a way that the probability of each world in the sequence dwarfs the combined probability of the remaining worlds in the sequence. That is, for each w_i, $\rho(w_i) \geq (1 - \delta)\rho(w_i \vee \ldots \vee w_m)$. Then for each a_k and b_k, we have $\rho(b_k/a_k) > 1 - \delta$. For suppose w_j is the nearest a_k-world to w^*. Then a_k holds at w_j, and holds at no w_i that is nearer than w_j to w^*. Moreover, since $a_k \mathbin{\square\!\!\rightarrow}_{\sigma} b_k$ holds at w^*, b_k holds at w_j. Thus the

a_k-worlds consist of some subset of $\{w_j, \ldots, w_m\}$, and the $a_k b_k$-worlds consist of at least w_j. Hence since $\rho(a_k b_k) \geqslant (1 - \delta)\rho(w_j \vee \ldots \vee w_m) \geqslant (1 - \delta)\rho(a_k)$, we have $\rho(a_k b_k)/\rho(a_k) \geqslant 1 - \delta$, or in other words, $\rho(b_k/a_k) \geqslant 1 - \delta$. The construction has succeeded, and the Theorem is proved.

We have shown that any s-consequence is a p-consequence and conversely: the Adams and Stalnaker logics are equivalent on their common domain. A striking aspect of both halves of the equivalence proof is that they draw on no important similarity of the pictures that motivate the two logics. Rather, each proof is based on a trick. The trick behind the proof that any s-consequence is a p-consequence is to find, for any probability measure ρ, a Van Fraassen s-function: a function σ which has the formal properties of a Stalnaker selection function, but is so far from reflecting any intuitive idea of 'nearness' of possible worlds that, for proposition a and world w in which a does not hold, it selects an a-world as formally 'nearest' to w at random. The trick behind the proof that any p-consequence is an s-consequence is to find, for any s-function σ, a probability measure ρ which corresponds in the needed way, but whose correspondance to σ is contrived and unnatural. We took an arbitrary world w^* in the intersection of a set of Stalnaker conditionals, ordered the antecedants by distance from w^*, and concentrated all the probability in the antecedant worlds nearest to w^*, in decreasing orders of magnitude as the antecedents grew more distant from w^*. The proofs in both directions, in short, match s-functions and probability measures in a contrived way.

Here, then, is our situation. We saw earlier that the superficial connection between the Stalnaker and Adams accounts – that both depend in some way on a 'nearest' revision – masked a fundamental disparateness of the accounts. Still, the two accounts yield the same logic in the domain on which they both pronounce. That suggests that in some deep way the two accounts indeed are connected. The proofs I have given, though, display the sameness of logics as resting not on a deep connection between the two accounts, but on contrivances. That does not preclude there being a deep connection which some other proof might display, but at this point we have no reason to regard the sameness of logics as anything but coincidence.

4. GRAMMATICALLY INDICATIVE AND SUBJUNCTIVE CONDITIONALS

I turn now to natural language. Semantically, I shall argue, conditionals are of two major kinds, which are alike only superficially. The Adams account applies to conditionals of one kind, which I shall call 'epistemic conditionals'; the Stalnaker account applied to conditionals of the other kind, which I shall call 'nearness conditionals'. To this semantical distinction there roughly corresponds a syntactical distinction. For the most part, what I shall call 'grammatically indicative' conditionals are epistemic conditionals and what I shall call 'grammatically subjunctive' conditionals are nearness conditionals. The grammatical distinction has little to do with the indicative and subjunctive moods, and I use the terms I do only for want of better.

As grammatical paradigms, take the pair:

(5)   If Oswald hadn't shot Kennedy, no one would have shot Oswald.

(6)   If Oswald didn't shoot Kennedy, no one shot Oswald.

(5) I shall treat as a paradigm of a grammatically subjunctive conditional; (6), as a paradigm of a grammatically indicative conditional. There are two grammatical differences between them, one in the antecedent and one in the consequent. In the antecedent, (5) uses the past perfect 'hadn't shot' where (6) uses the simple past 'didn't shoot'. In the consequent, (5) uses 'would have shot' where (6) uses 'shot'. Consider each of these in turn.

Although grammatically, only the antecedent of (6) is in the simple past, semantically both antecedents concern the simple past. The antecedent of (5) is thus grammatically prior to its time of reference. That can be seen most clearly from a variant of (5) with the same meaning and general grammatical form,

(7)   If Oswald hadn't shot Kennedy when he did, no one would have shot Oswald.

The situation posited in the antecedent of (7) is one in which Oswald didn't shoot Kennedy at the time when in actual fact he did. It is not one in which Oswald hadn't shot Kennedy at the time when in fact he did; that could only mean that he hadn't already shot Kennedy at the time when in fact he shot Kennedy, which is true – or at least would be true in fact, Oswald shot Kennedy only once. The antecedent of (7), though, clearly posits a contrary to fact situation in which Oswald didn't shoot Kennedy at the very time when in fact he shot Kennedy. I shall call this use of a grammatical tense prior to the one ordinarily appropriate for the time of the antecedent an *antecedent tense shift*.

The second feature by which (5) differs from (6) is in its use of 'would' in the consequent. It will turn out that grammatically, 'would' acts as the past tense of 'will', and so the feature of (5) to note is that there is a form of 'will' in the consequent.

I shall call a conditional with antecedent tense shift and a model auxilliary such as 'will' in the consequent *grammatically subjunctive,* and other conditionals *grammatically indicative.* The terms are unfortunate in a way, since the antecedent of a grammatically subjunctive conditional, we shall see, need not be in the subjunctive mood. I use these terms because I can find no simple, familiar terms to mark the systematic distinction I want to make which are not at least equally misleading.

Here are general directions for constructing conditionals with the features I have noted. Conditionals which are constructed in this way and which, as a result, exhibit antecedent tense shift, form, I shall claim, a significant grammatical class. Begin with a stem conditional, which, because its verbs lack tense and person, is not itself a piece of English.

> (8)　　*If he be upset, she comfort him.*

Optionally, either stem may be transformed into a perfect, progressive, or perfect progressive.

> (9)　　*If he have been upset, she have comforted him.*

> (10)　　*If he be upset, she be comforting him.*

> (11)　　*If he have been upset, she comfort him.*

A model auxilliary[9] such as 'will' is now applied to the verb stem of the consequent, so that (8)–(11) become

> (12)　　*If he be upset, she will comfort him.*

> (13)　　*If he have been upset, she will have comforted him.*

> (14)　　*If he be upset, she will be comforting him.*

> (15)　　*If he have been upset, she will comfort him.*

Finally, either present tense, with appropriate person, is applied to both clauses, or past tense is applied to both clauses. (12)–(15) now become

> (16)　　If he is upset, she will comfort him.

> (17*)　　If he was upset, she would comfort him.*

> (18)　　If he has been upset, she will have comforted him.

(19)   If he had been upset, she would have comforted him.

(20)   If he is upset, she will be comforting him.

(21*)   If he was upset, she would be comforting him.*

(22)   If he has been upset, she will comfort him.

(23)   If he had been upset, she would comfort him.

(17*) and (21*) are not securely part of standard English; their antecedent need to be in the subjunctive mood.

(17)   If he were upset, she would comfort him.

(21)   If he were upset, she would be comforting him.

Present antecedents in (16), (18), (20), and (22), on the other hand, seem quaint or worse in the subjunctive mood; see (12)–(15). The subjunctive mood is now vestigial in English, and applies to antecedents of what I am calling 'grammatically subjunctive conditionals' only in the past tense.

The instructions I have given allow thirty-two grammatically subjunctive 'will' conditionals to be constructed from the stem conditional (8). Optional features are past or present tense, perfect antecedent, progressive antecedent, perfect consequent, and progressive consequent. Perhaps not all thirty-two are easily interpretable, but I think that for each, some imaginable context can be found in which it can be read with antecedent tense shift.

The rules for the formation of indicative conditionals are much more flexible: a modal auxiliary in the consequent is optional, and past and present tenses can apply separately to antecedent and consequent. The following, for instance, are allowable.

(24)   If he was upset, she is comforting him.

(25)   If he is upset, she had been yelling at him.

(26)   If he was upset, she will learn about it.

These rules allow the construction both of sentence like (24)–(26) which cannot be constructed with the rules for subjunctive conditionals, and of sentences which can. In the case of sentences like (22), which can be constructed by either set of rules, the sentence is syntactically ambiguous: it is read as a grammatically subjunctive conditional if it is read as having antecedent tense shift, and otherwise it is read as a grammatically indicative conditional.

I propose that what I have been calling 'grammatically subjunctive conditionals' form a significant grammatical class. Grammatically subjunctive conditionals are those conditionals with the following two features.

(i)   A modal auxilliary is the stem verb of the consequent, its tense agreeing with that of the antecedent.

(ii)   Feature (i) induces an antecedent tense shift: the time to which the antecedent is understood as referring is after the time to which if would refer if uttered as an independent sentence . .

Indicative conditionals lack feature (ii) and may lack feature (i). Only grammatically subjunctive conditionals have antecedents in the subjunctive mood, but many have antecedents in the indicative mood. The next question is whether this grammatical distinction has any semantical consequences apart from the antecedent tense shift.

5. THE SEMANTICAL DISTINCTION

Sly Pete and Mr. Thomas Stone are playing poker aboard a Mississippi River boat. Both Pete and Stone are good poker players, and Pete, in addition, is unscrupulous. Stone has bet up to the limit for the hand, and it is now up to Pete to call or fold. Zack has seen Stone's hand, which is quite good, and signalled its contents to Pete. (Call this moment t_0). Stone, suspecting something, demands that the room be cleared. Five minutes later, Zack is standing by the bar, confident that the hand has been played out but ignorant of its outcome. (Call this moment t_1). He now entertains these two conditionals.

(27)   If Pete called, he won.

(28)   If Pete had called, he would have won.

At t_1, Zack accepts (27), because he knows that Pete is a crafty gambler who knew Stone's hand; thus Zack knows that Pete would not have called unless he had a winning hand. (28), on the other hand, Zack regards as probably false. For he knows that Stone's hand was quite good, and therefore regards it as unlikely that Pete had a winning hand. Thus he regards it as unlikely that if Pete had called, he would have won.

(27) is grammatically indicative, whereas (28) is grammatically subjunctive. To this grammatical difference, we have seen, there corresponds a semantical difference. What is it? Zack knows enough that were he to learn

that Pete had called and to learn nothing else, he would come to believe that Pete had won. In other words, because Zack believes that Pete would not have called unless he had a winning hand, Zack's conditional credence p (Pete won / Pete called) is close to one. (27), then, seems to fit Adams' theory: Zack's acceptance of (27) depends on the corresponding conditional probability's being high, the probability in question being Zack's credence.

F. P. Ramsey wrote in a footnote (1931, p. 247), "If two people are arguing 'If p will q?' and both are in doubt as to p, they are adding p hypothetically to their stock of knowledge and arguing on that basis about q .... We can say they are fixing their degrees of belief in q given p." This test – to see whether you accept q if p, add p hypothetically to your knowledge and note whether you now accept q – will be called the *Ramsey test*.[10] A conditional to which the Ramsey test applies will be called an *epistemic conditional*. A past grammatically indicative conditional like (27), it appears, is an epistemic conditional: it is accepted by anyone whose corresponding conditional credence is sufficiently high.

Although (27) has a clear acceptance condition, it does not have clear truth conditions. Suppose that in fact, as Zack suspects, Pete did not call, because he know he held a losing hand. It is not clear what then has to be true for (27) to be true.

(28), on the other hand, Zack regards as unlikely, because he thinks it unlikely that Pete had a winning hand. It seems, then, that Zack regards (28) as a proposition which is true or false according as Pete has a winning or a losing hand. Clearly his acceptance of (28) does not go by the Ramsey test, since (28) passes the test but Zack does not accept it. Whereas, then, (27) is an epistemic conditional which Zack does not in any obvious way treat as a proposition, the grammatically subjunctive conditional (28) is not an epistemic conditional, but is treated as epistemically equivalent to the proposition that Pete had a winning hand.

In many respects, at least, (28) fits Stalnaker's theory. (28) is treated as a proposition, whose truth or falsity depends on qualities of its subject matter, the game and its players, rather than the state of mind of the person who entertains it. Whether (28) is true depends, it seems reasonable to claim, on whether Pete wins in a world in which Pete calls, and which of all such worlds is nearest to the actual world by the following criteria: it is exactly like the actual world until it is time for Pete to call or fold; then it is like the actual world apart from whatever it is that constitutes Pete's decision to call or to fold, and from then on it develops in accordance with natural laws.[11] I do not mean to commit myself here to the details of Stalnaker's theory, but to note that the picture that guides Stalnaker applies without undue strain to past grammatically subjunctive conditionals like (28). I shall call any conditional to which such a picture applies a *nearness conditional.*

In the past tense, it seems from the example, a grammatically indicative conditional is an epistemic conditional and a grammatically subjunctive conditional is a nearness conditional. What of the future tense? For it, there is no obvious grammatical contrast to make; the salient relevant future conditional is

(29)   If Pete calls, he'll win.

This I have classified as grammatically subjunctive; semantically is it an epistemic conditional or a nearness conditional? Indeed can that distinction be made for the future at all?

Consider (29) as uttered by Zack at t_0, when Pete is about to fold or call. If (27) is an epistemic conditional, the Zack accepts it at t_0. For at t_0, he knows everything relevant that he knows at t_1: that Pete is a skilled player who knows Stone's hand, and thus would not call unless he had a winning hand. Thus Zack's conditional credence at t_0, ρ_0 (Pete will win / Pete will call), is close to one, and if he treats (29) as an epistemic conditinal, he accepts (29). If, on the other hand, (29) is a nearness conditional, which is true if an only if Pete has a winning hand, then Zack at t_0 regards (29) as unlikely. For he regards it as highly likely that Stone has the better hand.

We can, then, distinguish a reading of a future tense conditional as an epistemic conditional from a reading of it as a nearness conditional. To see whether you read (29) as an epistemic conditional or as a nearness conditional, put yourself in Zack's epistemic situations at t_0 and see whether you accept (29). If you do, you read it as an epistemic conditional; if you regard (29) as unlikely, then you read it as a nearness conditional.

My informal polls on whether Zack accepts (29) have been inconclusive, but most people I have asked think he does. Thus (29) seems to be read as an epistemic conditional, and thus semantically like the future of an indicative rather than a subjunctive conditional.

If Pete were to call, he would win.

If Pete called, he would win.

are generally treated as nearness conditionals: they are regarded as unlikely, given the information available to Zack at t_0, and as true if and only if Pete has a winning hand. A reading as an epistemic conditional is perhaps most securely elicited by the sentence

   If Pete's going to call, he'll win.

Grammatically indicative conditionals seem in general to be epistemic conditionals; these I shall call simply *indicative conditionals*. Grammatically subjunctive conditionals with 'would' are, I have argued, nearness conditionals, and it is these that I shall call *subjunctive conditionals*. Conditionals with antecedent tense shift and 'will' in the consequent I shall leave aside.

6. THOUGHT AND COMMUNICATION WITHOUT CONDITIONAL PROPOSITIONS

Nearness conditionals are propositions, whereas nothing so far in our account of epistemic conditionals requires that they be propositions. We have given acceptance conditions for them but not truth-conditions, and propositions need truth-conditions. Moreover, it looks difficult to interpret epistemic conditionals as propositions, for the most obvious approaches to doing so were ruled out by the results in Section 2. An epistemic conditional $a \to b$ is accepted by a person if and only if, where ρ is his credence measure, we have $\rho(b/a) \approx 1$. If we do suppose that $a \to b$ is a proposition, this amounts to the condition

$$\rho(a \to b) \approx 1 \quad \text{iff} \quad \rho(b/a) \approx 1.$$

If we suppose further that that is because always $\rho(a \to b) = \rho(b/a)$, then the results in Section 2 raise obstacles.

   None of this, however, need be disturbing. What a theory of indicative conditionals should do is to explain their role in thought and communication, and that task in no way demands that indicative conditionals be construed as propositions. To see this, let me propose a line of explanation that does without indicative conditional propositions.

   Take first thought, and consider belief in propositions. One can correctly be said to 'believe' or 'accept' a proposition, on this line of explanation, iff one's subjective probability for that proposition, or *credence* in it, is as close to one as matters for the purposes at hand.[12] For most purposes, for instance, I can be said to accept that my house will be standing in its usual place when I go home, but for purposes of explaining why I never allow my fire insurance to lapse, even for short periods, I cannot be said fully to accept that my house will not burn down before I go home, and hence cannot be said to accept that my house will be standing in its usual place when I go home.

Acceptance of an indicative conditional can be explained along the same lines, without invoking indicative conditional propositions. One accepts an indicative conditional iff one's corresponding conditional credence is as close to one as matters for the purposes at hand. (Such a high conditional credence I shall call a *conditional belief,* and I shall call sufficiently high conditional credence in b given a *belief in b given a.*

Take next communication, and consider first the communication of a proposition. In the standard, felicitous circumstances of communication, I accept a proposition and express it in a sentence, and my audience, hearing the sentence, comes to accept the proposition. That happens because I exploit certain conventions to get the audience to accept that I have the belief that I do. In felicitous cases, the audience trusts my sincerity and command of language, and for that reason it accepts, on account of my having uttered a sentence S in the circumstances, that I believe a certain proposition a. If the audience accepts a on my authority, it is because the audience supposes that I would not believe a unless I had adequate grounds for a, and takes my having adequate grounds for a as evidence sufficient to warrant accepting a.[14]

Now any such account of communication[15] – take it in your favorite version – will extend naturally to communication of conditional belief. In felicitous cases, I utter an indicative conditional, and thereby insure that the audience comes to accept that I have a certain conditional belief, belief in b given a. The audience does so because it trusts my sincerity and command of language. The audience then infers from my believing b given a that I have some good grounds for so believing, and takes that as a reason for itself believing b given a. Thus is my conditional belief communicated to them.

Conditional propositions, then, need play no role in an account of indicative conditionals. Conditional beliefs – states of high conditional credence – are just as much states of mind as are unconditional beliefs. There is no reason in what has been said here to suppose that a conditional belief constitutes an unconditional belief in a conditional proposition. We have, moreover, no reason to suppose that conditional beliefs must be communicated by means of conditional propositions: the devices which allow the communication of unconditional beliefs, we have seen, could just as well allow the communication of conditional beliefs. Conditional propositions, it seems, are superfluous in the communication of conditional beliefs.

7. PROPOSITIONAL THEORIES WITH CONDITIONAL NON-CONTRADICTION

Suppose we nevertheless do want to treat indicative conditionals as propositions. One way to do so is to adopt the theory that indicative conditionals are truth-functional; I shall discuss that theory in the next section. Other theories that treat conditionals as propositions – Stalnaker's, Lewis's, and Van Fraassen's (1976, p. 276, display line (18)) – share a law of *Conditional Non-contradiction*: that $a \rightarrow \bar{b}$ is inconsistent with $a \rightarrow b$. Now any theory with Conditional Non-contradiction confronts an anomaly which is illustrated by this version of the Sly Pete story.

Sly Pete and Mr. Stone are playing poker on a Mississippi riverboat. It is now up to Pete to call or fold. My henchman Zack sees Stone's hand, which is quite good, and signals its content to Pete. My henchman Jack sees both hands, and sees that Pete's hand is rather low, so that Stone's is the winning hand. At this point, the room is cleared. A few minutes later, Zack slips me a note which says "If Pete called, he won," and Jack slips me a note which says "If Pete called, he lost." I know that these notes both come from my trusted henchmen, but do not know which of them sent which note. I conclude that Pete folded.

If both these utterances express propositions, then I think we can see that both express true propositions. In the first place, both are assertable, given what their respective utterers know. Zack knows that Pete knew Stone's hand. He can thus appropriately assert "If Pete called, he won." Jack knows that Pete held the losing hand, and thus can appropriately assert "If Pete called, he lost." From this, we can see that neither is asserting anything false. For one sincerely asserts something false only when one is mistaken about something germane. In this case, neither Zack nor Jack has any relevant false beliefs. The relevant facts are these: (a) Pete had the losing hand, (b) he knew Stone's hand as well as his own, (c) he was disposed to fold on knowing that he had the losing hand, and (d) he folded. Zack knows (b) and (c), and he suspects (a) and therefore (d). Jack knows (a) and (c), and knowing Pete as he does, may well suspect (b) and therefore (d). Neither has any relevant false beliefs, and indeed both may well suspect the whole relevant truth. Neither, then, could sincerely be asserting anything false. Each is sincere, and so each, if he is asserting a proposition at all, is asserting a true proposition.

It follows that

(27)   If Pete called, he won

as uttered by Zack is consistent with

(30)   If Pete called, he didn't win

as uttered by Jack. For clearly as uttered by Jack,

(31)   If Pete called, he lost

entails (30) in any context, so that if (31) as uttered by Jack is true, then (30) is. Then since both (27) as uttered by Zack and (30) as uttered by Jack are true, they are consistent. The only apparent way to reconcile this with Conditional Non-contradiction is to suppose that the sentence "If Pete called, he won" as uttered by Zack expresses a different proposition from the one the same sentence would express if it were uttered by Jack.

That fits Stalnaker's contention that the selection function is pragmatically determined (1968, pp. 109–111), so that different contexts of utterence invoke different s-functions. If the context in which Zack passes his note invokes an s-function σ and the context in which Jack passes his note invoke a different s-function τ, then (27) and (30) may express Stalnaker conditionals of the form $a \boxbar\!\!\rightarrow_\sigma b$ and $a \boxbar\!\!\rightarrow_\tau \bar{b}$, so that even though Conditional Non-contradiction holds for any fixed s-function, (27) as uttered by Zack does not contradict (30) as uttered by Jack. That is to say, even though $a \boxbar\!\!\rightarrow_\sigma b$ contradicts $a \boxbar\!\!\rightarrow_\sigma \bar{b}$ and $a \boxbar\!\!\rightarrow_\tau b$ contradicts $a \boxbar\!\!\rightarrow_\tau \bar{b}$, (27) as uttered by Zack expresses $a \boxbar\!\!\rightarrow_\sigma b$ and (30) as uttered by Jack expresses $a \boxbar\!\!\rightarrow_\tau \bar{b}$, and these do not contradict each other.

The difference in contexts here, though, has a strange feature. Ordinarily when context resolves a pragmatic ambiguity, the features of the context that resolve it are common knowledge between speaker and audience. If the chairman of a meeting announces "Everyone has voted 'yes' on that motion", what the audience knows about the context allows it to judge the scope of 'everyone'. In Example 2, in contrast, whatever contextual differences between the utterances there may be, they are unknown to the audience. I, the audience, know exactly the same thing about the two contexts: that the sentence is the content of a note handed me by one of my henchmen. Whatever differences in the context make them invoke different s-functions is completely hidden from me, the intended audience.

That seems strange, for suppose it is so. I trust my henchmen and they are not contradicting each other. I presumably believe Zack's message "If Pete called, he won," and that constitutes believing a proposition c. Had Jack instead of Zack slipped me the message, I would have believed that message, but that would constitute believing a different proposition c'. Yet since I don't know which of them slipped me the message, there is no difference in my intrinsic mental state in the two cases: whether my mental state constitutes believing c or c' depends not on what that state is intrinsically like, but on who slipped me the note.

Perhaps, though, this feature of the context theory is not as bizarre as I have made it out, for perhaps sentences with indexicals have the very feature I have depicted. Suppose Zack or Jack, I know not which, slips me a note which says

I have swiped Mr. Stone's gold watch chain.

Perhaps this expresses the proposition

(32)   Zack has swiped Mr. Stone's gold watch chain

if the note came from Zack and

(33)   Jack has swiped Mr. Stone's gold watch chain

if the note came from Jack. I believe the message, but that constitutes believing neither (32) nor (33), but that the writer of the note swiped Stone's gold watch chain.

What is left of the contextual theory is that indicative conditionals act in an indexical-like way, where the workings of the indexical elements can depend on things the audience does not know. Now since the assertability of an indicative conditional depends on the utterer's credences, it seems reasonable to suppose that its propositional content too depends on the utterer's credences. Indeed we are driven to accepting this dependence if we want conditionals to be propositions to which the Ramsey test applies.[16] For suppose that where ρ is the utterer's credence measure, the indicative conditional $a \to b$ is assertable iff $\rho(b/a) \approx 1$ and it has a propositional content c such that $a \to b$ is assertable iff $\rho(c) \approx 1$. A variant of the Lewis proof will show that there is no ρ-independent propositional function $\to$ that satisfies these conditions.

(i) ab entails $a \to b$

(ii) $a \to b$ is inconsistent with $a \to \bar{b}$

(iii)   For all ρ such that $\rho(a) \neq 0$, $\rho(a \to b) \approx 1$ iff $\rho(b/a) \approx 1$.

For let $\rho(b/a) \approx 1$ and $\rho(a) \approx 0$. Then by (iii), $\rho(a \rightarrow b) \approx 1$, and hence by (ii), $\rho(a \rightarrow \bar{b}) \approx 0$. Obtain ρ' by conditionalizing on $a\bar{b}$; then $\rho'(\bar{b}/a) = 1$, but since $\rho(a\bar{b}) \approx 1, \rho'(a \rightarrow \bar{b})$ remains close to zero.

That leaves only one way for indicative conditionals to be treated as propositions which satisfy Conditional Non-contradiction. Let the propositional content of an indicative conditional depend on the utterer's epistemic state, and do so in such a way that the proposition is accepted by the utterer if and only if the utterer's corresponding conditional credence is sufficiently high. We might, in other words, have a three-place function $\rightarrow$ which yields a proposition $a \rightarrow_\rho b$ as a function of propositions a and b and probability measure ρ, with $\rho(a \rightarrow_\rho b) \approx 1$ when and only when $\rho(b/a) \approx 1$. We can interpret the indicative conditional as a propositional function satisfying Conditional Non-contradiction only at the cost of such radical dependence of the utterer's epistemic state.

8. EMBEDDING AND THE TRUTH-FUNCTIONAL THEORY

Why might we even want indicative conditionals to be propositions? Lewis (1976, p. 305) offers a reason: we then have an account of embedded indicative conditionals – of sentences of such forms as

$$a \rightarrow (b \rightarrow c), (a \rightarrow b) \rightarrow c, a \,\&\, (b \rightarrow c), a \vee (b \rightarrow c), \text{ and } - (a \rightarrow b).$$

Now only a truth-functional theory will account for such embeddings straightforwardly. Here and from now on, let $\rightarrow$ be the indicative conditional connective. It seems[17] to be a logical truth that $a \rightarrow (b \rightarrow c)$ is equivalent to $ab \rightarrow c$. For instance

> If they were outside, then if it rained they got wet

seems to be equivalent to

> If they were outside and it rained, then they got wet.

From this equivalence and a few quite weak additional assumptions, though, it follows that $\rightarrow$, if it is a propositional function at all, is the truth-functional connective $\supset$. The additional assumptions are just that any conditional $a \rightarrow b$ is false in all $a\bar{b}$ worlds, and that if a entails b then $a \rightarrow b = t$. Our assumptions, then, are these: for all non-empty propositions a, b, and c,

(i) $[a \rightarrow (b \rightarrow c)] = (ab \rightarrow c)$

(ii) $(a \rightarrow b) \subseteq (a \supset b)$

(iii)   If $a \subseteq b$, then $(a \to b) = t$.

Given (ii), it remains to be shown that $(a \supset b) \subseteq (a \to b)$. Consider the sentence

(34) $(a \supset b) \to (a \to b)$.

By (i), this is $[(a \supset b) \& a] \to b$, or equivalently, $ab \to b$. By (iii), this is t. By (ii), (34) entails

$$(a \supset b) \supset (a \to b),$$

and to say that t entails a truth-functional conditional is to say that its antecedent entails its consequent. That completes the proof.

If, then, we want the indicative conditional to be a propositional function, and to account in that way for our readings of embedded indicative conditionals, then the function must be $\supset$. That a non-propositional theory of indicative conditionals fails to account for some embeddings, though, may be a strength. Many embeddings of indicative conditionals, after all, seem not to make sense. Suppose I tell you, of a conference you don't know much about,

(35)   If Kripke was there if Strawson was, then Anscomb was there.

Do you know what you have been told? On the truth functional theory, you have been told that either Strawson was there and Kripke wasn't, or Anscomb was there. That seems surprising, though perhaps that is because some feature of conversational implicatures keeps the iterated conditional from being assertable.

Some iterated conditionals of the same form do seem to be assertable, but the way they are read is at odds with the truth-functional theory – at odds with it in a way for which conversational implicatures cannot account. Take, for instance, the conditional $(d \to b) \to f$,

(36)   If the cup broke if dropped, then it was fragile.

That seems assertable by someone who knows that the cup was being held at a moderate height over a carpeted floor, even if he gives rather low credence to the cup's being dropped or to its being fragile. Suppose, though, that in fact, things are as he thinks likely: the cup was not dropped and was not fragile. Then interpreted truth-functionally, (36) is false. For $(d \supset b) \supset f$ is equivalent to $d\bar{b} \vee f$, and neither disjunct is true. The speaker, indeed, gives low credence to both disjuncts, and hence gives low credence to the disjunction. (36), then, is assertable, even though on a truth-functional interpretation it gets low credence. Conversational implicatures will not save the truth-functional theory from this anomaly. Conversational implicatures, after all, explain only why a sentence believed true may not be appropriately assertable, whereas this is a case of a sentence which is appropriately assertable, but according to the truth-functional theory, false.[18]

The advantage of the truth-functional theory over Adams' theory, according to Lewis, is that it accounts for embedded conditionals. We have seen that it does so incorrectly, and so the alleged advantage turns out to yield a reason for rejecting the truth-functional theory of indicative conditionals.

9. AN ASSESSMENT

If the truth-functional theory will not handle embedding, will a propositional theory with Conditional Non-contradiction? We learned in Section 7 that at best, such a theory will do the job only if the propositional function represented by $\rightarrow$ depends on the utterer's epistemic state. We have now learned that even that kind of dependence will not suffice, for as we saw at the beginning of the last section, if in a fixed epistemic context, $\rightarrow$ represents a fixed propositional function, then to account for some of the behaviour of the indicative conditional, we must conclude that $\rightarrow$ is truth-functional. Since we know that the indicative conditional is not truth-functional, we can eliminate that possibility.

One other possibility remains: that $\rightarrow$ always represents a propositional function, but that what that function is depends not only on the utterer's epistemic state, but on the place of the connective in the sentence. In $a \rightarrow (b \rightarrow c)$, for instance, we might suppose that the two different arrows represent two different propositional functions. Nothing we have seen rules that out.

The pursuit of such a theory, though, has now lost its advantage. A theory of indicative conditionals as propositions was supposed to give, at no extra cost, a general theory of sentences with indicative conditional components: simply add the theory of conditionals to our extant theory of the ways truth-conditions of sentences depend on the truth-conditions of their components. The alternative was to develop a new theory to account for each way indicative conditionals might be embedded in longer sentences, and that seemed costly. Now it turns out that for each way indicative conditonals might be embedded in longer sentences, a propositional theory will have to account for their propositional content, and do so in a way that is sensitive to the place of each indicative conditional in its sentence. In $a \rightarrow (b \rightarrow c)$, the right and left arrows must be treated separately. What must be done with the left and right arrow in $(a \rightarrow b) \rightarrow c$ or with the arrows in $a \& (b \rightarrow c)$ and $a \vee (b \rightarrow c)$ we do not yet know. Thus, for instance, no account of sentences of the form $(a \rightarrow b) \rightarrow c$ will fall out of a simple general account of indicative conditionals as propositions; rather the account of indicative conditionals itself will have to confront separately the way left-embedded arrows work. A propositional theory would not save labor; instead it would demand all the labor that would have to be done without it.

What, then, of the alternative: to deal *ad hoc* with each kind of embedding without treating indicative conditionals as propositions? Here I think the prospects are not so bleak as might be supposed. In the first place, some sentences with indicative conditional components, such as (35), make no sense. An *ad hoc* treatment is more likely to account for this fact than is a theory which systematically assigns truth-conditions to every sentence. In the second place, various *ad hoc* accounts do turn out to work. Sentences with the apparent form $a \rightarrow (b \rightarrow c)$ can be read as really having the form $ab \rightarrow c$. A sentence with the apparent form $(a \rightarrow b) \& (c \rightarrow d)$ can be read as expressing a combination of two conditional beliefs, belief in b given a and belief in d given c.

More difficult is the conditional $(d \rightarrow b) \rightarrow f$,

(36)   If the cup broke if dropped, then it was fragile.

I think, though, that an account can be given that explains why this sentence seems to make sense and others of the same form do not. Consider first the antecedent $d \rightarrow b$, "If the cup was dropped, it broke." Such an indicative conditional may have an *obvious basis*: a proposition c such that it is presupposed, for both utterer and audience, that he will believe the consequent given the antecedent iff he believes c. The obvious basis for the conditional $d \rightarrow b$ is c,

(37)   The cup was disposed to break on being dropped.

c might hold either because the cup was fragile, or because it was being held over an especially hard floor or at an especially great height. Now when a conditional of the form $(d \rightarrow b) \rightarrow f$ is understandable, I propose, it is because the antecedent $d \rightarrow b$ has an obvious basis c and its obvious basis is understood in its place. The compound conditional $(d \rightarrow b) \rightarrow f$, then, is ordinarily understood as $c \rightarrow f$; (36) is read as

(38)   If the cup was disposed to break on being dropped, then it was fragile.

This is indeed assertable by someone who knows that the cup was being held at a moderate height above a carpeted floor and does not know whether it was fragile. For were he to learn that it was disposed to break on being dropped, be would come to believe it fragile.

The obvious basis for an indicative conditional is not synonymous with the conditional itself. 'The cup broke if dropped' is assertable by me if I believe this:

(i)   The cup was being held by someone who would not have dropped it unless it was highly fragile, but who might well have dropped it if it was highly fragile.

It is also assertable by me if I believe this:

(ii)   A trusted assistant was under order to inform me if the cup dropped without breaking, and not to bother me otherwise. The cup may have been dropped, for all I know, but I have not been informed that it dropped without breaking.

Neither of these entails that the cup was disposed to break on dropping but belief in either would give me grounds for asserting that the cup broke if dropped.

On the account I am giving, understanding (36) as (38) depends on contextual prosuppositions that might have been absent. Suppose possibility (i) is taken seriously: both utterer and audience take it as something they might well come to believe, and there is vivid common awareness of the possibility. Then the presuppositions that make (37) the obvious basis for $d \rightarrow b$ have broken down. Thus on the account I am giving, (38) ceases to gloss (36), and indeed it becomes unclear how (36) is to be interpreted. That seems to me to be what indeed would happen in that case. The same goes for a case in which (ii) is taken seriously.

Perhaps, then, what is explainable about sentences with indicative conditional components is explainable in an *ad hoc* way, without the invocation of indicative conditional propositions. The alternative, propositional account may require at least as much *ad hocum,* while rendering strange and mysterious the central fact about indicative conditionals: that their assertability goes with high conditional credence.

10. CONCLUSION

None of the arguments I have given for denying that indicative conditionals express propositions apply to subjunctive conditionals. In the first place, the

Ramsey test does not apply to subjunctive conditionals, and so no problem of reconciling a propositional account of subjunctive conditionals with the Ramsey test arises. In the second, place, subjunctive conditionals embed. Take, for instance, (35), and render the antecedant subjunctive.

> If Kripke would have been there if Strawson had been, then Anscomb was there.

The result is Delphic but not incomprehensible; with ingenuity, we can imagine circumstances that would make this assertable. Finally, with subjunctive conditionals, it is often possible to give at least a rough account of their truth conditions. "If Pete had called, he would have won" is true if normal conditions of play prevailed and Pete had a winning hand, and false if normal conditions of play prevailed and Pete has a losing hand.

There is a clear need for both kinds of conditionals. Epistemic conditionals prepare us for the acquisition of new information, and nearness conditionals help us express our understanding of what depends on what in the world. It is not surprising, then, that we should have linguistic devices for both jobs. The surprise, once we realize how disparate the jobs are, should be that similar linguistic devices do both jobs. The devices are similar not, as far as I have been able to discover, because of any deep connection between the two jobs, but for these reasons. First, belief in a nearness conditional is often grounds for conditional belief: belief in a nearness conditional $a \mathbin{\square\!\!\rightarrow} b$ is grounds for belief in b given a in the frequent circumstances that the proposition $a \mathbin{\square\!\!\rightarrow} b$ is epistemically independent of its antecedent a. (See Lewis, 1976, p. 309) Thus it is easy to conflate the job a nearness conditional does with the expression of conditional belief.[19] Second, in their common domain, the two functions have the same logic. That surprising fact itself, though, as far as I have been able to discover, manifests no deep connnection between the two function, but only the possibility of tricks.

Indicative and subjunctive conditionals, I conclude, have distinct jobs, and do them in ways that have little important in common.

APPENDIX: EQUIVALENCE PROOFS

We note first that the relation of s-consequence is invariant when possible worlds are more closely subdivided. Let $\mathscr{F}$ be a field of sets with universal set t, let t^* be a set, and let $\mathscr{F}^*$ be a set of subsets of t^*. Where h is a function from t^* onto t, we define what it is for h to 'extend' $\mathscr{F}$ to $\mathscr{F}^*$. For any w, let

$$w^* = \{v \mid v \in t^* \& h(v) = w\},$$

and for any $a \in \mathscr{F}$, let

$$a^* = \{v \mid v \in t^* \& h(v) \in a\},$$

the h-correspondant of a. The function h extends field of sets $\mathscr{F}$ to $\mathscr{F}^*$ iff (i) h is a function whose domain is all of t^* and whose range is all of t, (ii) $\mathscr{F}^*$ is a field of sets, and (iii) for every $a \in \mathscr{F}$, $a^* \in \mathscr{F}^*$. Where $a \to b$ is a conditional A of $\mathscr{F}$, A^* is its h-correspondant $a^* \to b^*$.

THEOREM 1. *Let $\mathscr{F}$ be a field of sets, and let h extend $\mathscr{F}$ to $\mathscr{F}^*$. Let $A_1, \ldots, A_n$ be conditionals of $\mathscr{F}$, and let $A_1^*, \ldots, A_n^*$ be their h-correspondants. Then $\{A_1^*, \ldots, A_n^*\}$ is s-consistent iff $\{A_1, \ldots, A_n\}$ is s-consistent.*

THEOREM 2. *A finite set of conditionals is s-consistent in the strong sense iff it is s-consistent.*

Theorem 2 is proved in passing, since like Theorem 1, it follows from the three lemmas that are stated and proved below.

NOTATION: Let $N = \{1, \ldots, n\}$, and let $A_i = (a_i \to b_i)$ and $A_i^* = (a_i^* \to b_i^*)$ for each $i \in N$. Let $\mathscr{A} = \{A_1, \ldots, A_n\}$ and $\mathscr{A}^* = \{A_1^*, \ldots, A_n^*\}$.

LEMMA 1. *Let $\mathscr{F}$ consist of all subsets of a finite set t, and suppose $\mathscr{A}$ is s-consistent. Then $\mathscr{A}^*$ is s-consistent in the strong sense.*

Proof: We are given that for some s-function σ for $\mathscr{F}$ and world $u \in t$, $\sigma(a_i, u) \in b_i$ for all $i \in \{1, \ldots, n\}$. Construct an s-function τ for $\mathscr{F}^*$ as follows. Let $\prec$ be an arbitrary well-ordering of t^*, and for any non-empty subset x of t^*, let $\chi(x)$ be the $\prec$-first member of x. For any proposition $p \in \mathscr{F}^*$, let $p' = \{w \mid w \in t \& w^* \cap p \neq f\}$ and for any world $v \in t^*$, let $\tau(p, v) = v$ if $v \in p$, and otherwise, let

(39) $\quad \tau(p, v) = \chi(p \cap \sigma(p', h(v))^*).$

Note that $p \cap \sigma(p', h(v))^*$ is non-empty, since by (S1), $\sigma(p', h(v)) \in p'$, and p' is the set of worlds w such that $p \cap w^* \neq f$.

Clearly τ satisfies (S1) and (S2). To check (S3), suppose $p, q \in t^*$, $p \subseteq q$, and $\tau(q, v) \in p$. Now if $v \in q$, then by (S2) $\tau(q, v) = v$; thus $v \in p$ and by (S2), $\tau(p, v) = v = \tau(q, v)$. Suppose $v \notin q$, so that $v \notin p$. Thus by (39),

$$\chi(q \cap \sigma(q', h(v))^*) \in p,$$

and so where $w = \sigma(q', h(v))$, w^* intersects p and $w \in p' \subseteq q'$. Therefore by (S3), $w = \sigma(p', h(v))$. Since $\chi(qw^*) \in pw^*$ and $pw^* \subseteq qw^*$, from the construction of χ we have $\chi(qw^*) = \chi(pw^*)$, and so (S3) is satisfied. Thus τ is an s-function.

τ is an internal s-function for $\mathscr{F}^*$. For let $p \to q$ be a conditional of $\mathscr{F}^*$. The proof is this, where the range of the variable v is t^*.

$$p \,\square\!\!\rightarrow_\tau q \;=\; \{v \mid \tau(p, v) \in q\}$$
$$=\; pq \vee \{v \mid v \notin p \,\&\, \tau(p, w) \in q\}$$
$$=\; pq \vee \{v \mid v \notin p \,\&\, \chi(p \cap \sigma(p', h(v))^*) \in q\}$$
$$=\; pq \vee \bar{p}c$$

where $c = \{v \mid \chi(p \cap \sigma(p', h(v))^*) \in q\}$. Now $\chi(p \cap \sigma(p', h(v))^*)$ depends on v only through its dependence on $h(v)$; thus for any $w \in t$, $\chi(p \cap \sigma(p', h(v))^*) \in q$ either throughout w^* or nowhere in w^*. Thus c is the union of a finite number of propositions of $\mathscr{F}^*$ of the form w^*, and is thus itself a proposition. Therefore $pq \vee \bar{p}c$ is a proposition, and τ is an internal s-function.

Finally, let $\hat{u} \in u^*$. Then for each conditional $a \to b \in \mathscr{A}$, $\hat{u} \in (a^* \,\square\!\!\rightarrow_\tau b^*)$, and so $\mathscr{A}^*$ is s-consistent. For $u \in (a \,\square\!\!\rightarrow_\sigma b)$; hence either (i) $u \in ab$ or (ii) $u \notin a$ and $\sigma(a, u) \in b$. In case (i), $\hat{u} \in a^*b^*$, and so $\hat{u} \in (a^* \,\square\!\!\rightarrow_\tau b^*)$. In case (ii), $\hat{u} \notin a^*$, an so by (39),

$$\tau(a^*, \hat{u}) \;=\; \chi(a^* \cap \sigma(a^{*\prime}, h(\hat{u}))^*)$$
$$=\; \chi(a^* \cap \sigma(a, u)^*).$$

Since $\sigma(a, u) \in ab$, $a^* \cap \sigma(a, u)^* \subseteq a^*b^*$, and so $\chi(a^* \cap \sigma(a, u)^*) \in b^*$. Thus $\tau(a^*, \hat{u}) \in b^*$, and $\hat{u} \in (a^* \,\square\!\!\rightarrow_\tau b^*)$. That completes the proof of the Lemma.

LEMMA 2. If $\mathscr{A}^*$ is s-consistent, then $\mathscr{A}$ is s-consistent.

Proof: Suppose $\mathscr{A}^*$ is s-consistent. Then for some s-function τ for $\mathscr{F}^*$ and world $\hat{u} \in t^*$, $\tau(a_i^*, \hat{u}) \in b_i^*$ for each $i \in N$. Let $u = h(\hat{u})$, and for each $a \in \mathscr{F}$, let $\sigma(a, u) = h(\tau(a^*, \hat{u}))$. For worlds of $\mathscr{F}$ other than u, take an arbitrary s-function σ' for $\mathscr{F}$, and let $\sigma(a, w) = \sigma'(a, w)$. Then σ is an s-function. For from (S1), $\tau(a^*, \hat{u}) \in a^*$; therefore $h(\tau(a^*, \hat{u})) \in a$, satisfying (S1). From (S2), if $u \in a$, $\hat{u} \in a^*$; hence $\tau(a^*, \hat{u}) = \hat{u}$ and $h(\tau(a^*, \hat{u})) = u$, satisfying (S2). If $a \subseteq b$ and $\sigma(b, u) \in a$, i.e. $h(\tau(b^*, \hat{u}))$, then $\tau(b^*, \hat{u}) \in a^*$.

and $a^* \subseteq b^*$. Hence by (S3), $\tau(b^*, \hat{u}) = \tau(a^*, \hat{u})$ and $\sigma(b, u) = \sigma(a, u)$. For $w \neq u$, (S1)–(S3) follow directly from their satisfaction by σ'.

Now from the way b^* is defined, we have that for any conditional $a \to b$ of $\mathcal{F}$, $\tau(a^*, \hat{u}) \in b^*$ iff $h(\tau(a^*, \hat{u})) \in b$. This last is just $\sigma(a, u) \in b$, and so we see that for any conditionals A of $\mathcal{F}$, $u \in A^\sigma$ iff $\hat{u} \in A^{*\tau}$. Therefore $u \in A_1^\sigma, \ldots, u \in A_n^\sigma$, and $\mathscr{A}$ is s-consistent.

LEMMA 3. There exists a function g, a field of sets $\mathcal{F}''$ consisting of all subsets of a finite set t'', and conditionals $A_1'', \ldots, A_n''$ of $\mathcal{F}''$ such that (i) g extends $\mathcal{F}''$ to $\mathcal{F}$, and (ii) $A_1, \ldots, A_n$ are g-correspondants of $A_1'', \ldots, A_n''$ respectively.

Proof: Take the set $\{a_1, \ldots, a_n, b_1, \ldots, b_n\}$, and let $\mathcal{F}'$ be its *Boolean closure*: its closure under the operations of union and complementation. $\mathcal{F}'$ is finite, and hence has a finite number of atoms: non-empty sets $\in \mathcal{F}'$ which partition t, none of which has a proper non-empty subset $\in \mathcal{F}'$. Let t'' be the set of atoms of $\mathcal{F}'$, and let $\mathcal{F}''$ be the set of all subsets of t''. Then where g takes each world of t into the atom of $\mathcal{F}'$ of which it is a member, g extends $\mathcal{F}''$ to $\mathcal{F}$. For each $i \in N$, A_i is the g-correspondant of a conditional A_i'' of $\mathcal{F}''$, for each a_i or b_i is the union of a finite number of atoms $w_1, \ldots, w_m$ of $\mathcal{F}'$, and hence is the g-correspondant of proposition $\{w_1, \ldots, w_m\}$ of $\mathcal{F}''$.

Proof of Theorem 1: The composition of g and h extends $\mathcal{F}''$ to $\mathcal{F}^*$. From Lemmas 1 and 2, we have that $\mathscr{A}$ is s-consistent iff $\mathscr{A}''$ is s-consistent, and $\mathscr{A}^*$ is s-consistent iff $\mathscr{A}''$ is s-consistent. Therefore $\mathscr{A}^*$ is s-consistent iff $\mathscr{A}$ is s-consistent.

Proof of Theorem 2: Suppose $\mathscr{A}$ is s-consistent. Since g extends $\mathcal{F}''$ to $\mathcal{F}$, $\mathscr{A}''$ is s-consistent by Lemma 3. Thus by Lemma 1, since $\mathscr{A}''$ is s-consistent, $\mathscr{A}$ is s-consistent in the strong sense. Therefore if $\mathscr{A}$ is s-consistent, then it is s-consistent in the strong sense. The converse is trivial.

Here is the part of the Van Fraassen result needed for the equivalence proof.

THEOREM 3 (Van Fraassen). *Let $\mathcal{F}''$ be a finite field of sets. Then there are h and $\mathcal{F}^*$ such that h extends $\mathcal{F}''$ to $\mathcal{F}^*$, and for every probability measure ρ'' on $\mathcal{F}''$, there are an internal s-function τ for $\mathcal{F}^*$ and a probability measure ρ^* on $\mathcal{F}^*$ such that*

(i)   *For every $a'' \in \mathcal{F}''$, $\rho^*(a^*) = \rho''(a'')$*

(ii)   *For every $a'', b'' \in \mathcal{F}''$ with $\rho''(a'') \neq 0$, $\rho^*(a^* \,\square\!\!\rightarrow_\tau b^*) = \rho''(b''/a'')$.*

THEOREM 4. *Let $A_1, \ldots, A_n$, C be conditionals of a field of sets $\mathscr{F}$, and let $\mathscr{A} = \{A_1, \ldots, A_n\}$. Then if C is an s-consequence of $\mathscr{A}$, then C is a p-consequence of $\mathscr{A}$.*

Proof: Let $A_i = (a_i \to b_i)$ for each $i \in N = \{1, \ldots, n\}$, and let $C = (c \to d)$. Let C be an s-consequence of $\mathscr{A}$. Given $\epsilon > 0$, let $\delta = \epsilon/n$. Suppose ρ is a probability measure on $\mathscr{F}$ with $\rho(a_i) > 0$ for each $i \in N$, $\rho(c) > 0$, and $\rho(b_i/a_i) \geqslant 1 - \delta$ for all $i \in N$; we are to prove that $\rho(d/c) \geqslant 1 - \epsilon$. Let $\mathscr{F}''$ be constructed from the conditionals $A_1, \ldots, A_n$, C as in Lemma 3, and for each proposition a'' of $\mathscr{F}''$, let $\rho''(a'') = \rho(a)$ where a is its g-correspondant in $\mathscr{F}$. Let h extend $\mathscr{F}''$ to $\mathscr{F}^*$ as in Theorem 3. Then by Theorem 1, C^* is an s-consequence of $\{A_1^*, \ldots, A_n^*\}$, and so $\{A_1^{*\tau}, \ldots, A_n^{*\tau}\}$ entails $C^{*\tau}$. Therefore since for each $i \in N$,

$$\rho^*(A_i^{*\tau}) = \rho''(b_i''/a_i'') = \rho(b_i/a_i) > 1 - \delta,$$

we have $1 - \epsilon < \rho^*(C^{*\tau}) = \rho''(d''/c'') = \rho(d/c)$, and the proof is complete.

THEOREM 5. *Any finite s-consistent set of conditionals is p-consistent.*

Proof: Let $\{a_1 \to b_1, \ldots, a_n \to b_n\}$ be s-consistent. Then for some s-function σ, the set

$$\{a_1 \mathbin{\Box\!\!\to}_\sigma b_1, \ldots, a_n \mathbin{\Box\!\!\to}_\sigma b_n\}$$

is consistent. Hence we may let w^* be a world, fixed for the rest of the proof, such that all of $a_1 \mathbin{\Box\!\!\to}_\sigma b_1, \ldots, a_n \mathbin{\Box\!\!\to}_\sigma b_n$ hold at w^*.

Define $\sigma^*(a_i) = \sigma(w^*, a_i)$. We weakly order the set $\{a_1, \ldots, a_n\}$ of antecedants by 'distance' from w^* as follows: $a_i \precsim a_j$ iff $\sigma^*(a_i \vee a_j) = \sigma^*(a_i)$. Let $a_i \approx a_j$ iff both $a_i \precsim a_j$ and $a_j \precsim a_i$. The relation $\precsim$ is connected and transitive, and $a_i \approx a_j$ iff $\sigma^*(a_i) = \sigma^*(a_j)$.

(i) $a_i \precsim a_j$ or $a_j \precsim a_i$, and if $a_i \approx a_j$, then $\sigma^*(a_i) = \sigma^*(a_j)$.

Proof: Let $v = \sigma^*(a_i \vee a_j)$. Then by definition, $a_i \precsim a_j$ iff $\sigma^*(a_i) = v$ and $a_j \precsim a_i$ iff $\sigma^*(a_j) = v$. It follows that if $a_i \approx a_j$, then $\sigma^*(a_i) = \sigma^*(a_j)$. Now by (S1), $v \in a_i \vee a_j$. If $v \in a_i$, then by (S3), $\sigma^*(a_i) = v$, and hence $a_i \precsim a_j$. Similarly, if $v \in a_j$, then $\sigma^*(a_j) = v$ and $a_j \precsim a_i$. We have seen that either $\sigma^*(a_i) = v$ or $\sigma^*(a_j) = v$. Hence if $\sigma^*(a_i) = \sigma^*(a_j)$, then both are v, and $a_i \approx a_j$.

(ii) Let $a_i \precsim a_j$ and $a_j \precsim a_k$. Then $a_i \precsim a_k$.

Proof: Let $u = \sigma^*(a_i \vee a_j \vee a_k)$. *Case* 1. $u \in a_i$. Then by (S3), $\sigma^*(a_i \vee a_k) = u$ and $\sigma^*(a_i) = u$; thus $a_i \precsim a_k$. *Case* 2. $u \in a_j$. Then by (S3), $\sigma^*(a_j) = u$, $\sigma^*(a_i \vee a_j) = u$, and hence $a_j \precsim a_i$. Since by hypothesis $a_i \precsim a_j$, we have $a_i \approx a_j$, and by (i), $\sigma^*(a_i) = \sigma^*(a_j) = u$, and so $u \in a_i$, and this reduces to Case 1. *Case* 3. $u \in a_k$. Then by similar reasoning, since $a_j \lesssim a_k$, we have $u \in a_j$, and this reduces to Case 2. That proves (ii).

Let $W = \{\sigma^*(a_1), \ldots, \sigma^*(a_n)\}$. Then the relation $\lesssim$ induces a linear ordering on w, again by 'distance' from w^*. Where $w, w' \in W$, let $w \prec w'$ iff for some a_i and a_j, $w = \sigma^*(a_i)$, $w' = \sigma^*(a_j)$, and $a_i \lesssim a_j$ but not $a_j \lesssim a_i$. Since, as we have seen, for any $w \in W$, the set $\{a \mid \sigma^*(a) = w\}$ is an equivalence class under $\approx$, the relation $\prec$ is a linear ordering. Let $w_1, \ldots, w_m$ be the members of W, with $w_1 \prec \ldots \prec w_m$.

(iii) If $\sigma^*(a_k) = w_j$ and $i < j$, then $w_i \in a_k$.

Proof: Since $i < j$, by the way the w_i's are indexed, $w_i \neq w_j$, and for some a_l, $w_i = \sigma^*(a_l)$, $a_l \lesssim a_k$, and not $a_k \lesssim a_l$. By the definition of '$\lesssim$', this means $w_i = \sigma^*(a_l \vee a_k)$. Suppose $w_i \in a_k$. Then by (S3), $w_i = \sigma(a_k)$, and $a_k \lesssim a_l$, contradicting what was said earlier. Thus $w_i \notin a_k$.

The proof of the Theorem is this. Let $\rho(w_m) = \delta^m$, and for each $i < m$, let $\rho(w_i) = \delta^i(1 - \delta)$. For any proposition x, then, we let $\rho(x) = \Sigma_{i=0}^{m} \rho_x(w_i)$, where $\rho_x(w_i) = \rho(w_i)$ if $w_i \in x$ and $\rho_x(w_i) = 0$ if $w_i \notin x$. Define $r_i = w_i \vee w_{i+1} \vee \ldots \vee w_m$; then $\rho(r_i) = \delta^i$, so that in particular $\rho(W) = \rho(r_0) = 1$.

Now let $w_j = \sigma^*(a_k)$. Since by (iii), for $i < j$, $w_i \notin a_k$, we have $\rho(a_k) \leqslant \rho(r_j) = \delta^j$. Now we assumed at the outset that $a_k \mathbin{\square\!\!\!\rightarrow}_\sigma b_k$ holds at w^*, and that is to say that $\sigma^*(a_k) \in b_k$; thus $w_j \in a_k b_k$. Thus $\rho(a_k b_k) \geqslant \rho(w_j) = \delta^i(1 - \delta)$. Therefore $\rho(b_k/a_k) = \rho(a_k b_k)/\rho(a_k) \geqslant 1 - \delta$. For arbitrary δ, we have shown how to construct a probability measure ρ such that $\rho(b_k/a_k) > 1 - \delta$ for all $k = 1, \ldots, n$. Thus $\{a_1 \rightarrow b_1, \ldots, a_n \rightarrow b_n\}$ is p-consistent.

University of Michigan

NOTES

[1] Adams himself argues, as I will later, that indicative conditionals are not propositions in the sense in which I am using the term (1975, pp. 5–6).

[2] See Lewis (1976, p. 311) for more discussion of the contrast between these two kinds of minimal revision.

[3] I get this use of Venn diagrams from Adams (1975).

[4] Stalnaker proposes this requirement (1970, p. 75), but later rejects it (see Stalnaker, 1976).

[5] Lewis (1976, pp. 300–303). Stalnaker has offered an interpretation of this possible divergence, which is developed in Gibbard and Harper (1978). Examples in which one's credence in a subjunctive conditional is not equal to one's corresponding conditional credence are given in Section 5 of this paper.

[6] Stalnaker's result, unlike Lewis's, depends on the conditional's having the Stalnaker logic; Van Fraassen (1976, pp. 286–291) has shown that a weaker logic of conditionals is compatible with the supposition that → is a propositional function such that, for some fixed ρ, $\rho(a \to b) = \rho(b/a)$ whenever $\rho(a) \neq 0$.

[7] I refer here to what Van Fraassen calls 'Stalnaker–Bernoulli models' (1976, pp. 279–282, 293–295). Here I state only those aspects of Van Fraassen's results that are needed in this paper; Van Fraassen's results are not, for instance, restricted to finite fields of sets.

[9] The modal auxilliaries are 'will', 'shall', 'can', 'may', and 'must'; I treat the first four as having the respective past tenses 'would', 'should', 'could', and 'might'. (See Chomsky, 1957, p. 40.) The past tense of 'must' is lacking, and 'should' as a synonym for 'ought to' does not act as the past tense of 'shall'. Only examples with 'will' and its past tense 'would' are given here.

[8] These rules, indeed, are sound for Lewis's semantics (1973). It follows that on the domain to which the Adams logic applies, the Lewis and Stalnaker logics are equivalent. That should not be surprising: The axiom by which the Lewis and Stalnaker logics differ, conditional excluded middle, says in effect that $-(a \to b)$ is equivalent to $a \to \bar{b}$. The constraint it imposes, then, bears solely on negations of conditionals and constructions that involve them, all of which include embedded conditionals.

[10] Harper (1976a, p. 97; 1976b) uses the term this way.

[11] This is the interpretation adopted in Gibbard and Harper (1978, p. 127). Lewis (1979) offers a more general account of truth conditions for such conditionals, and tries to show that the account given here is a consequence of his account and some deep contingent facts that, at least in part, constitute the direction of time.

[12] For a contrasting treatment of acceptance, see Harper (1976a, especially pp. 77–81).

[13] Adams (1976) proposes an alternative to the account of subjunctive conditionals I have given. On that account, they act much like epistemic conditionals, but the relevant conditional probability is not one's conditional credence but a 'prior' conditional probability. In his book (1975, pp. 129–133) he gives a counterexample to that theory, and proposes a more complex probabilistic account of subjunctive conditionals. Skyrms (this volume) proposes an account of subjunctive conditionals which I think correct: that a subjunctive conditional is accepted iff the subjectively expected value of the corresponding *propensity* is sufficiently high. On this view as I would express it, subjunctive conditionals involve conditional propositions, but those propositions involve objective chance: they take the form "If a had obtained at t, the chance, as of t, of b would have been α." Sobel (1978) calls the subjectively expected value of this propensity α the 'probable chance' of b given a. Here I endorse the nearness account only as an approximation to Skyrms' view that a subjunctive conditional is accepted iff the corresponding expectation is sufficiently high. Roughly, the expectation is high iff one puts high credence in a proposition: that the chance, as of t, with which b would obtain if a did is high. Thus to adopt the Skyrms account is roughly to treat subjunctive conditionals as propositions.

[14] Additional factors may be at work in determining what the audience can conclude from the fact that I uttered S: my audience may suppose that even if I believed a, I would not have said S unless certain other conditions also obtained – say, that I do not accept another proposition c. Thus we have conversation implicatures; see Grice (c. 1969).

[15] The classical account along these lines is in Grice (1957).

[16] Stalnaker (1975) treats indicative conditionals as context dependent propositions, and says "The most important element of a context, I suggest, is the common knowledge, or presumed common knowledge and common assumption of the participants in the discourse." He calls this presumed common ground the 'presuppositions' of the speaker. The example shows that if an indicative conditional utterance of the form 'If a then b' expresses a proposition, what proposition it expresses depends on more than a, b, and the speaker's presuppositions. Indeed it seems that the crucial aspect of the context that makes the utterance express the proposition it does is the speaker's conditional credence in b on a, and if that were presupposed, the utterance of the conditional would be pointless.

[17] Adams (1975, p. 33) regards the equivalence of $a \rightarrow (b \rightarrow c)$ with $ab \rightarrow c$ as problematical, since with *modus ponens*, it allows us to infer $b \rightarrow a$ from a. His argument is this: If the equivalence claim is correct, then $a \rightarrow (b \rightarrow a)$ is equivalent to $ab \rightarrow a$, which is a logical truth. Hence from a and a logical truth, we get $b \rightarrow a$ by *modus ponens*. My intuition are that sentences of the form $a \rightarrow (b \rightarrow a)$ indeed are logical truths, and are accepted even by someone for whom a is assertable and $b \rightarrow a$ is not. I am prepared to assert

> Andrew Jackson was President in 1836,

and I am not prepared to assert

> Even if Andrew Jackson died in 1835, he was President in 1836.

Nevertheless, the following strikes me as something which I accept as a logical truth.

> If Andrew Jackson was President in 1836, then even if he died in 1835, he was president in 1836.

[18] Adams (1975, pp. 31–37) gives a number of examples in which the truth-functional theory seems to fail for embedded 'will' conditionals. Here is one, adapted to the past tense.

> If switches A and B were both on, the motor was running. Therefore, either if switch A was on the motor was running or if switch B was on the motor was running.

I do not see how the apparent fallaciousness of this inference could be explained away with conversational implicatures.

[19] Adams (1976) argues that the logics of indicative and subjunctive conditionals are isomorphic (pp. 6–16), and notes that that is what would be expected on his 'prior probability' representation, according to which indicative and subjunctive conditionals have similar semantics. The equivalence of the Adams and restricted Stalnaker logics shown in Section 3 shows that there is an alternative explanation for a logical isomorphism of indicative and subjunctive conditionals.

BIBLIOGRAPHY

Adams, E. W., *The Logic of Conditionals*, D. Reidel, Dordrecht, Holland, 1975.
Adams, E. W., 'Prior Probabilities and Counterfactual Conditionals', in W. L Harper and

C. A. Hooker (eds.), *Foundations of Probability Theory, Statistical Inference, and Statistical Theories of Science,* Volume I. D. Reidel, Dordrecht, Holland, 1976, pp. 1–21.

Chomsky, N., *Syntactic Structures,* Mouton, The Hague, 1957.

Gibbard, A. and Harper, W. L., 'Counterfactuals and Two Kinds of Expected Utility', C. A. Hooker, J. J. Leach, and E. F. McClennen (eds.), *Foundations and Applications of Decision Theory,* Volume I, D. Reidel, Dordrecht, Holland, 1978, pp. 125–162.

Gibbard, A., 'Chance Conditionals and Beliefs about Influence', Duplicated typescript, University of Michigan, 1978.

Grice, H. P., 'Meaning', *Pilosophical Review* **67** (1957), 377–388.

Grice, H. P., *William James Lectures,* Duplicated typescript (c. 1969).

Harper, W. L., 1976a, 'Rational Belief Change, Popper Functions, and Counterfactuals', in W. L. Harper and C. A. Hooker (eds.), *Foundations of Probability Theory, Statistical Inference, and Statistical Theories of Science,* Volume I, D. Reidel, Dordrecht, Holland, 1976, pp. 73–112.

Harper, W. L., 1976b, 'Ramsey Test Conditionals and Iterated Belief Change (A Response to Stalnaker)', in W. L. Harper and C. A. Hooker (eds.), *Foundations of Probability Theory, Statistical Inference, and Statistical Theories of Science,* Volume I, D. Reidel, Dordrecht, Holland, 1976, p. 117.

Kyburg, H. E., *Probability and Inductive Logic,* Macmillan, London, 1970.

Lewis, D., *Counterfactuals,* Harvard Univeristy Press, Cambridge, Mass., 1973.

Lewis, D., 'Probabilities of Conditionals and Conditional Probabilities', *Philosophical Review* **85** (1976), 297–315; also this volume, pp. 129–147.

Lewis, D., 'Counterfactual Dependence and Time's Arrows', *Noûs* **13** (1979), 455–476.

Ramsey, F. P., 'General Propositions and Causality', *The Foundations of Mathematics and Other Logical Essays,* Routledge and Kegan Paul, London, 1931.

Skyrms, B., 'The Prior Propensity Account of Subjunctive Conditionals', this volume, pp. 259–265.

Sobel, J. H., *Choice, Chance, and Action: Newcomb's Problem Resolved,* Duplicated typescript, University of Toronto, 1978.

Stalnaker, R., 'A Theory of Conditionals', *Studies in Logical Theory, American Philosophical Quarterly Monograph Series,* No. 2, Blackwell, Oxford, 1968, pp. 98–112.

Stalnaker, R., 'Probability and Conditionals', *Philosophy of Science* **37** (1970), 64–80; also this volume, pp. 107–128.

Stalnaker, R., 'Indicative Conditionals', *Philosophia* **5** (1975), 269–286; also this volume, pp. 193–210.

Stalnaker, R., 'Letter to Van Fraassen', in W. L. Harper and C. A. Hooker (eds.), *Foundations of Probability Theory, Statistical Inference, and Statistical Theories of Science,* Volume I, D. Reidel, Dordrecht, Holland, 1976, pp. 302–306.

Van Fraassen, B. C., 'Probabilities of Conditonals', in W. L. Harper and C. A. Hooker (eds.), *Foundations of Probability Theory, Statistical Inference, and Statistical Theories of Science,* Volume I, D. Reidel, Dordrecht, Holland, 1976, pp. 261–308.

JOHN L. POLLOCK

INDICATIVE CONDITIONALS AND CONDITIONAL PROBABILITY

It has been suggested repeatedly in the literature of conditionals that there is an intimate connection between conditionals and conditional probability. Let us symbolize the indicative conditional as $\ulcorner P \mathscr{I} \mathscr{C} Q \urcorner$. Then the simplest proposal of this nature is:

(1)   One is justified in believing $\ulcorner P \mathscr{I} \mathscr{C} Q \urcorner$ iff $\mathrm{prob}(Q/P)$ is high.

'probability' here is taken to be 'degree of rational belief'. The degree to which one is justified in believing $\ulcorner P \mathscr{I} \mathscr{C} Q \urcorner$ would seem to be just $\mathrm{prob}(P \mathscr{I} \mathscr{C} Q)$, so the force of (1) is:

(2) $\mathrm{prob}(P \mathscr{I} \mathscr{C} Q)$ is high iff $\mathrm{prob}(Q/P)$ is high.

It seems eminently reasonable to suppose that what makes (2) true is that $\mathrm{prob}(P \mathscr{I} \mathscr{C} Q)$ is determined by $\mathrm{prob}(Q/P)$. At the very least:

(3) $\mathrm{prob}(P \mathscr{I} \mathscr{C} Q)$ is a monotonic increasing function of $\mathrm{prob}(Q/P)$.

The most plausible initial candidate for such a function is the identity function, which gives us the Stalnaker Hypothesis:

(SH) $\mathrm{prob}(P \mathscr{I} \mathscr{C} Q) = \mathrm{prob}(Q/P)$.

'prob' represents the degree of belief one could rationally have under the present circumstances. Let 'prob_R' represent the degree of belief one could rationally have if the proposition R were added to one's evidence. It has generally been accepted that:

(CH) $\mathrm{prob}_R(P) = \mathrm{prob}(P/R)$, provided $\mathrm{prob}(R) \neq 0$.

This is the Conditionalization Hypothesis. (CH) is really just the claim that the normal definition of conditional probability captures what it is supposed to capture. I will provisionally assume (CH). Its use at least appears unsuspicious in the examples to follow.

(3) and (SH) are supposed to be general principles not dependent upon our particular state of knowledge. Accordingly, they are defensible only insofar as the following stronger principles are defensible:

249

W. L. Harper, R. Stalnaker, and G. Pearce (eds.), Ifs, 249–252.
Copyright © 1980 by D. Reidel Publishing Company.

(3⁺)   ($\exists f$) {f is a monotonic increasing function &
       ($\forall R$) [$\text{prob}_R(P\mathscr{IC}Q) = f(\text{prob}_R(Q/P))$]};

(SH⁺)   ($\forall R$) [$\text{prob}_R(P\mathscr{IC}Q) = \text{prob}_R(Q/P)$].

The triviality results of Lewis and Stalnaker show that we cannot maintain both (SH⁺) and (CH). This suggests rejecting (SH⁺) and retreating to the weaker (3⁺). However, Gibbard has given a representation of Lewis' triviality result in terms of Venn diagrams, and further reflection on those diagrams enables us to see that (3⁺) must also be rejected if we are to maintain (CH). To see this, suppose that F is related to ⌜$P\mathscr{IC}Q$⌝ and P and Q as indicated in the following diagram:

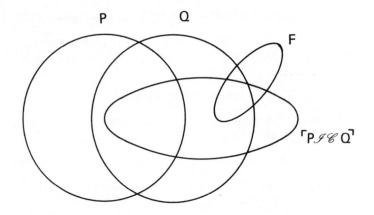

Let $R = \ulcorner{\sim}F\urcorner$. Assuming (CH), it follows from the probability calculus that $\text{prob}_R(Q/P) = \text{prob}(Q/P \& R)$. As P entails R, $\text{prob}(Q/P \& R) = \text{prob}(Q/P)$, so $\text{prob}_R(Q/P) = \text{prob}(Q/P)$. But $\text{prob}_R(P\mathscr{IC}Q) < \text{prob}(P\mathscr{IC}Q)$. Thus not only is (3⁺) false; $\text{prob}_R(P\mathscr{IC}Q)$ is not any function at all of $\text{prob}_R(Q/P)$.

In an attempt to salvage something like principle (1) in light of the foregoing, Gibbard has suggested that indicative conditionals do not express propositions. They play a different linguistic role, related to 'conditional assertion' or 'conditional belief', however that is to be made out. This has the effect that indicative conditionals are not included in the domain of the probability function, and hence the preceding reasoning is blocked. On this view, $\text{prob}(Q/P)$ provides the measure of the reasonableness of the conditional belief expressed by ⌜$P\mathscr{IC}Q$⌝, but that reasonableness cannot be expressed as ⌜$\text{prob}(P\mathscr{IC}Q)$⌝.

To test Gibbard's suggestion, we must ask whether anything which intuitively counts for or against an indicative conditional affects the conditional
probability in the same manner. The preceding diagrammatic example suggests
where to look for a concrete counterexample to this claim. If the claim is
false, we ought to be able to show that it is false by finding an R of the sort
required by that example. Such an R would diminish the reasonableness of
the conditional while leaving the conditional probability unchanged. A
concrete example of this sort is as follows.[1] Suppose we know of a vase
which was included in a certain shipment of vases. Seventy-five percent (or
however high we want to make this percentage) of the vases in the shipment
were ceramic and highly fragile, and the other 25% were plastic and virtually
unbreakable. We know of this shipment that every ceramic vase which was
dropped broke, and none of the plastic vases broke. Furthermore, we know
that when the shipment reached its destination, all broken vases and all
plastic vases were discarded, and of the discarded vases, 75% were plastic.
This completes our initial background information regarding the shipment.
On the basis of this information, we can reasonably believe that if the vase
was dropped, it broke. (If a probability of 75% is considered inadequate for
this, we can adjust the probability upwards as necessary).

Suppose we are now informed that the vase under consideration was discarded. This is the proposition R. As 75% of the discarded vases were plastic,
this makes it unreasonable to believe that if the vase was dropped then it
broke. Thus R makes the indicative conditional less reasonable. However,
given our background information, the vase's being dropped entails that it
was discarded. If it was dropped and ceramic, then it broke and was discarded
for that reason; if it was dropped and not ceramic, then it was plastic and was
discarded for that reason. Thus letting B be the proposition that the vase
broke and D be the proposition that it was dropped, $\mathrm{prob}(B/D) =
\mathrm{prob}(B/D\,\&\,R)$, and assuming (CH), $\mathrm{prob}_R(B/D) = \mathrm{prob}(B/D\,\&\,R)$. Conse-
quently, although R is relevant to the reasonableness of the conditional,
$\mathrm{prob}_R(B/D) = \mathrm{prob}(B/D)$. Thus we have a counterexample to the hypothesis
that the reasonableness of an indicative conditional is determined by the conditional probability.

The preceding counterexample does not depend upon supposing that
indicative conditionals express propositions. However, the only reason for
denying that supposition was to salvage some version of principle (1). We
have now seen that such a denial is not efficacious in the desired salvage, so
the most reasonable move would seem to be to return to the traditional
supposition that conditionals do express propositions. On that supposition, it

seems intuitively reasonable to make some further claims about the preceding counterexample. Let $\ulcorner C(x) \urcorner$ symbolize $\ulcorner x$ is ceramic $\urcorner$. Then for each member of the shipment, it seems that we have:

(4) $C(x) \equiv [D(x) \mathscr{I}\mathscr{C} B(x)]$.

Accordingly,

(5) $\mathrm{prob}_R (D \mathscr{I}\mathscr{C} B) = \mathrm{prob}_R (C) = 0.25$

whereas, assuming that $\mathrm{prob}(D/C) = \mathrm{prob}(D/\sim C)$,

(6) $\mathrm{prob}_R (B/D) = \mathrm{prob}(B/D) = \mathrm{prob}(C) = 0.75$.

Thus we have a concrete example in which the probability of the conditional and the conditional probability diverge. There seems to be no way to object to the use of (CH) in (6), so this example fortifies the triviality results by showing that they cannot be avoided by rejecting (CH). Even if (CH) sometimes fails (and I see no reason to think that it does), it surely holds in this case, and that suffices to show that the probability of the conditional cannot generally be identified with the conditional probability.

University of Arizona

NOTE

[1] I am indebted to Keith Lehrer for help in the formulation of this example.

ALLAN GIBBARD

INDICATIVE CONDITIONALS AND CONDITIONAL PROBABILITY: REPLY TO POLLOCK

In 'Indicative Conditionals and Conditional Probability' (this volume, pp. 249–252), Pollock constructs an intriguing situation to serve as a counter-example to the 'Ramsey test' thesis. Here by the *Ramsey test thesis*, I mean the thesis that, in whatever ways the acceptability, assertability, and the like of a proposition depend on its subjective probability, the acceptability, assertability, and the like of an indicative conditional $A \rightarrow B$ depend on the corresponding subjective conditional probability $\rho(B/A)$. (I shall use '$\rightarrow$' as a symbol for the indicative conditional connective, and otherwise follow Pollock's notation. The systematic development of the Ramsey test thesis was the work of Ernest Adams, 1975). I think that I can give an argument to show that in Pollock's example, contrary to what he judges, if $B \rightarrow D$ is acceptable before one learns R, it is acceptable after one learns R.

To make the example work at all, of course, we have to set the proportion of ceramic vases high enough to make it reasonable to assert at the outset, on the basis of frequency information, that the vase was ceramic. Suppose this proportion is 95%, and adjust the example accordingly. Now to fill in the example, we have to suppose that I, the utterer, regard the droppings as random. Otherwise I might deny "If the vase was dropped, it broke" on such grounds as this: "If the vase was dropped, it was probably plastic, for no one would have been careless enough to drop a ceramic vase. Therefore probably if the vase was dropped, it didn't break – because it was plastic." Suppose, then, that I regard the droppings as random, and suppose further, to avoid statistical complications, that the shipment is large, so that in effect I accept that 95% of the dropped vases were ceramic.

A large part of the force that Pollock's example will have for a reader may come from a failure to realize the following fact: that even if 95% of the discarded vases are plastic, 95% of the discarded vases that were dropped are ceramic. This diagnosis will be confirmed if Pollock's conclusion that $\rho_R(B/D) = \rho(B/D)$ comes as a surprise to the reader – as I think it does. Now if this diagnosis is correct, then our natural assessments of reasonable subjective probabilities in the case are wrong. Hence on the Ramsey test thesis, our natural judgments of the reasonable assertability of the indicative conditional should be correspondingly wrong.

253

W. L. Harper, R. Stalnaker, and G. Pearce (eds.), Ifs, 253–256

When intuitions are befuddled by a complex situation, it should often be possible to present an *argument* to the correct conclusion. I propose this argument.

(1)   This vase was discarded.

(2)   Of the discarded vases that had been dropped, 95% were ceramic.

(3)   Therefore if this vase was dropped, then it was ceramic.

(4)   If this vase was dropped and it was ceramic, then it broke.

(5)   Therefore if this vase was dropped, it broke.

Here (1) is given and (2) has been shown. (3) begs the point at issue, since it uses 95% conditional probability as sufficient for reasonable assertability. It does seem to me, though, that a person might well find the step from (1) and (2) to (3) intuitively convincing, and the intuition in question is a simple one – more on this later. (4) is given, and the derivation of (5) from (3) and (4) is sanctioned by disparately based logics of conditionals, including Stalnaker's, Lewis's, and Adams'. Where $C =$ The vase was ceramic, the argument takes the form

(3) $D \to C$

(4) $DC \to B$

(5) $D \to B.$

Part of our tendency to regard R as evidence against $D \to B$, I have suggested, stems from confusion about conditional probabilities in a complex situation. Another part of the tendency, I suppose, stems from a tendency to regard $D \to B$ as a proposition which is true if and only if the vase was fragile. There is a tendency, in other words, implicitly to accept the *dispositional thesis*: A statement of the form $D(x) \to B(x)$ expresses a property true of those and only those things that are disposed to be B on being D. For me, at least, all of the appeal of Pollock's (4) stems from the appeal of the dispositional thesis – although I am not sure whether he wants to accept the dispositional thesis. Now I think that the tendency to accept the dispositional thesis should be regarded as a temptation rather than a guide. To accept the dispositional thesis is to go against other strong tendencies we have in our evaluations of indicative conditionals, and I suspect it will be more in accord with the central tendencies of our thought to reject the dispositional thesis than to accept it. Suppose, for instance, I knew that only plastic vases had been dropped. It seems to me that I could then say, "If this vase was dropped, it was plastic" without supposing that I am talking about a vase that was disposed to be plastic on being dropped. On the dispositional thesis, to do so is incoherent. True, from the dispositional statement one can normally infer the indicative conditional, but if we suppose that that is because the two are the same and we retain our ordinary judgments of indicative conditionals, we will face contradictions. It seems to me, then, that we should reject judgments that seem to derive their plausibility from the dispositional thesis as based on a *dispositional fallacy*.

My hunch, then, is not that our judgments of indicative conditionals clearly and consistently conform to the Ramsey test thesis, but that the preponderance of them do, and that no alternative account of indicative conditionals will systematize so much of our usage. If that is so, then it would seem a good idea to use the indicative conditional consistently as a Ramsey test conditional, and regard any reasoning with indicative conditionals that would be invalid for Ramsey test conditionals as fallacious.

My argument (1)–(5) pits the intuition that (1) and (2) warrant (3) against Pollock's intuition that R warrants the denial of $B \rightarrow D$. Now we cannot simply trust either intuition, for we have mistaken intuitions about all sorts of things. In Pollock's example, for instance, I would intuitively expect $\rho_R(B/D)$ to be lower than $\rho(B/D)$, but Pollock shows that the two are equal. (Tversky and Kahneman (1978) have done a psychological study of fallacies involving conditional probabilities, which seem to be systematic and pervasive.) It seems that virtually whenever we get clear on how a subject matter works, we find that some of our previous untutored intuitions went wrong. The traditional way of dealing with this problem is by reasoning: by putting together intuitions about simple matters to form conclusions about complex matters. That is what I have tried to do. It seems to me that a judgment of whether (1) and (2) warrant (3) is simpler than our evaluation of the intricately wrought example Pollock gives, and hence should be given more credence – especially when we consider that one's judgment of the complex situation Pollock constructs may be infected with the dispositional thesis, which is incompatible with going from "Only plastic vases were dropped" to "If this vase was dropped, then it is plastic."

University of Michigan

BIBLIOGRAPHY

Ernest Adams, *The Logic of Conditionals*, *An Application of Probability to Deductive Logic*. D. Reidel, Dordrecht, Holland, 1975.

Amos Tversky and Daniel Kahneman, 'Causal Schemas in Judgments Under Uncertainty', in M. Fishbein (ed.), *Progress in Social Psychology*, Lawrence Erlbaum Associates, Hillsdale, 1978.

PART 6

CHANCE, TIME, AND THE SUBJUNCTIVE CONDITIONAL

BRIAN SKYRMS

THE PRIOR PROPENSITY ACCOUNT OF SUBJUNCTIVE CONDITIONALS[1]

I agree with Ernest Adams and Brian Ellis that assertability of uniterated indicative conditionals goes by epistemic conditional probability. It might be thought that subjunctives are mere stylistic variants of indicatives, the counterfactual being used only to convey the extra information that we are in a counterfactual belief state. There are striking examples which argue that this is not always the case.

An unknown sample is placed in the flame and burns green. We will maintain "If it *is* a sodium salt it didn't burn yellow" since it didn't, but will be quite confident in asserting "If it *had been* a sodium salt, it would have burned yellow." Or consider Adams' example: we will follow epistemic conditional probability in asserting: "If Oswald didn't kill Kennedy, then someone else did" but we will not follow it far enough to assert: "If Oswald hadn't killed Kennedy, then someone else would have." The imprint of any stable connection between the antecedent and consequent of the conditional on the probability distribution has been swamped by our quite certain knowledge about the truth value of the consequent in the absence of equally definitive knowledge of the antecedent.

A natural suggestion is that in evaluating the subjunctive, we should not look at the conditional probability in our *present* probability distribution, but rather look at the conditional probability in a prior distribution in which the perturbing knowledge of the truth value of the consequent is suppressed. In our belief state prior to putting the sample in the flame – that is, prior to conditionalizing on our observation that the sample didn't burn yellow – the probability of it burning yellow conditional on it being sodium is indeed high. If we imagine a belief state which includes our best knowledge of the events leading up to the assassination but suppressing our knowledge that it did in fact take place, we will find that in that distribution our conditional probabilities coincide with our degree of confidence in asserting the corresponding counterfactual.

Adams makes just this suggestion in 'Counterfactual Conditions and Prior Probabilities'. He there points out that in simple urn examples, where it is quite clear what the appropriate prior is, the account gives just the right

259

W. L. Harper, R. Stalnaker, and G. Pearce (eds,), Ifs, 259–265 results:

Two urns, *A* and *B* are filled with black and white balls, urn *A* containing 0.1% white
and 99.9% black, urn *B* containing 50% of each color. One urn is selected at random and
placed before an 'observer' who draws one ball at random from it. This ball proves to be
white.

The probability that a black ball *would have been* drawn if urn *A had* been
selected is 0.999 while the probability that a black ball *was* drawn if urn *A*
was selected is zero.

But although the epistemic prior probability account of counterfactuals
does a credible job on these examples, it cannot be quite right. Consider
Adams' counterexample to his own theory:

Imagine the following situation. We have just entered a room and are standing in front
of a metal box with two buttons marked '*A*' and '*B*' and a light, which is off at the
moment, on its front panel. Concerning the light we know the following. It may go on
in one minute, and whether it does or not depends on what combinations of buttons *A*
and *B*, if either, have been pushed a short while before, prior to our entering the room.
If exactly one of the two buttons has been pushed then the light will go on, but if either
both buttons or neither button has been pushed then it will stay off. We think it highly
unlikely that either button has been pushed, but if either or both were pushed then they
were pushed independently, the chances of *A*'s having been pushed being 1 in a thousand,
while the chances of *B*'s having been pushed is a very remote 1 in a million. In the circumstances we think there is only a very small chance of 1 000 999 in one billion (about
1 in a thousand) that the light will go on, but a high probability of 999 in a thousand
that *if B was pushed, the light will go on.*

Now suppose that to our surprise the light does go on, and consider what we would
infer in consequence. Leaving out numerical probabilities for the moment, we would no
doubt conclude that the light probably lit because *A* was pushed and *B* wasn't, and not
because *B* was pushed and *A* wasn't. Therefore, since *A* was probably the button pushed,
if B had been pushed the light wouldn't have gone on, for then both buttons would have
been pushed. *The point here is that the counterfactual would be affirmed a posteriori in
spite of the fact that the corresponding indicative was very improbable a priori*, because
its contrary "if *B* was pushed then the light will go on" had a probability of 0.999 *a
priori.*

My suggestion is that the prior probability account can be saved by one small
change – the probabilities involved are the prior *propensities* rather than the
prior epistemic probabilities. In cases where the correct propensities are
known with certainty,[2] the two accounts coincide. But, if we do not know
for certain the values of the prior propensities, we may have to do with a
weighted average – the expected prior propensities. The weights in this
average will be epistemic probabilities, and we should use the best ones available – for this job – the *posterior* epistemic probabilities. I will use PR for

epistemic probabilities and pr for propensities. I will superscript i and f for prior (initial) and posterior (final) respectively. Let the double arrow, $\Rightarrow$, symbolize the subjunctive conditional, and BAV be 'Basic Assertability Value'. Then the theory that I am suggesting can be succinctly expressed thusly:

$$*** \qquad \text{BAV}\,(p \Rightarrow q) = \sum_j \text{PR}^f[\text{pr}_j^i] \cdot \text{pr}_j^i\,(q \text{ given } p).$$

where the pr_j^i's are the appropriate prior propensity distributions.

Let us analyze Adams' example according to this theory. Letting $A = A$ pushed; $B = B$ was pushed: $L = $ the light goes on, the counterfactual to be analyzed is $B \Rightarrow L$. The relevant prior propensities depend on whether A was pushed or not, and are gotten by conditionalizing out on these two alternatives: $\text{pr}_A^i(p) = \text{PR}^i(p \text{ given } A)$; $\text{pr}_{-A}^i(p) = \text{PR}^i(p \text{ given } -A)$. The values of $\text{PR}^f[\text{pr}_A^i] = \text{PR}^f(A)$ and $\text{PR}_A^f[\text{pr}_{-A}^i] = \text{PR}^f(-A)$ are gotten by Bayes' theorem. The happy result is that $\text{PR}^f(B \Rightarrow L)$ as defined by $***$ is appropriately small as desired, as a consequence of the high posterior epistemic probability of the prior propensities associated with A being pushed. (Adams, in fact, gives an '*ad hoc* two factor model' for his example which is equivalent to the analysis forthcoming from the prior propensity approach. But from this point of view, Adams' model is not *ad hoc* but entirely natural.) Where p and q lie on the future we can usually take $\text{PR}^i = \text{PR}^f$. In other cases, like the ones just cited, there is a natural choice for the appropriate prior. But it should come as no surprise that choice of the appropriate prior may be less than routine. All the vagueness and ambiguities of the subjunctive still exist.[3]

Let me sketch[4] how this analysis works on a few familiar examples:

1. *Fischer*: Suppose that smoking and lung cancer are effects of a common genetic cause. There are two prior propensity distributions corresponding to having the gene or not. A man refrains from smoking and remains healthy. His epistemic probability of not having the smoking-cancer gene and the corresponding prior propensity is therefore high. He may therefore justifiably assert, "Had I smoked, I would still have avoided lung cancer."

2. *Lewis*: Suppose that jogging does not strengthen the heart, but rather people with strong hearts tend to jog. There are n prior propensity distributions corresponding to prior strength of heart. Within each prior propensity distribution, jogging and heart attacks are statistically independent. We may fairly say to the jogging enthusiast who has a strong healthy heart, "You would have escaped heart attack, even if you hadn't jogged."

3. *Spence*: College graduates perform better in certain jobs, not because of what they learn in college, but because college screens out people who lack the skills or motivation to do well. Analysis follows the pattern of the foregoing. We may say: "She would have done just as well if she hadn't gone to college."

4. *Newcomb*: Prof. L faces two boxes, one transparent and one opaque. The transparent box can be seen to contain $ 1000. He may take either the opaque box and all its contents, or both boxes and all their contents. A predictor who is very good at predicting the choice that people made has put $ 1 000000 under the opaque box if he predicted that Prof. L would take only the opaque box and nothing if he predicted that Prof. L would take both. The predictor is not only very good, but also very good for those who take only the opaque box, in the following sense: Of those who take the opaque box only, his percentage of correct predictions is very high. Likewise for those who take both boxes. In these circumstances, Prof. L takes only the opaque box. Under it he finds $ 1 000 000. "I'm rich," he cries. "You would have been $ 1000 richer if you had taken both boxes," Prof. G remarks.

Here again, we have two prior propensity distributions, PR_1 and PR_2 corresponding to whether the predictor predicted a choice of only the opaque or both boxes respectively:

PR_1 (1 000 000 given take only the opaque box) = 1

PR_1 (1 001 000 given both boxes) = 1

PR_2 (0 given only the opaque box) = 1

PR_2 (1000 given both boxes) = 1

so Prof. G's remarks are correct.

5. *Nozick–Schlesinger–Gettier*: The set up is as in the Newcomb case, except that Prof. L has a confederate who has peeked under the opaque box, and knows what money is there. Prof. L is disposed to take only the opaque box, but his confederate signals him to take both boxes. He does, and finds nothing under the opaque box together with the thousand from under the transparent box. Later, when he and his confederate are alone, he is furious: "You fool," he says. "If you hadn't signalled me, I'd have a million dollars now!"

The analysis of example 4 shows that Prof. L's outburst is quite unwarranted. Indeed, his confederate would be warranted in the following sharp retort: "If I hadn't signalled you, you would have nothing now. And you shouldn't need a confederate know that no matter what's under the opaque box, you'll be $ 1000 richer taking both boxes than you would be taking only the opaque one."

6. *Looking Backward*: A: "What is that red light for?" B: "If that light had gone on half an hour ago, we would all be dead now. It signals nuclear attack, and only gives us 5 minutes to retaliate before we are incinerated." A: "But since there was no attack launched, if the light had gone on, it would have been a spurious signal, and we would still be alive." B: "Ha, ha! Der Professor strikes again!"

Notice that A and B here decompose into possible propensity distributions in different ways. For B there are two possible propensity distributions corresponding to whether the warning system is working properly or not. The epistemic posterior probability is high that the system is working properly. Within the corresponding propensity distribution, the probability of being dead now conditional on the light going on half an hour ago is high. For A, on the other hand, the two possible propensity distributions correspond to whether an attack has been launched (say by an hour ago) or not. The posterior probability is high that an attack has not been launched. Within the corresponding propensity distribution, the probability of being dead now conditional on the light going on half an hour ago is low.

The Stalnaker conditional has the property that $p \& q$ entails $p \Rightarrow q$, so that $\Pr(p \Rightarrow q) > \Pr(p \& q)$. The prior propensity account of the subjunctive does not satisfy the analogous principle. It need not be the case that BAV $(p \Rightarrow q) > \Pr(p \& q)$. In this respect, the prior propensity account is more similar to Lewis' system with weak centering, than to Stalnaker's original treatment.

There is another way in which this account differs from the classical Stalnaker account with respect to chance processes. Suppose I flip a coin and it comes up heads (or particle decays in one of several possible ways). Consider the conditional: "If I had flipped the coin a millisecond ago, it would have come up heads." (If the particle had decayed a millisecond ago, it would have decayed in the same way.) On the idea of similarity of possible worlds in which the antecedent is true, these conditionals should be as strongly assertable as any: making the consequent true picks up similarity at no cost. But intuitively, this counterfactual is not so unproblematic. The prior propensity account which gives the counterfactual a basic assertability value of 1/2 seems more in line with ordinary usage.[5]

NOTES

[1] I discuss this, and related matters in greater detail in *Causal Necessity*, Yale University Press, New Haven, 1980; and in 'Randomness and Physical Necessity' (forthcoming).

[2] As in Adams' urn example, where the relevant propensity is known to be 0.999.

[3] Choice of the appropriate prior may depend on the individuation of the relevant chance process. A slot machine comes up 3 pears. Would we say 'If the left wheel had come up apple, the other two would (still) have come up pears"? A coin is flipped twice and comes up tails both times. Would we say, "If it had come up heads on toss 1, it (still) would have come up tails on toss 2"? It depends on whether we think of the play of the slot (the double coin flip) as a single, or multiple chance process. This explains the 'spinner' example in Adams, pp. 132–133.

[4] As these are only sketches, I rely on the reader to fill in the details that he finds necessary in a sympathetic manner.

[5] The preceding sorts of examples might be accommodated within the general Stalnaker approach by considering Stalnaker conditionals with propensity-attributing consequents:

If p then $pr(q) = a$.

Then I could fairly say of the coin that if it had been flipped a millisecond ago, then the probability of it coming up heads would have been 1/2; and it need not be the case that for any true statements, p: q; if p then $pr(q) = 1$. This, however, would still leave the conditional with unqualified consequent "If this coin had been flipped a millisecond ago, it would have come up heads" assertable alongside the conditional with probabilistic consequent: "... would have had a 50/50 chance of coming up heads". The proponent of the Stalnaker approach might deal with this remaining deviation from usage, by maintaining that we blur the distinctions between probabilistic consequents and take as the assertability value of a subjunctive with an unqualified consequent, the expected value of the probability of the consequent of the true conditional with probabilistic consequent. The story would then be that a conditional with unqualified consequent is not being used literally, but rather doing useful duty as a carrier for an expectation. But this is not *my* story, and I won't attempt to complete it here.

BIBLIOGRAPHY

Adams, E. W., 'Prior Probabilities and Counterfactual Conditionals', in W. L. Harper and C. A. Hooker (eds.), *Foundations of Probability Theory, Statistical Inference, and Statistical Theories of Science*, Volume I, D. Reidel, Dordrecht, Holland, 1976, pp. 1–21.

Adams, E. W., *The Logic of Conditionals. An Application of Probability to Deductive Logic*, D. Reidel, Dordrecht, Holland, 1975.

Ellis, B., 'Probability Logic', manuscript, 1968.

Lewis, D., *Counterfactuals*, Harvard University Press, Cambridge, 1974.

Skyrms, B., *Causal Necessity*, Yale University Press, New Haven, 1980.
Skyrms, B., 'Randomness and Physical Necessity', Pittsburgh Lecture in the Philosophy of Science, forthcoming.
Stalnaker, R. C., 'A Theory of Conditionals', in Rescher (ed.), *Studies in Logical Theory*, Blackwell, London, 1968.

DAVID LEWIS

A SUBJECTIVIST'S GUIDE TO OBJECTIVE CHANCE*

INTRODUCTION

We subjectivists conceive of probability as the measure of reasonable partial belief. But we need not make war against other conceptions of probability, declaring that where subjective credence leaves off, there nonsense begins. Along with subjective credence we should believe also in objective chance. The practice and the analysis of science require both concepts. Neither can replace the other. Among the propositions that deserve our credence we find, for instance, the proposition that (as a matter of contingent fact about our world) any tritium atom that now exists has a certain chance of decaying within a year. Why should we subjectivists be less able than other folk to make sense of that?

Carnap (1945) did well to distinguish two concepts of probability, insisting that both were legitimate and useful and that neither was at fault because it was not the other. I do not think Carnap chose quite the right two concepts, however. In place of his 'degree of confirmation' I would put *credence* or *degree of belief*; in place of his 'relative frequency in the long run' I would put *chance* or *propensity*, understood as making sense in the single case. The division of labor between the two concepts will be little changed by these replacements. Credence is well suited to play the role of Carnap's probability$_1$, and chance to play the role of probability$_2$.

Given two kinds of probability, credence and chance, we can have hybrid probabilities of probabilities. (Not 'second order probabilities', which suggests one kind of probability self-applied.) Chance of credence need not detain us. It may be partly a matter of chance what one comes to believe, but what of it? Credence about chance is more important. To the believer in chance, chance is a proper subject to have beliefs about. Propositions about chance will enjoy various degrees of belief, and other propositions will be believed to various degrees conditionally upon them.

As I hope the following questionnaire will show, we have some very firm and definite opinions concerning reasonable credence about chance. These opinions seem to me to afford the best grip we have on the concept of

267

W. L. Harper, R. Stalnaker. and G. Pearce (eds.), Ifs, 267–297.

chance. Indeed, I am led to wonder whether anyone *but* a subjectivist is in a position to understand objective chance!

QUESTIONNAIRE

First question. A certain coin is scheduled to be tossed at noon today. You are sure that this chosen coin is fair: it has a 50% chance of falling heads and a 50% chance of falling tails. You have no other relevant information. Consider the proposition that the coin tossed at noon today falls heads. To what degree should you now believe that proposition?

Answer. 50%, of course.

(Two comments. (1) It is abbreviation to speak of the coin as fair. Strictly speaking, what you are sure of is that the entire 'chance set-up' is fair: coin, tosser, landing surface, air, and surroundings together are such as to make it so that the chance of heads is 50%. (2) Is it reasonable to think of coin-tossing as a genuine chance process, given present-day scientific knowledge? I think so: consider, for instance, that air resistance depends partly on the chance making and breaking of chemical bonds between the coin and the air molecules it encounters. What is less clear is that the toss could be designed so that you could reasonably be sure that the chance of heads is 50% exactly. If you doubt that such a toss could be designed, you may substitute an example involving radioactive decay.)

Next question. As before, except that you have plenty of seemingly relevant evidence tending to lead you to expect that the coin will fall heads. This coin is known to have a displaced center of mass, it has been tossed 100 times before with 86 heads, and many duplicates of it have been tossed thousands of times with about 90% heads. Yet you remain quite sure, despite all this evidence, that the chance of heads this time is 50%. To what degree should you believe the proposition that the coin falls heads this time?

Answer. Still 50%. Such evidence is relevant to the outcome by way of its relevance to the proposition that the chance of heads is 50%, not in any other way. If the evidence somehow fails to diminish your certainty that the coin is fair, then it should have no effect on the distribution of credence about outcomes that accords with that certainty about chance. To the extent that uncertainty about outcomes is based on certainty about their chances, it is a stable, resilient sort of uncertainty – new evidence won't get rid of it. (The term 'resiliency' comes from Skyrms (1977); see also Jeffrey (1965), §12.5.)

Someone might object that you could not reasonably remain sure that the coin was fair, given such evidence as I described and no contrary evidence that I failed to mention. That may be so, but it doesn't matter. Canons of reasonable belief need not be counsels of perfection. A moral code that forbids all robbery may also prescribe that if one nevertheless robs, one should rob only the rich. Likewise it is a sensible question what it is reasonable to believe about outcomes if one is unreasonably stubborn in clinging to one's certainty about chances.

Next question. As before, except that now it is afternoon and you have evidence that became available after the coin was tossed at noon. Maybe you know for certain that it fell heads; maybe some fairly reliable witness has told you that it fell heads; maybe the witness has told you that it fell heads in nine out of ten tosses of which the noon toss was one. You remain as sure as ever that the chance of heads, just before noon, was 50%. To what degree should you believe that the coin tossed at noon fell heads?

Answer. Not 50%, but something not far short of 100%. Resiliency has its limits. If evidence bears in a direct enough way on the outcome – a way which may nevertheless fall short of outright implication – then it may bear on your beliefs about outcomes otherwise than by way of your beliefs about the chances of the outcomes. Resiliency under all evidence whatever would be extremely unreasonable. We can only say that degrees of belief about outcomes that are based on certainty about chances are resilient under *admissible* evidence. The previous question gave examples of admissible evidence; this question gave examples of inadmissible evidence.

Last question. You have no inadmissible evidence; if you have any relevant admissible evidence, it already has had its proper effect on your credence about the chance of heads. But this time, suppose you are not sure that the coin is fair. You divide your belief among three alternative hypotheses about the chance of heads, as follows.

You believe to degree 27% that the chance of heads is 50%.

You believe to degree 22% that the chance of heads is 35%.

You believe to degree 51% that the chance of heads is 80%.

Then to what degree should you believe that the coin falls heads?

Answer. $(27\% \times 50\%) + (22\% \times 35\%) + (51\% \times 80\%)$; that is, 62%. Your degree of belief that the coin falls heads, conditionally on any one of the hypotheses about the chance of heads, should equal your unconditional degree of belief if you were sure of that hypothesis. That in turn should equal the chance of heads according to the hypothesis: 50% for the first hypothesis, 35% for the second, and 80% for the third. Given your degrees of belief that the coin falls heads, conditionally on the hypotheses, we need only apply the standard multiplicative and additive principles to obtain our answer.

THE PRINCIPAL PRINCIPLE

I have given undefended answers to my four questions. I hope you found them obviously right, so that you will be willing to take them as evidence for what follows. If not, do please reconsider. If so, splendid – now read on.

It is time to formulate a general principle to capture the intuitions that were forthcoming in our questionnaire. It will resemble familiar principles of direct inference except that (1) it will concern chance, not some sort of actual or hypothetical frequency, and (2) it will incorporate the observation that certainty about chances – or conditionality on propositions about chances – makes for resilient degrees of belief about outcomes. Since this principle seems to me to capture all we know about chance, I call it

THE PRINCIPAL PRINCIPLE Let C be any reasonable initial credence function. Let t be any time. Let x be any real number in the unit interval. Let X be the proposition that the chance, at time t, of A's holding equals x. Let E be any proposition compatible with X that is admissible at time t. Then

$$C(A/XE) = x.$$

That will need a good deal of explaining. But first I shall illustrate the principle by applying it to the cases in our questionnaire.

Suppose your present credence function is $C(-/E)$, the function that comes from some reasonable initial credence function C by conditionalizing on your present total evidence E. Let t be the time of the toss, noon today, and let A be the proposition that coin tossed today falls heads. Let X be the proposition that the chance at noon (just before the toss) of heads is x. (In our questionnaire, we mostly considered the case that x is 50%.) Suppose that nothing in your total evidence E contradicts X; suppose also that it is not yet noon, and you have no foreknowledge of the outcome, so everything that is included in E is entirely admissible. The conditions of the Principal Principle are met. Therefore $C(A/XE)$ equals x. That is to say that x is your present degree of belief that the coin falls heads, conditionally on the proposition that its chance of falling heads is x. If in addition you are sure that the chance of heads is x – that is, if $C(X/E)$ is one – then it follows also that x is your present unconditional degree of belief that the coin falls heads. More generally, whether or not you are sure about the chance of heads, your unconditional degree of belief that the coin falls heads is given by summing over alternative hypotheses about chance:

$$C(A/E) = \Sigma_x C(X_x/E)C(A/X_xE) = \Sigma_x C(X_x/E)x,$$

where X_x, for any value of x, is the proposition that the chance at t of A equals x.

Several parts of the formulation of the Principal Principle call for explanation and comment. Let us take them in turn.

THE INITIAL CREDENCE FUNCTION C

I said: let C be any reasonable initial credence function. By that I meant, in part, that C was to be a probability distribution over (at least) the space whose points are possible worlds and whose regions (sets of worlds) are propositions. C is a non-negative, normalized, finitely additive measure defined on all propositions.

The corresponding conditional credence function is defined simply as a quotient of unconditional credences:

$$C(A/B) =_{df} C(AB)/C(B).$$

I should like to assume that it makes sense to conditionalize on any but the empty proposition. Therefore I require that C is *regular*: $C(B)$ is zero, and $C(A/B)$ is undefined, only if B is the empty proposition, true at no worlds. You may protest that there are too many alternative possible worlds to permit regularity. But that is so only if we suppose, as I do not, that the values of the function C are restricted to the standard reals. Many propositions must have infintesimal C-values, and $C(A/B)$ often will be defined as a quotient of infinitesimals, each infinitely close but not equal to zero. (See Bernstein and Wattenberg (1969).) The assumption that C is regular will prove convenient, but it is not justified only as a convenience. Also it is required as a condition of reasonableness: one who started out with an irregular credence function (and who then learned from experience by conditionalizing) would stubbornly refuse to believe some propositions no matter what the evidence in their favor.

In general, C is to be reasonable in the sense that if you started out with it as your initial credence function, and if you always learned from experience by conditionalizing on your total evidence, then no matter what course of experience you might undergo your beliefs would be reasonable for one who had undergone that course of experience. I do not say what distinguishes a reasonable from an unreasonable credence function to arrive at after a given course of experience. We do make the distinction, even if we cannot analyze it; and therefore I may appeal to it in saying what it means to require that C be a reasonable initial credence function.

I have assumed that the method of conditionalizing is *one* reasonable way to learn from experience, given the right initial credence function. I have not assumed something more controversial: that it is the *only* reasonable way. The latter view may also be right (the cases where it seems wrong to conditionalize may all be cases where one departure from ideal rationality is needed to compensate for another) but I shall not need it here.

(I said that C was to be a probability distribution over *at least* the space of worlds; the reason for that qualification is that sometimes one's credence might be divided between different possibilities within a single world. That is the case for someone who is sure what sort of world he lives in, but not at all sure who and when and where in the world he is. In a fully general treatment of credence it would be well to replace the worlds by something like the 'centered worlds' of Quine (1969), and the propositions by something corresponding to properties. But I shall ignore these complications here.)

THE REAL NUMBER x

I said: let x be any real number in the unit interval. I must emphasize that 'x' is a quantified variable; it is not a schematic letter that may freely be replaced by terms that designate real numbers in the unit interval. For fixed A and t, 'the chance, at t, of A's holding' is such a term; suppose we put it in for the variable x. It might seem that for suitable C and E we have the following: if X is the proposition that the chance, at t, of A's holding equals the chance, at t, of A's holding – in other words, if X is the necessary proposition – then

$$C(A/XE) = \text{the chance, at } t, \text{ of } A\text{'s holding.}$$

But that is absurd. It means that if E is your present total evidence and $C(-/E)$ is your present credence function, then if the coin is in fact fair – whether or not you think it is! – then your degree of belief that it falls heads is 50%. Fortunately, that absurdity is not an instance of the Principal Principle. The term 'the chance, at t, of A's holding' is a non-rigid designator; chance being a matter of contingent fact, it designates different numbers at different worlds. The context 'the proposition that . . .', within which the variable 'x' occurs, is intensional. Universal instantiation into an intensional context with a non-rigid term is a fallacy. It is the fallacy that takes you, for instance, from the true premise 'For any number x, the proposition that x is nine is non-contingent' to the false conclusion 'The proposition that the number of planets is nine is non-contingent'. See Jeffrey (1970) for discussion of this point in connection with a relative of the Principal Principle.

I should note that the values of 'x' are not restricted to the standard reals in the unit interval. The Principal Principle may be applied as follows: you are sure that some spinner is fair, hence that it has infinitesimal chance of coming to rest at any particular point; therefore (if your total evidence is admissible) you should believe only to an infinitesimal degree that it will come to rest at any particular point.

THE PROPOSITION X

I said: let X be the proposition that the chance, at time t, of A's holding equals x. I emphasize that I am speaking of objective, single-case chance – not credence, not frequency. Like it or not, we have this concept. We think that a coin about to be tossed has a certain chance of falling heads, or that a radioactive atom has a certain chance of decaying within the year, quite regardless of what anyone may believe about it and quite regardless of whether there are any other similar coins or atoms. As philosophers we may well find the concept of objective chance troublesome, but that is no excuse to deny its existence, its legitimacy, or its indispensability. If we can't understand it, so much the worse for us.

Chance and credence are distinct, but I don't say they are unrelated. What is the Principal Principle but a statement of their relation? Neither do I say that chance and frequency are unrelated, but they are distinct. Suppose we have many coin-tosses with the same chance of heads (not zero or one) in each case. Then there is some chance of getting any frequency of heads whatever; and hence some chance that the frequency and the uniform single-case chance of heads may differ, which could not be so if these were one and the same thing. Indeed the chance of difference may be infinitesimal if there are infinitely many tosses, but that is still not zero. Nor do hypothetical frequencies fare any better. There is no such thing as *the* infinite sequence of outcomes, or *the* limiting frequency of heads, that *would* eventuate if some particular coin-toss were somehow repeated forever. Rather there are countless sequences, and countless frequencies, that *might* eventuate and would have some chance (perhaps infinitesimal) of eventuating. (See Jeffrey (1977), Skyrms (1977), and the discussion of 'might' counterfactuals in Lewis (1973).)

Chance is not the same thing as credence or frequency; this is not yet to deny that there might be some roundabout way to analyze chance in terms of credence or frequency. I would only ask that no such analysis be accepted unless it is compatible with the Principal Principle. We shall consider how this requirement bears on the prospects for an analysis of chance, but without settling the question of whether such an analysis is possible.

I think of chance as attaching in the first instance to propositions: the chance of an event, an outcome, etc. is the chance of truth of the proposition that holds at just those worlds where that event, outcome, or whatnot occurs. (Here I ignore the special usage of 'event' to simply mean 'proposition'.) I have foremost in mind the chances of truth of propositions about localized matters of particular fact – a certain toss of a coin, the fate of a certain tritium atom on a certain day – but I do not say that those are the only propositions to which chance applies. Not only does it make sense to speak of the chance that a coin will fall heads on a particular occasion; equally it makes sense to speak of the chance of getting exactly seven heads in a particular sequence of eleven tosses. It is only caution, not any definite reason to think otherwise, that stops me from assuming that chance of truth applies to any proposition whatever. I shall assume, however, that the broad class of propositions to which chance of truth applies is closed under the Boolean operations of conjunction (intersection), disjunction (union), and negation (complementation).

We ordinarily think of chance as time-dependent, and I have made that dependence explicit. Suppose you enter a labyrinth at 11:00 a.m., planning to choose your turn whenever you come to a branch point by tossing a coin. When you enter at 11:00, you may have a 42% chance of reaching the center by noon. But in the first half hour you may stray into a region from which it is hard to reach the center, so that by 11:30 your chance of reaching the center by noon has fallen to 26%. But then you turn lucky; by 11:45 you are not far from the center and your chance of reaching it by noon is 78%. At 11:49 you reach the center; then and forevermore your chance of reaching it by noon is 100%.

Sometimes, to be sure, we omit reference to a time. I do not think this means that we have some timeless notion of chance. Rather, we have other ways to fix the time than by specifying it explicitly. In the case of the labyrinth we might well say (before, after, or during your exploration) that your chance of reaching the center by noon is 42%. The understood time of reference is the time when your exploration begins. Likewise we might speak simply of the chance of a certain atom's decaying within a certain year, meaning the chance at the beginning of that year. In general, if A is the proposition that something or other takes place within a certain interval beginning at time t, then we may take a special interest in what I shall call the *endpoint chance* of A's holding: the chance at t, the beginning of the interval in question. If we speak simply of the chance of A's holding, not mentioning a time, it is this endpoint chance – the chance at t of A's holding – that we are likely to mean.

Chance also is world-dependent. Your chance at 11:00 of reaching the center of the labyringth by noon depends on all sorts of contingent features of the world: the structure of the labyrinth and the speed with which you can walk through it, for instance. Your chance at 11:30 of reaching the center by noon depends on these things, and also on where in the labyrinth you then are. Since these things vary from world to world, so does your chance (at either time) of reaching the center by noon. Your chance at noon of reaching the center by noon is one at the worlds where you have reached the center; zero at all others, including those worlds where you do not explore the labyrinth at all, perhaps because you or it do not exist. (Here I am speaking loosely, as if I believed that you and the labyrinth could inhabit several worlds at once. See Lewis (1968) for the needed correction.)

We have decided this much about chance, at least: it is a function of three arguments. To a proposition, a time, and a world it assigns a real number. Fixing the proposition A, the time t, and the number x, we have our proposition X; it is the proposition that holds at all and only those worlds w such that this function assigns to A, t, and w the value x. This is the proposition that the chance, at t, of A's holding is x.

THE ADMISSIBLE PROPOSITION E

I said: let E be any proposition that is admissible at time t. Admissible propositions are the sort of information whose impact on credence about outcomes comes entirely by way of credence about the chances of those outcomes. Once the chances are given outright, conditionally or unconditionally, evidence bearing on them no longer matters. (Once it is settled that the suspect fired the gun, the discovery of his fingerprint on the trigger adds nothing to the case against him.) The power of the Principal Principle depends entirely on how much is admissible. If nothing is admissible it is vacuous. If everything is admissible it is inconsistent. Our questionnaire suggested that a great deal is admissible, but we saw examples also of inadmissible information. I have no definition of admissibility to offer, but must be content to suggest sufficient (or almost sufficient) conditions for admissibility. I suggest that two different sorts of information are generally admissible.

The first sort is historical information. If a proposition is entirely about matters of particular fact at times no later than t, then as a rule that proposition is admissible at t. Admissible information just before the toss of a coin, for example, includes the outcomes of all previous tosses of that coin and others like it. It also includes every detail – no matter how hard it might be to discover – of the structure of the coin, the tosser, other parts of the set-up, and even anything nearby that might somehow intervene. It also includes a great deal of other information that is completely irrelevant to the outcome of the toss.

A proposition is *about* a subject matter – about history up to a certain time, for instance – if and only if that proposition holds at both or neither of any two worlds that match perfectly with respect to that subject matter. (Or we can go the other way: two worlds match perfectly with respect to a subject matter if and only if every proposition about that subject matter holds at both or neither.) If our world and another are alike point for point, atom for atom, field for field, even spirit for spirit (if such there be) throughout the past and up until noon today, then any proposition that distinguishes the two cannot be entirely about the respects in which there is no difference. It cannot be entirely about what goes on no later than noon today. That is so even if its linguistic expression makes no overt mention of later times; we must beware lest information about the future is hidden in the predicates, as in 'Fred was mortally wounded at 11:58'. I doubt that any linguistic test of aboutness will work without circular restrictions on the language used. Hence it seems best to take either 'about' or 'perfect match with respect to' as a primitive.

Time-dependent chance and time-dependent admissibility go together. Suppose the proposition A is about matters of particular fact at some moment or interval t_A, and suppose we are concerned with chance at time t. If t is later than t_A, then A is admissible at t. The Principal Principle applies with A for E. If X is the proposition that the chance at t of A equals x, and if A and X are compatible, then

$$1 = C(A/XA) = x.$$

Put contrapositively, this means that if the chance at t of A, according to X, is anything but one, then A and X are incompatible. A implies that the chance at t of A, unless undefined, equals one. What's past is no longer

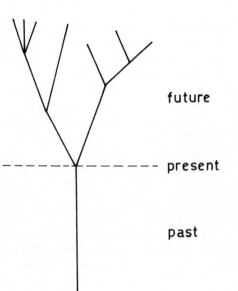

future

present

past

chancy. The past, unlike the future, has no chance of being any other way than the way it actually is. This temporal asymmetry of chance falls into place as part of our conception of the past as 'fixed' and the future as 'open' – whatever that may mean. The asymmetry of fixity and of chance may be pictured by a tree. The single trunk is the one possible past that has any present chance of being actual. The many branches are the many possible futures that have some present chance of being actual. I shall not try to say here what features of the world justify our discriminatory attitude toward past and future possibilities, reflected for instance in the judgment that historical information is admissible and similar information about the future is not. But I think they are contingent features, subject to exception and absent altogether from some possible worlds.

That possibility calls into question my thesis that historical information is invariably admissible. What if the commonplace *de facto* asymmetries between past and future break down? If the past lies far in the future, as we are far to the west of ourselves, then it cannot simply be that propositions about the past are admissible and propositions about the future are not. And if the past contains seers with foreknowledge of what chance will bring, or time travellers who have witnessed the outcome of coin-tosses to come, then patches of the past are enough tainted with futurity so that historical information about them may well seem inadmissible. That is why I qualified my claim that historical information is admissible, saying only that it is so 'as a rule'. Perhaps it is fair to ignore this problem in building a case that the Principal Principle captures our common opinions about chance, since those opinions may rest on a naive faith that past and future cannot possibly get mixed up. Any serious physicist, if he remains at least open-minded both about the shape of the cosmos and about the existence of chance processes, ought to do better. But I shall not; I shall carry on as if historical information is admissible without exception.

Besides historical information, there is at least one other sort of admissible information: hypothetical information about chance itself. Let us return briefly to our questionnaire and add one further supposition to each case. Suppose you have various opinions about what the chance of heads would be under various hypotheses about the detailed nature and history of the chance set-up under consideration. Suppose further that you have similar hypothetical opinions about other chance set-ups, past, present, and future. (Assume that these opinions are consistent with your admissible historical information and your opinions about chance in the present case.) It seems quite clear to me – and I hope it does to you also – that these added opinions do not change anything. The correct answers to the questionnaire are just as before. The added opinions do not bear in any overly direct way on the future outcomes of chance processes. Therefore they are admissible.

We must take care, though. Some propositions about future chances do reveal inadmissible information about future history, and these are inadmissible. Recall the case of the labyrinth: you enter at 11:00, choosing your turns by chance, and hope to reach the center by noon. Your subsequent chance of success depends on the point you have reached. The proposition that at 11:30 your chance of success has fallen to 26% is not admissible information at 11:00; it is a giveaway about your bad luck in the first half hour. What is admissible at 11:00 is a conditional version: if you were to reach a certain point at 11:30, your chance of success would then be 26%. But even some conditionals are tainted: for instance, any conditional that could yield inadmissible information about future chances by *modus ponens* from admissible historical propositions. Consider also the truth-functional conditional that if history up to 11:30 follows a certain course, then you will have a 98% chance of becoming a monkey's uncle before the year is out. This conditional closely resembles the denial of its antecedent, and is inadmissible at 11:00 for the same reason.

I suggest that conditionals of the following sort, however, are admissible; and indeed admissible at all times. (1) The consequent is a proposition about chance at a certain time. (2) The antecedent is a proposition about history up to that time; and further, it is a complete proposition about history up to that time, so that it either implies or else is incompatible with any other proposition about history up to that time. It fully specifies a segment, up to the given time, of some possible course of history. (3) The conditional is made from its consequent and antecedent not truth-functionally, but rather by means of a strong conditional operation of some sort. This might well be the counterfactual conditional of Lewis (1973); but various rival versions would serve as well, since many differences do not matter for the case at hand. One feature of my treatment will be needed, however: if the antecedent of one of our conditionals holds at a world, then both or neither of the conditional and its consequent hold there.

These admissible conditionals are propositions about how chance depends (or fails to depend) on history. They say nothing, however, about how history chances to go. A set of them is a theory about the way chance works. It may or may not be a complete theory, a consistent theory, a systematic theory, or a credible theory. It might be a miscellany of unrelated propositions about what the chances would be after various fully specified particular courses of events. Or it might be systematic, compressible into generalizations to the effect that after any course of history with property J there would follow a chance distribution with property K. (For instance, it might say that any coin with a certain structure would be fair.) These generalizations are universally quantified conditionals about single-case chance; if lawful, they are probabilistic laws in the sense of Railton (1978). (I shall not consider here what would make them lawful; but see Lewis (1973), §3.3, for a treatment that could cover laws about chance along with other laws.) Systematic theories of chance are the ones we can express in language, think about, and believe to substantial degrees. But a reasonable initial credence function does not reject any possibility out of hand. It assigns some non-zero credence to any consistent theory of chance, no matter how unsystematic and incompressible it is.

Historical propositions are admissible; so are propositions about the dependence of chance on history. Combinations of the two, of course, are also admissible. More generally, we may assume that any Boolean combination of propositions admissible at a time also is admissible at that time. Admissibility consists in keeping out of a forbidden subject matter – how the chance processes turned out – and there is no way to break into a subject matter by making Boolean combinations of propositions that lie outside it.

There may be sorts of admissible propositions besides those I have considered. If so, we shall have no need of them in what follows.

This completes an exposition of the Principal Principle. We turn next to an examination of its consequences. I maintain that they include all that we take ourselves to know about chance.

THE PRINCIPLE REFORMULATED

Given a time t and world w, let us write P_{tw} for the *chance distribution* that obtains at t and w. For any proposition A, $P_{tw}(A)$ is the chance, at time t and world w, of A's holding. (The domain of P_{tw} comprises those propositions for which this chance is defined.)

Let us also write H_{tw} for the *complete history* of world w up to time t: the conjunction of all propositions that hold at w about matters of particular fact no later than t. H_{tw} is the proposition that holds at exactly those worlds that perfectly match w, in matters of particular fact, up to time t.

Let us also write T_w for the *complete theory of chance* for world w: the conjunction of all the conditionals from history to chance, of the sort just considered, that hold at w. Thus T_w is a full specification, for world w, of the way chances at any time depend on history up to that time.

Taking the conjunction $H_{tw}T_w$, we have a proposition that tells us a great deal about the world w. It is nevertheless admissible at time t, being simply a giant conjunction of historical propositions that are admissible at t and conditionals from history to chance that are admissible at any time. Hence the Principal Principle applies:

$$C(A/XH_{tw}T_w) = x$$

when C is a reasonable initial credence function, X is the proposition that the chance at t of A is x, and $H_{tw}T_w$ is compatible with X.

Suppose X holds at w. That is so if and only if x equals $P_{tw}(A)$. Hence we can choose such an X whenever A is in the domain of P_{tw}. $H_{tw}T_w$ and X both hold at w, therefore they are compatible. But further, $H_{tw}T_w$ implies X. The theory T_w and the history H_{tw} together are enough to imply all that is true (and contradict all that is false) at world w about chances at time t. For consider the strong conditional with antecedent H_{tw} and consequent X. This conditional holds at w, since by hypothesis its antecedent and consequent hold there. Hence it is implied by T_w, which is the conjunction of all conditionals of its sort that hold at w; and this conditional and H_{tw} yield X

by *modus ponens*. Consequently the conjunction $XH_{tw}T_w$ simplifies to $H_{tw}T_w$. Provided that A is in the domain of P_{tw}, so that we can make a suitable choice of X, we can substitute $P_{tw}(A)$ for x, and $H_{tw}T_w$ for $XH_{tw}T_w$, in our instance of the Principal Principle. Therefore we have

THE PRINCIPAL PRINCIPLE REFORMULATED. Let C be any reasonable initial credence function. Then for any time t, world w, and proposition A in the domain of P_{tw}

$$P_{tw}(A) = C(A/H_{tw}T_w).$$

In words: the chance distribution at a time and a world comes from any reasonable initial credence function by conditionalizing on the complete history of the world up to the time, together with the complete theory of chance for the world.

This reformulation enjoys less direct intuitive support than the original formulation, but it will prove easier to use. It will serve as our point of departure in examining further consequences of the Principal Principle.

CHANCE AND THE PROBABILITY CALCULUS

A reasonable initial credence function is, among other things, a probability distribution: a non-negative, normalized, finitely additive measure. It obeys the laws of mathematical probability theory. There are well-known reasons why that must be so if credence is to rationalize courses of action that would not seem blatantly unreasonable in some circumstances.

Whatever comes by conditionalizing from a probability distribution is itself a probability distribution. Therefore a chance distribution is a probability distribution. For any time t and world w, P_{tw} obeys the laws of mathematical probability theory. These laws carry over from credence to chance via the Principal Principle. We have no need of any independent assumption that chance is a kind of probability.

Observe that although the Principal Principle concerns the relationship between chance and credence, some of its consequences concern chance alone. We have seen two such consequences. (1) The thesis that the past has no present chance of being otherwise than it actually is. (2) The thesis that chance obeys the laws of probability. More such consequences will appear later.

CHANCE AS OBJECTIFIED CREDENCE

Chance is an objectified subjective probability in the sense of Jeffrey (1965), §12.7. Jeffrey's construction (omitting his use of sequences of partitions, which is unnecessary if we allow infinitesimal credences) works as follows. Suppose given a partition of logical space: a set of mutually exclusive and jointly exhaustive propositions. Then we can define the *objectification* of a credence function, with respect to this partition, at a certain world, as the probability distribution that comes from the given credence function by conditionalizing on the member of the given partition that holds at the given world. Objectified credence is credence conditional on the truth – not the whole truth, however, but exactly as much of it as can be captured by a member of the partition without further subdivision of logical space. The member of the partition that holds depends on matters of contingent fact, varying from one world to another; it does not depend on what we think (except insofar as our thoughts are relevant matters of fact) and we may well be ignorant or mistaken about it. The same goes for objectified credence.

Now consider one particular way of partitioning. For any time t, consider the partition consisting of the propositions $H_{tw}T_w$ for all worlds w. Call this the *history-theory partition* for time t. A member of this partition is an equivalence class of worlds with respect to the relation of being exactly alike both in respect of matters of particular fact up to time t and in respect of the dependence of chance on history. The Principal Principle tells us that the chance distribution, at any time t and world w, is the objectification of any reasonable credence function, with respect to the history-theory partition for time t, at world w. Chance is credence conditional on the truth – *if* the truth is subject to censorship along the lines of the history-theory partition, and *if* the credence is reasonable.

Any historical proposition admissible at time t, or any admissible conditional from history to chance, or any admissible Boolean combination of propositions of these two kinds – in short, any sort of admissible proposition we have considered – is a disjunction of members of the history-theory partition for t. Its borders follow the lines of the partition, never cutting between two worlds that the partition does not distinguish. Likewise for any proposition about chances at t. Let X be the proposition that the chance at t of A is x, let Y be any member of the history-theory partition for t, and let C be any reasonable initial credence function. Then, according to our reformulation of the Principal Principle, X holds at all worlds in Y if $C(A/Y)$

equals x, and at no worlds in Y otherwise. Therefore X is the disjunction of all members Y of the partition such that $C(A/Y)$ equals x.

We may picture the situation as follows. The partition divides logical space into countless tiny squares. In each square there is a black region where A holds and a white region where it does not. Now blur the focus, so that divisions within the squares disappear from view. Each square becomes a grey patch in a broad expanse covered with varying shades of grey. Any maximal region of uniform shade is a proposition specifying the chance of A. The darker the shade, the higher is the uniform chance of A at the worlds in the region. The worlds themselves are not grey – they are black or white, worlds where A holds or where it doesn't – but we cannot focus on single worlds, so they all seem to be the shade of grey that covers their region. Admissible propositions, of the sorts we have considered, are regions that may cut across the contours of the shades of grey. The conjunction of one of these admissible propositions and a proposition about the chance of A is a region of uniform shade, but not in general a maximal uniform region. It consists of some, but perhaps not all, the members Y of the partition for which $C(A/Y)$ takes a certain value.

We derived our reformulation of the Principal Principle from the original formulation, but have not given a reserve derivation to show the two formulations equivalent. In fact the reformulation may be weaker, but not in any way that is likely to matter. Let C be a reasonable initial credence function; let X be the proposition that the chance at t of A is x; let E be admissible at t (in one of the ways we have considered) and compatible with X. According to the reformulation, as we have seen, XE is a disjunction of incompatible propositions Y, for each of which $C(A/Y)$ equals x. If there were only finitely many Y's, it would follow that $C(A/XE)$ also equals x. But the implication fails in certain cases with infinitely many Y's (and indeed we would expect the history-theory partition to be infinite) so we cannot quite recover the original formulation in this way. The cases of failure are peculiar, however, so the extra strength of the original formulation in ruling them out seems unimportant.

KINEMATICS OF CHANCE

Chance being a kind of probability, we may define conditional chance in the usual way as a quotient (leaving it undefined if the denominator is zero):

$$P_{tw}(A/B) =_{\mathrm{df}} P_{tw}(AB)/P_{tw}(B).$$

To simplify notation, let us fix on a particular world – ours, as it might be – and omit the subscript 'w'; let us fix on some particular reasonable initial credence function C, it doesn't matter which; and let us fix on a sequence of times, in order from earlier to later, to be called 1, 2, 3, ... (I do not assume they are equally spaced.) For any time t in our sequence, let the proposition I_t be the complete history of our chosen world in the interval from time t to time $t + 1$ (including $t + 1$ but not t). Thus I_t is the set of worlds that match the chosen world perfectly in matters of particular fact throughout the given interval.

A complete history up to some time may be extended by conjoining complete histories of subsequent intervals. H_2 is $H_1 I_1$, H_3 is $H_1 I_1 I_2$, and so on. Then by the Principal Principle we have:

$$P_1(A) = C(A/H_1 T),$$
$$P_2(A) = C(A/H_2 T) = C(A/H_1 I_1 T) = P_1(A/I_1),$$
$$P_3(A) = C(A/H_3 T) = C(A/H_1 I_1 I_2 T) = P_2(A/I_2)$$
$$= P_1(A/I_1 I_2),$$

.
.
.

and in general

$$P_{t+n+1}(A) = P_t(A/I_t \ldots I_{t+n}).$$

In words: a later chance distribution comes from an earlier one by conditionalizing on the complete history of the interval in between.

The evolution of chance is parallel to the evolution of credence for an agent who learns from experience, as he reasonably might, by conditionalizing. In that case a later credence function comes from an earlier one by conditionalizing on the total increment of evidence gained in the interval in between. For the evolution of chance we simply put the world's chance distribution in place of the agent's credence function, and the totality of particular fact about a time in place of the totality of evidence gained at that time.

In the interval from t to $t + 1$ there is a certain way that the world will in fact develop: namely, the way given by I_t. And at t, the last moment before the interval begins, there is a certain chance that the world will develop in that way: $P_t(I_t)$, the endpoint chance of I_t. Likewise for a longer interval, say from time 1 to time 18. The world will in fact develop in the way given by $I_1 \ldots I_{17}$, and the endpoint chance of its doing so is $P_1(I_1 \ldots I_{17})$. By definition of conditional chance

$$P_1(I_1 \ldots I_{17}) = P_1(I_1) \cdot P_1(I_2/I_1) \cdot P_1(I_3/I_1 I_2) \ldots P_1(I_{17}/I_1 \ldots I_{16}),$$

and by the Principal Principle, applied as above,

$$P_1(I_1 \ldots I_{17}) = P_1(I_1) \cdot P_2(I_2) \cdot P_3(I_3) \ldots P_{17}(I_{17}).$$

In general, if an interval is divided into subintervals, then the endpoint chance of the complete history of the interval is the product of the endpoint chances of the complete histories of the subintervals.

Earlier we drew a tree to represent the temporal asymmetry of chance. Now we can embellish the tree with numbers to represent the kinematics of chance. Take time 1 as the present. Worlds – those of them that are compatible with a certain common past and a certain common theory of chance – lie along paths through the tree. The numbers on each segment give the endpoint chance of the course of history represented by that segment, for any world that passes through that segment. Likewise, for any path consisting of several segments, the product of numbers along the path gives the endpoint chance of the course of history represented by the entire path.

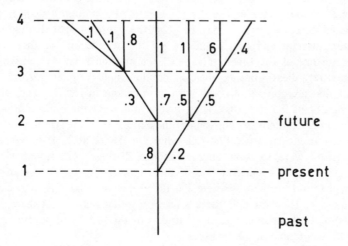

CHANCE OF FREQUENCY

Suppose that there is to be a long sequence of coin tosses under more or less standardized conditions. The first will be in the interval between time 1 and time 2, the second in the interval between 2 and 3, and so on. Our chosen world is such that at time 1 there is no chance, or negligible chance, that the planned sequence of tosses will not take place. And indeed it does take place. The outcomes are given by a sequence of propositions $A_1, A_2, \ldots$. Each A_t states truly whether the toss between t and $t + 1$ fell heads or tails. A conjunction $A_1 \ldots A_n$ then gives the history of outcomes for an initial segment of the sequence.

The endpoint chance $P_1(A_1 \ldots A_n)$ of such a sequence of outcomes is given by a product of conditional chances. By definition of conditional chance,

$$P_1(A_1 \ldots A_n) = P_1(A_1) \cdot P_1(A_2/A_1) \cdot P_1(A_3/A_1A_2) \ldots$$
$$\cdot P_1(A_n/A_1 \ldots A_{n-1}).$$

Since we are dealing with propositions that give only incomplete histories of intervals, there is no general guarantee that these factors equal the endpoint chances of the A's. The endpoint chance of A_2, $P_2(A_2)$, is given by $P_1(A_2/I_1)$; this may differ from $P_1(A_2/A_1)$ because the complete history I_1 includes some relevant information that the incomplete history A_1 omits about chance occurrences in the first interval. Likewise for the conditional and endpoint chances pertaining to later intervals.

Even though there is no general guarantee that the endpoint chance of a sequence of outcomes equals the product of the endpoint chances of the individual outcomes, yet it may be so if the world is right. It may be, for instance, that the endpoint chance of A_2 does not depend on those aspects of the history of the first interval that are omitted from A_1 – it would be the same regardless. Consider the class of all possible complete histories up to time 2 that are compatible both with the previous history H_1 and with the outcome A_1 of the first toss. These give all the ways the omitted aspects of the first interval might be. For each of these histories, some strong conditional holds at our chosen world that tells what the chance at 2 of A_2 would be if that history were to come about. Suppose all these conditionals have the same consequent: whichever one of the alternative histories were to come about, it would be that X, where X is the proposition that the chance at 2 of A_2 equals x. Then the conditionals taken together tell us that the endpoint chance of A_2 is independent of all aspects of the history of the first interval except the outcome of the first toss.

In that case we can equate the conditional chance $P_1(A_2/A_1)$ and the endpoint chance $P_2(A_2)$. Note that our conditionals are of the sort implied by T, the complete theory of chance for our chosen world. Hence A_1, H_1, and T jointly imply X. It follows that A_1H_1T and XA_1H_1T are the same proposition. It also follows that X holds at our chosen world, and hence that x equals $P_2(A_2)$. Note also that A_1H_1T is admissible at time 2. Now, using the Principal Principle first as reformulated and then in the original formulation, we have

$$P_1(A_2/A_1) = C(A_2/A_1H_1T) = C(A_2/XA_1H_1T) = x = P_2(A_2).$$

If we also have another such battery of conditionals to the effect that the endpoint chance of A_3 is independent of all aspects of the history of the first two intervals except the outcomes A_1 and A_2 of the first two tosses, and another battery for A_4, and so on, then the multiplicative rule for endpoint chances follows:

$$P_1(A_1 \ldots A_n) = P_1(A_1) \cdot P_2(A_2) \cdot P_3(A_3) \ldots P_n(A_n).$$

The conditionals that constitute the independence of endpoint chances mean that the incompleteness of the histories $A_1, A_2, \ldots$ doesn't matter. The missing part wouldn't make any difference.

We might have a stronger form of independence. The endpoint chances might not depend on *any* aspects of history after time 1, not even the outcomes of previous tosses. Then conditionals would hold at our chosen world to the effect that if any complete history up to time 2 which is compatible with H_1 were to come about, it would be that X (where X is again the proposition that the chance at 2 of A_2 equals x). We argue as before, leaving out A_1: T implies the conditionals, H_1 and T jointly imply X, H_1T and XH_1T are the same, X holds, x equals $P_2(A_2)$, H_1T is admissible at 2; so, using the Principal Principle in both formulations, we have

$$P_1(A_2) = C(A_2/H_1T) = C(A_2/XH_1T) = x = P_2(A_2).$$

Our strengthened independence assumption implies the weaker independence assumption of the previous case, wherefore

$$P_1(A_2/A_1) = P_2(A_2) = P_1(A_2).$$

If the later outcomes are likewise independent of history after time 1, then we have a multiplicative rule not only for endpoint chances but also for unconditional chances of outcomes at time 1:

$$P_1(A_1 \ldots A_n) = P_1(A_1) \cdot P_1(A_2) \cdot P_1(A_3) \ldots P_1(A_n).$$

Two conceptions of independence are in play together. One is the familiar probabilistic conception: A_2 is independent of A_1, with respect to the chance distribution P_1, if the conditional chance $P_1(A_2/A_1)$ equals the unconditional chance $P_1(A_2)$; equivalently, if the chance $P_1(A_1A_2)$ of the conjunction equals the product $P_1(A_1) \cdot P_1(A_2)$ of the chances of the conjuncts. The other conception involves batteries of strong conditionals with different antecedents and the same consequent. (I consider this to be *causal* independence, but that's another story.) The conditionals need not have anything to do with probability; for instance, my beard does not depend on my politics since I would have such a beard whether I were Republican, Democrat, Prohibitionist, Libertarian, Socialist Labor, or whatever. But one sort of consequent that can be independent of a range of alternatives, as we have seen, is a consequent about single-case chance. What I have done is to use the Principal Principle to parlay battery-of-conditionals independence into ordinary probabilistic independence.

If the world is right, the situation might be still simpler; and this is the case we hope to achieve in a well-conducted sequence of chance trials. Suppose the history-to-chance conditionals and the previous history of our chosen world give us not only independence (of the stronger sort) but also uniformity of chances: for any toss in our sequence, the endpoint chance of heads on that toss would be h (and the endpoint chance of tails would be $1 - h$) no matter which of the possible previous histories compatible with H_1 might have come to pass. Then each of the A_t's has an endpoint chance of h if it specifies an outcome of heads, $1 - h$ if it specifies an outcome of tails. By the multiplicative rule for endpoint chances,

$$P_1(A_1 \ldots A_n) = h^{fn} \cdot (1 - h)^{(n-fn)}$$

where f is the frequency of heads in the first n tosses according to $A_1 \ldots A_n$.

Now consider any other world that matches our chosen world in its history up to time 1 and in its complete theory of chance, but not in its sequence of outcomes. By the Principal Principle, the chance distribution at time 1 is the same for both worlds. Our assumptions of independence and uniformity apply to both worlds, being built into the shared history and theory. So all goes through for this other world as it did for our chosen world. Our calculation of the chance at time 1 of a sequence of outcomes, as a function of the uniform single-case chance of heads and the length and frequency of heads in the sequence, goes for any sequence, not only for the sequence $A_1, A_2, \ldots$ that comes about at our chosen world.

Let F be the proposition that the frequency of heads in the first n tosses is

f. F is a disjunction of propositions each specifying a sequence of n outcomes with frequency f of heads; each disjunct has the same chance at time 1, under our assumptions of independence and uniformity; and the disjuncts are incompatible. Multiplying the number of these propositions by the uniform chance of each, we get the chance of obtaining some or other sequence of outcomes with frequency f of heads:

$$P_1(F) = \frac{n! \cdot h^{fn} \cdot (1-h)^{(n-fn)}}{(fn)! \cdot (n-fn)!}.$$

The rest is well known. For fixed h and n, the right hand side of the equation peaks for f close to h; the greater is n, the sharper is the peak. If there are many tosses, then the chance is close to one that the frequency of heads is close to the uniform single-case chance of heads. The more tosses, the more stringent we can be about what counts as 'close'. That much of frequentism is true; and that much is a consequence of the Principal Principle, which relates chance not only to credence but also to frequency.

On the other hand, unless h is zero or one, the right hand side of the equation is non-zero. So, as already noted, there is always some chance that the frequency and the single-case chance may differ as badly as you please. That objection to frequentist analyses also turns out to be a consequence of the Principal Principle.

EVIDENCE ABOUT CHANCES

To the subjectivist who believes in objective chance, particular or general propositions about chances are nothing special. We believe them to varying degrees. As new evidence arrives, our credence in them should wax and wane in accordance with Bayesian confirmation theory. It is reasonable to believe such a proposition, like any other, to the degree given by a reasonable initial credence function conditionalized on one's present total evidence.

If we look at the matter in closer detail, we find that the calculations of changing reasonable credence involve *likelihoods*: credences of bits of evidence conditionally upon hypotheses. Here the Principal Principle may act as a useful constraint. Sometimes when the hypothesis concerns chance and the bit of evidence concerns the outcome, the reasonable likelihood is fixed, independently of the vagaries of initial credence and previous evidence. What is more, the likelihoods are fixed in such a way that observed frequencies tend to confirm hypotheses according to which these frequencies differ not too much from uniform chances.

To illustrate, let us return to our example of the sequence of coin tosses. Think of it as an experiment, designed to provide evidence bearing on various hypotheses about the single-case chances of heads. The sequence begins at time 1 and goes on for at least n tosses. The evidence gained by the end of the experiment is a proposition F to the effect that the frequency of heads in the first n tosses was f. (I assume that we use a mechanical counter that keeps no record of individual tosses. The case in which there is a full record, however, is little different. I also assume, in an unrealistic simplification, that no other evidence whatever arrives during the experiment.) Suppose that at time 1 your credence function is $C(-/E)$, the function that comes from our chosen reasonable initial credence function C by conditionalizing on your total evidence E up to that time. Then if you learn from experience by conditionalizing, your credence function after the experiment is $C(-/FE)$. The impact of your experimental evidence F on your beliefs, about chances or anything else, is given by the difference between these two functions.

Suppose that before the experiment your credence is distributed over a range of alternative hypotheses about the endpoint chances of heads in the experimental tosses. (Your degree of belief that none of these hypotheses is correct may not be zero, but I am supposing it to be negligible and shall accordingly neglect it.) The hypotheses agree that these chances are uniform, and each independent of the previous course of history after time 1; but they disagree about what the uniform chance of heads is. Let us write G_h for the hypothesis that the endpoint chances of heads are uniformly h. Then the credences $C(G_h/E)$, for various h's, comprise the *prior distribution* of credence over the hypotheses; the credences $C(G_h/FE)$ comprise the *posterior distribution*; and the credences $C(F/G_hE)$ are the likelihoods. Bayes' Theorem gives the posterior distribution in terms of the prior distribution and the likelihoods:

$$C(G_h/FE) = \frac{C(G_h/E) \cdot C(F/G_hE)}{\Sigma_h \left[C(G_h/E) \cdot C(F/G_hE) \right]}.$$

(Note that 'h' is a bound variable of summation in the denominator of the right hand side, but a free variable elsewhere.) In words: to get the posterior distribution, multiply the prior distribution by the likelihood function and renormalize.

In talking only about a single experiment, there is little to say about the prior distribution. That does indeed depend on the vagaries of initial credence and previous evidence.

Not so for the likelihoods. As we saw in the last section, each G_h implies a proposition X_h to the effect that the chance at 1 of F equals x_h, where x_h is given by a certain function of h, n, and f. Hence G_hE and X_hG_hE are the same proposition. Further, G_hE and X are compatible (unless G_hE is itself impossible, in which case G_h might as well be omitted from the range of hypotheses). E is admissible at 1, being about matters of particular fact – your evidence – at times no later than 1. G_h also is admissible at 1. Recall from the last section that what makes such a proposition hold at a world is a certain relationship between that world's complete history up to time 1 and that world's history-to-chance conditionals about the chances that would follow various complete extensions of that history. Hence any member of the history-theory partition for time 1 either implies or contradicts G_h; G_h is therefore a disjunction of conjunctions of admissible historical propositions and admissible history-to-chance conditionals. Finally, we supposed that C is reasonable. So the Principal Principle applies:

$$C(F/G_hE) = C(F/X_hG_hE) = x_h.$$

The likelihoods are the endpoint chances, according to the various hypotheses, of obtaining the frequency of heads that was in fact obtained.

When we carry the calculation through, putting these implied chances for the likelihoods in Bayes' theorem, the results are as we would expect. An observed frequency of f raises the credences of the hypotheses G_h with h close to f at the expense of the others; the more sharply so, the greater is the number of tosses. Unless the prior distribution is irremediably biased, the result after enough tosses is that the lion's share of the posterior credence will go to hypotheses putting the single-case chance of heads close to the observed frequency.

CHANCE AS A GUIDE TO LIFE

It is reasonable to let one's choices be guided in part by one's firm opinions about objective chances or, when firm opinions are lacking, by one's degrees of belief about chances. *Ceteris paribus*, the greater chance you think a lottery ticket has of winning, the more that ticket should be worth to you and the more you should be disposed to chose it over other desirable things. Why so?

There is no great puzzle about why credence should be a guide to life. Roughly speaking, what makes it be so that a certain credence function is *your* credence function is the very fact that you are disposed to act in more or less the ways that it rationalizes. (Better: what makes it be so that a certain reasonable initial credence function and a certain reasonable system of basic intrinsic values are both yours is that you are disposed to act in more or less the ways that are rationalized by the pair of them together, taking into account the modification of credence by conditionalizing on total evidence; and further, you would have been likewise disposed if your life history of experience, and consequent modification of credence, had been different; and further, no other such pair would fit your dispositions more closely.) No wonder your credence function tends to guide your life. If its doing so did not accord to some considerable extent with your dispositions to act, then it would not be your credence function. You would have some other credence function, or none.

If your present degrees of belief are reasonable – or at least if they come from some reasonable initial credence function by conditionalizing on your total evidence – then the Principal Principle applies. Your credences about outcomes conform to your firm beliefs and your partial beliefs about chances. Then the latter guide your life because the former do. The greater chance you think the ticket has of winning, the greater should be your degree of belief that it will win; and the greater is your degree of belief that it will win, the more, *ceteris paribus*, it should be worth to you and the more you should be disposed to choose it over other desirable things.

PROSPECTS FOR AN ANALYSIS OF CHANCE

Consider once more the Principal Principle as reformulated:

$$P_{tw}(A) = C(A/H_{tw}T_w).$$

Or in words: the chance distribution at a time and a world comes from any reasonable initial credence function by conditionalizing on the complete history of the world up to the time, together with the complete theory of chance for the world.

Doubtless it has crossed your mind that this has at least the form of an analysis of chance. But you may well doubt that it is informative as an analysis; that depends on the distance between the analysandum and the concepts employed in the analysans.

Not that it has to be informative *as an analysis* to be informative. I hope I have convinced you that the Principal Principle is indeed informative, being rich in consequences that are central to our ordinary ways of thinking about chance.

There are two different reasons to doubt that the Principal Principle qualifies as an analysis. The first concerns the allusion in the analysans to reasonable initial credence functions. The second concerns the allusion to complete theories of chance. In both cases the challenge is the same: could we possibly get any independent grasp on this concept, otherwise than by way of the concept of chance itself? In both cases my provisional answer is: most likely not, but it would be worth trying. Let us consider the two problems in turn.

It would be natural to think that the Principal Principle tells us nothing at all about chance, but rather tells us something about what makes an initial credence function be a reasonable one. To be reasonable is to conform to objective chances in the way described. Put this strongly, the response is wrong: the Principle has consequences, as we noted, that are about chance and not at all about its relationship to credence. (They would be acceptable, I trust, to a believer in objective single-case chance who rejects the very idea of degree of belief.) It tells us more than nothing about chance. But perhaps it is divisible into two parts: one part that tells us something about chance, another that takes the concept of chance for granted and goes on to lay down a criterion of reasonableness for initial credence.

Is there any hope that we might leave the Principal Principle in abeyance, lay down other criteria of reasonableness that do not mention chance, and get a good enough grip on the concept that way? It's a lot to ask. For note that just as the Principal Principle yields some consequences that are entirely about chance, so also it yields some that are entirely about reasonable initial credence. One such consequence is as follows. There is a large class of propositions such that if Y is any one of these, and C_1 and C_2 are any two reasonable initial credence functions, then the functions that come from C_1 and C_2 by conditionalizing on Y are exactly the same. (The large class is, of course, the class of members of history-theory partitions for all times.) That severely limits the ways that reasonable initial credence functions may differ, and so shows that criteria adequate to pick them out must be quite strong. What might we try? A reasonable initial credence function ought to (1) obey the laws of mathematical probability theory; (2) avoid dogmatism, at least by never assigning zero credence to possible propositions and perhaps also by never assigning infinitesimal credence to certain kinds of possible propositions; (3) make it possible to learn from experience by having a built-in bias in favor of worlds where the future in some sense resembles the past; and perhaps (4) obey certain carefully restricted principles of indifference, thereby respecting certain symmetries. Of these, criteria (1)–(3) are all very well, but surely not yet strong enough. Given C_1 satisfying (1)–(3), and given any proposition Y that holds at more than one world, it will be possible to distort C_1 very slightly to produce C_2, such that $C_1(-/Y)$ and $C_2(-/Y)$ differ but C_2 also satisfies (1)–(3). It is less clear what (4) might be able to do for us. Mostly that is because (4) is less clear *simpliciter*, in view of the fact that it is not possible to obey too many different restricted principles of indifference at once and it is hard to give good reasons to prefer some over their competitors. It also remains possible, of course, that some criterion of reasonableness along different lines than any I have mentioned would do the trick.

I turn now to our second problem: the concept of a complete theory of chance. In saying what makes a certain proposition be the complete theory of chance for a world (and for any world where it holds), I gave an explanation in terms of chance. Could these same propositions possibly be picked out in some other way, without mentioning chance?

The question turns on an underlying metaphysical issue. A broadly Humean doctrine (something I would very much like to believe if at all possible) holds that all the facts there are about the world are particular facts, or combinations thereof. This need not be taken as a doctrine of analyzability, since some combinations of particular facts cannot be captured in any finite way. It might be better taken as a doctrine of supervenience: if two worlds match perfectly in all matters of particular fact, they match perfectly in all other ways too – in modal properties, laws, causal connections, chances, ... It seems that if this broadly Humean doctrine is false, then chances are a likely candidate to be the fatal counterinstance. And if chances are not supervenient on particular fact, then neither are complete theories of chance. For the chances at a world are jointly determined by its complete theory of chance together with propositions about its history, which latter plainly are supervenient on particular fact.

If chances are not supervenient on particular fact, then neither chance itself nor the concept of a complete theory of chance could possibly be analyzed in terms of particular fact, or of anything supervenient thereon. The only hope for an analysis would be to use something in the analysans which is itself not supervenient on particular fact. I cannot say what that something might be.

How might chance, and complete theories of chance, be supervenient on particular fact? Could something like this be right: the complete theory of chance for a world is that one of all possible complete theories of chance that somehow best fits the global pattern of outcomes and frequencies of outcomes? It could not. For consider any such global pattern, and consider a time long before the pattern is complete. At that time, the pattern surely has some chance of coming about and some chance of not coming about. There is surely some chance of a very different global pattern coming about; one which, according to the proposal under consideration, would make true some different complete theory of chance. But a complete theory of chance is not something that could have some chance of coming about or not coming about. By the Principal Principle,

$$P_{tw}(T_w) = C(T_w/H_{tw}T_w) = 1.$$

If T_w is something that holds in virtue of some global pattern of particular fact that obtains at world w, this pattern must be one that has no chance at any time (at w) of not obtaining. If w is a world where many matters of particular fact are the outcomes of chance processes, then I fail to see what kind of global pattern this could possibly be.

But there is one more alternative. I have spoken as if I took it for granted that different worlds have different history-to-chance conditionals, and hence different complete theories of chance. Perhaps this is not so: perhaps all worlds are exactly alike in the dependence of chance on history. Then the complete theory of chance for every world, and all the conditionals that comprise it, are necessary. They are supervenient on particular fact in the trivial way that what is non-contingent is supervenient on anything – no two worlds differ with respect to it. Chances are still contingent, but only because they depend on contingent historical propositions (information about the details of the coin and tosser, as it might be) and not also because they depend on a contingent theory of chance. Our theory is much simplified if this is true. Admissible information is simply historical information; the history-theory partition at t is simply the partition of alternative complete histories up to t; for any reasonable initial credence function C

$$P_{tw}(A) = C(A/H_{tw}),$$

so that the chance distribution at t and w comes from C by conditionalizing on the complete history of w up to t. Chance is reasonable credence conditional on the whole truth about history up to a time. The broadly Humean doctrine is upheld, so far as chances are concerned: what makes it true at a time and a world that something has a certain chance of happening is something about matters of particular fact at that time and (perhaps) before.

What's the catch? For one thing, we are no longer safely exploring the consequences of the Principal Principle, but rather engaging in speculation. For another, our broadly Humean speculation that history-to-chance conditionals are necessary solves our second problem by making the first one worse. Reasonable initial credence functions are constrained more narrowly than ever. Any two of them, C_1 and C_2, are now required to yield the same function by conditionalizing on the complete history of any world up to any time. Put it this way: according to our broadly Humean speculation (and the Principal Principle) if I were perfectly reasonable and knew all about the course of history up to now (no matter what that course of history actually is, and no matter what time is now) then there would be only one credence function I could have. Any other would be unreasonable.

It is not very easy to believe that the requirements of reason leave so little leeway as that. Neither is it very easy to believe in features of the world that are not supervenient on particular fact. But if I am right, that seems to be the choice. I shall not attempt to decide between the Humean and the anti-Humean variants of my approach to credence and chance. The Principal Principle doesn't.

Princeton University

NOTE

* I am grateful to several people for valuable discussions of this material; especially John Burgess, Nancy Cartwright, Richard Jeffrey, Peter Railton, and Brian Skyrms. I am also much indebted to Mellor (1971), which presents a view very close to mine; exactly how close I am not prepared to say.

BIBLIOGRAPHY

Bernstein, Allen R. and Wattenberg, Frank, 'Non-Standard Measure Theory', in *Applications of Model Theory of Algebra, Analysis, and Probability*, ed. by W. Luxemburg, Holt, Reinhart, and Winston, 1969.
Carnap, Rudolf, 'The Two Concepts of Probability', *Philosophy and Phenomenological Research* 5 (1945), 513–32.
Jeffrey, Richard C., *The Logic of Decision*, McGraw-Hill, 1965.
Jeffrey, Richard C., review of articles by David Miller *et al.*, *Journal of Symbolic Logic* 35 (1970), 124–27.
Jeffrey, Richard C., 'Mises Redux', in *Basic Problems in Methodology and Linguistics: Proceedings of the Fifth International Congress of Logic, Methodology and Philosophy of Science*, Part III, ed. by R. Butts and J. Hintikka, D. Reidel, Dordrecht, Holland, 1977.
Lewis, David, 'Counterpart Theory and Quantified Modal Logic', *Journal of Philosophy* 65 (1968), 113–26.
Lewis, David, *Counterfactuals*, Blackwell, 1973.

Mellor, D. H., *The Matter of Chance*, Cambridge University Press, 1971.

Quine, W. V. 'Propositional Objects', in *Ontological Relativity and Other Essays*, Columbia University Press, 1969.

Railton, Peter, 'A Deductive-Nomological Model of Probabilistic Explanation', *Philosophy of Science* **45** (1978), 206–26.

Skyrms, Brian, 'Resiliency, Propensities, and Causal Necessity', *Journal of Philosophy* **74** (1977), 704–13.

RICHMOND H. THOMASON AND ANIL GUPTA

A THEORY OF CONDITIONALS IN THE CONTEXT
OF BRANCHING TIME*

1. INTRODUCTION

In Stalnaker [9] and in Stalnaker and Thomason [10], a theory of conditionals is presented that involves a "selection function". Intuitively, the value of the function at a world is the world as it would be if a certain formula (the antecedent of a conditional) were true.

In these two papers, the notion of a possible world is left entirely blank and abstract; worlds are simply treated as points. This approach has the advantage of generality, but could also be misleading. Clearly, in a situation in which there are likenesses among possible worlds the selection function will be affected. Suppose, for instance, that we can speak of those worlds that are like w and those that are unlike it. Then the function should not choose a world unlike w when one like w would do as well. The moral of this is that if we pass to a logical theory in which "possible worlds" are given a certain amount of structure, we can't expect the logic of conditionals to remain unaffected – for this structure may provide some purchase on world similarity.[1]

In this paper we want to explore one of the most pervasive and important cases of this sort: the interaction of conditionals with tense. This interaction can be rather intricate in even the most commonplace examples of conditionals: consider, for instance, the following two.

(1.1)   You'll lose this match if you lose this point.

(1.2)   If he loves her then he will marry her.

We believe there is a difference in logical form here: (1.1) has the form $FQ > FR$ (or perhaps the form $F(Q > FR)$), while (1.2) has the form $Q > FR$.[2]

Or consider the following pair.

(1.3)   If Max missed the train he would have taken the bus.

(1.4)   Max took the bus if he missed the train.

We believe that the form of (1.3) is $P(Q > FR)$ (so that in this sentence the

299

W. L. Harper, R. Stalnaker, and G. Pearce (eds.), Ifs, 299–322.

word 'would' is the past tense of 'will'), while that of (1.4) is $PQ > PR$. (Thus, it is not, on our view, a matter of the form of (1.4) that one would normally expect the bus-catching to have followed the train-missing if (1.4) and its antecedent are true; (1.4) has the same form as 'He took the bus if he came here from the bus station'.)

These sentences help to bring out a fundamental point: if you approach tense logic with the theory of conditionals in mind that we have just sketched, it's natural to want a logic capable of dealing with temporal structures that branch towards the future. Take example (1.1), for instance, and imagine you are evaluating it at a moment of time with a single past but various possible futures, some in which you win this point and some in which you lose it. We want the selection function to choose a future course of events (or *scenario*, or *history*) in which you lose this point; the truth of (1.1) depends on whether you lose the match on this scenario. Other examples, like (1.3), can lead to scenarios that might have occurred, but didn't. Given a form like $P(Q > FR)$, we are led to a *past* moment, at which we choose a scenario in which Q is true. Now, it may be that there is no such scenario if we confine ourselves to what actually happened. In the example under consideration, this will be the case if Max didn't in fact miss the train. In this case, we want the selection function to choose a scenario that might have been actualized, but wasn't.

So we will draw on logical work concerning branching temporal structures. A number of logics are discussed in Prior [6], but the version of tense logic that we will use is that of Thomason [13].

2. THE FIRST THEORY

We will begin by developing a theory of tense and conditionals which is a simple adaptation of Stalnaker's theory. The difficulties that this first theory encounters will motivate some of the central ideas of the second and the third theories that we present later on.

In Stalnaker's theory a conditional $A > B$ is true at a world w if and only if B is true at a world w' determined by A and w. Intuitively, w' is the world at which A is true (i.e., w' is an A-*world*), and which is the closest possible A-world to w.[3] Formally, the theory posits a function s which for each antecedent A and world w picks out a world $w' = s(A, w)$.[4] Then clause (2.1) gives the truth conditions of a conditional formula.

(2.1) $A > B$ is true at w if and only if B is true at $s(A, w)$.

Constraints imposed on the s-function will determine the logical properties of conditionals; in Stalnaker [9] and in Stalnaker and Thomason [10] such constraints are elaborated, and the corresponding properties explored. These constraints can be understood as reflecting the idea that the A-world is the "closest" one in which A is true.

When time is brought in to the picture, worlds give way to evolving histories. Thus, the truth value of a sentence – and you should now think of sentences as including tensed ones, whose truth conditions may refer to past and future moments – will depend on a history h and a moment i along h. That is, formulas are evaluated at moment-history pairs $\langle i, h \rangle$. Further, in assigning a truth value to $A > B$, you do not merely want to consider the closest A-world. Rather you want to consider the closest moment-history pair at which A is true. Thus we want the Stalanker function s to take as arguments a formula A and a moment-history pair $\langle i, h \rangle$ and to yield as value the closest pair $\langle i', h' \rangle$ to $\langle i, h \rangle$ at which A is true. Conditionals will then be interpreted by (2.2).

(2.2) $A > B$ is true at $\langle i, h \rangle$ if and only if B is true at $\langle i', h' \rangle$, where $\langle i', h' \rangle = s(A, \langle i, h \rangle)$.

Rule (2.2) brings time into the picture, so that the elements we are selecting are complex. This means that we must look more closely at closeness. As a first step, we would assume that i' is an "alternative present" to i: 'If I were in Rome . . .' amounts to 'If I were in Rome now . . .', 'If I had been born in Wales . . .' to 'If it were true now that I had been born in Wales . . .', and so forth.[5]

Second, we wish to make a claim: closeness among moment-history pairs conforms to the following condition, the condition of *Past Predominance*.

(2.3)   In determining how close $\langle i_1, h_1 \rangle$ is to $\langle i_2, h_2 \rangle$ (where i_1 and i_2 are alternative presents to one another), past closeness predominates over future closeness; that is, the portions of h_1 and h_2 not after [6] i_1 and i_2 predominate over the rest of h_1 and h_2.[7]

This informal principle is to be interpreted as strongly as possible: if h_3 up to i_3 is even a little closer to h_1 up to i_1 than is h_2 up to i_2, then $\langle i_3, h_3 \rangle$ is closer to $\langle i_1, h_1 \rangle$ than $\langle i_2, h_2 \rangle$ is, even if h_2 after i_2 is much closer to h_1 after i_1, than is h_3 after i_3. Any gain with respect to the past counts more than even the largest gain with respect to the future. Our formal theory will incorporate the hypothesis that (2.3) is correct.

Contrast this hypothesis of Past Predominance with the notion that neither the past nor the future predominates in evaluating conditionals. Call this, the most natural rival of (2.3), the *Overall Similarity Theory*. The two differ in important logical ways. For one thing, the Overall Similarity Theory allows considerations of time and tense to influence the truth conditions of $A > B$ only insofar as they affect the truth conditions of the antecedent A and the consequent B. But on our theory they can influence the truth conditions directly. Thus, when A and B are eternal, $A > B$ is eternal on the Overall Similarity Theory, but need not be on ours.[8]

For another thing, since they motivate different constraints on the ordering of histories and hence on the s-functions, the two theories yield different principles of interaction between tenses and conditionals. This point is most easily seen in connection with a language having metric tense operators F^n and P^n. (F^n, for instance, may be read "it will be the case n minutes hence that.") The Overall Similarity Theory will validate the following distribution principles.

(2.4) $F^n(A > B) \supset (F^nA > F^nB)$

(2.5) $P^n(A > B) \supset (P^nA > P^nB)$

(2.6) $(F^nA > F^nB) \supset F^n(A > B)$

(2.7) $(P^nA > P^nB) \supset P^n(A > B)$

Take (2.4), for instance. Suppose $F^n(A > B)$ is true at $\langle i_1, h \rangle$; this means that $A > B$ is true at $\langle i_2, h \rangle$, where i_2 is the moment n minutes further along h than i_1. We can safely assume that A is true at $\langle i_2', h' \rangle$ for some i_2' copresent with i_2 and h' containing i_2', for otherwise (2.4) is vacuously true. So we conclude that at the closest pair $\langle i_2^*, h^* \rangle$, A and B both are true. On the Overall Similarity Theory, h* will be the history most resembling h overall, among those histories that meet the condition that A be true on them at the moment copresent with i_2. But then h* is also the closest history to h *overall* among those that meet the condition that F^nA be true on them at the moment copresent with i_1. Thus $F^nA > F^nB$ is true at $\langle i, h \rangle$, since B is true at $\langle i_2', h' \rangle$. Similar arguments yield the validity of (2.5)–(2.7) on the Overall Similarity Theory.

But none of these four formulas is valid on our proposal. Again, we take (2.4). This may fail to be true at $\langle i_1, h \rangle$ because the closest history to h, given what has happened before i_1, need not be the same as the closest history to h, given what has happened before i_2. Readers who are not content with this informal account can add metric tenses to the formal language we interpret below, and show that (2.4)–(2.7) are indeed falsifiable.

Differences like this emerge also with ordinary, nonmetric tense operators. The following four formulas, for example, are valid on the Overall Similarity Theory but invalid on the Past Predominance Theory.

(2.8) $G(A > B) \supset (FA > FB)$

(2.9) $H(A > B) \supset (PA > PB)$

(2.10) $(FA > G(A \supset B)) \supset F(A > B)$

(2.11) $(PA > H(A \supset B)) \supset P(A > B)$

(G and H are understood respectively as "It will always be the case that" and "It has always been the case that.")

These considerations show that the difference between Overall Similarity and Past Predominance is substantive; it affects the logic of tenses and conditionals.[9] Why do we choose the latter logic? Firstly, because we are not persuaded that (2.4)–(2.11) are logical truths. Consider (2.7). Imagine that David and Max have been playing a simple coin-tossing and betting game. Max flips the coin. Unknown to David, Max has two coins, one with heads on each side and one with tails on each side. If David bets tails, Max flips the first coin; if he bets heads, Max flips the second. Two minutes ago David bet that the coin would come up heads on the next flip. Max flipped the coin and it came up tails. Now the following can be said truly (say, by someone who does not know which way David bet).

(2.12)   If two minutes ago David bet tails then he wins now.

So, it seems that formula

(2.13) $(P^2 Q > P^2 R)$

is true, where Q stands for the sentence 'David bet tails' and R for the sentence 'David wins now'. But the formula

(2.14) $P^2(Q > R)$

is false. If David had bet tails two minutes ago, he would still have lost. So we have a situation in which (2.7) is false.

Secondly, Past Predominance explains our intuition about the truth conditions of English conditionals better than the Overall Similarity Theory. Consider the following variant of an example of Kit Fine's. (See Fine [3].)

(2.15)   If button A is pushed within a minute, there will be a nuclear holocaust.

Imagine ways in which the button may be hooked up, or may fail to be hooked up, to a doomsday device. In some of these, (2.15) is true, and in others false. And among the former cases, we can well imagine ones in which the button is not pushed, and no holocaust occurs; say that one of these cases involves a certain moment i and history h. The Overall Similarity Theory has difficulties with such cases. For a history h' in which the button is disconnected and no holocaust occurs when the button is pressed is much more similar to h, overall, than a history h" in which there is a holocaust when the button is pressed. But if h' is used to evaluate (2.15), the sentence is false. Moreover, (2.16) is true.

(2.16)   If button A is pushed within a minute, it is already disconnected.

Here, and in other cases too numerous to mention,[10] Overall Similarity would be hard put to explain our intuitions about truth. On the other hand, Past Predominance fits these intuitions. A hypothetical disconnecting of a button that is already connected counts for more than any hypothetical change regarding what *will* happen.

Thirdly, Past Predominance makes possible an approach for explaining differences between indicative and subjunctive conditionals. We wish to suggest (tentatively) that some examples that have been contrasted simply along the indicative-subjunctive dimension also involve scope differences with respect to tenses. For instance, consider Ernest Adams' lovely pair of examples (Adams [1], p. 90.)

(2.17)   If Oswald didn't shoot Kennedy then Kennedy is alive today.

(2.18)   If Oswald hadn't shot Kennedy then Kennedy would be alive today.

Our proposal is that (2.17) should be formalized as

(2.19) $PQ > R,$

while (2.18) should be formalized as

(2.20) $P(Q > R).$

In (2.19) and (2.20), Q stands for 'Oswald doesn't shoot Kennedy' and R for the eternal sentence 'Kennedy is alive today', and we understand the past tense operator P to be relativized to an indexically specified interval of time. The difference in the truth conditions of the two sentences arises because (2.19) requires us to maximize closeness to the present moment while (2.20) requires us to maximize closeness only up to some past moment. (2.20), and hence (2.18), are true because at some past moment the corresponding indicative $Q > R$, i.e., 'If Oswald doesn't shoot Kennedy then Kennedy will be alive . . .', is true. More generally, we want to propose (tentatively) that a subjunctive asserts that the corresponding indicative sentence *was* true in some contextually determined interval of time.[11]

We now put these ideas to work by sketching a formal interpretation of a propositional language $\mathscr{L}$ with the usual truth-functional connectives, the conditional $>$, the past and future tense operators P and F, and the "settledness" operator L. As far as tenses and settledness are concerned, we adopt the theory developed in Thomason [13].

Model structures consist of a nonempty set $\mathscr{K}$ of moments (to be thought of as world states, and not to be confused with clock times[12]), and two relations $<$ and $\simeq$ on $\mathscr{K}$. The relation $<$ orders members of $\mathscr{K}$ into a treelike structure, the branches of which give various possible courses of events. We impose the following conditions on $<$: (1) it is transitive; and (2) if $i_1, i_2 < i$ then $i_1 < i_2$ or $i_1 = i_2$ or $i_2 < i_1$.[13] The relation $\simeq$ relates "copresent" moments. Branches (i.e., maximal chains with respect to $<$) through a moment i give the various possible courses of history at i. We let $\mathscr{H}_i$ be the set of all branches passing through i. The following conditions ensure that $\simeq$ and $<$ interact in the proper way: (4) $\simeq$ is an equivalence relation; (5) if $i_1 \simeq i_2$, then $i_1 \not< i_2$; (6) If $i_1 \simeq i_2$ and $i_3 \simeq i_4$ and $i_1 < i_3$, then $i_4 \not< i_2$.

Models for $\mathscr{L}$ involve a valuation function V and two Stalnaker functions, s_1 and s_2. These can be thought of as two components of a single Stalnaker function. Intuitively, s_1 gives us the closest moment at which a condition is true, and s_2 provides a history through this moment. Formally, s_1 is a function which takes a formula A and a moment i as arguments and yields a moment $s_1(A, i)$. And s_2 is a function which takes a formula A, a moment i and a history h (it is required that $i \in h$) as arguments and yields a history $s_2(A, i, h)$, which we require to be a member of $\mathscr{H}_{s_1(A, i)}$.

Note that s_1 depends only on i and A, while s_2 depends on h as well. Now it is clear that s_2 should depend on h, for we want $s_2(A, i, h) = h$ when A is true on h at i. (If this principle is given up, *modus ponens* will become invalid.) On the other hand, s_1 does not depend on h, because of Past Predominance. This principle dictates that any gain in past similarity offsets any loss in future similarity. Thus if h, $h' \in \mathscr{H}_i$ then $s_2(A, i, h)$ and $s_2(A, i, h')$ must have the same past, and therefore $s_1(A, i, h) = s_1(A, i, h')$. Which is to say that s_1 depends only on A and i.

Lastly the valuation V gives the truth value of each atomic formula at each moment. We define by the usual recursion the truth value that V gives to a formula A on a history h and moment i, namely, the value $V_i^h(A)$. The recursion clauses for truth-functional connectives and the tense operators are standard. For conditionals and settledness we have the following natural clauses.

(2.21) $V_i^h(A > B) = T$ iff $V_{s_1(A, i)}^{s_2(A, i, h)}(B) = T$.

(2.22) $V_i^h(LA) = T$ iff $V_i^{h'}(A) = T$ for all $h' \in \mathscr{H}_i$.

We understand a conditional $A > B$ to be true when the antecedent A is impossible. There are several equally effective ways of achieving this formally; we suppose that one such way has been adopted.

The final part of the theory consists of constraints on the Stalnaker functions s_1 and s_2. These constraints ensure, among other things, that the conditionals of $\mathscr{L}$ are interpreted in accordance with (2.3), the Principle of Past Predominance. We will not state the constraints formally, for they are analogous to Conditions (i)–(vi) and (ix), stated in Section 3.

We turn now to the breakdown of this theory. One intimation of the problem is that no reasonable conditions on s_1 and s_2 will ensure that the following inferences are valid.

(Edelberg)   From $L\sim A$ and $L(A > B)$ to infer $A > L(A \supset B)$.

(Weak Edelberg) From $L\sim A$, $A > LA$ and $L(A > B)$ to infer $A > LB$.[14]

These examples involve a claim to the effect that a "counterfactural" (where "factual" is taken to mean "settled") conditional is settled. Before going any further, we want to point out that we do make claims of the form $L(A > B)$ – even when it is understood that $L\sim A$ is true – and that, in some cases at least, we feel that such claims have definite truth values.

Suppose, for instance, that you're waiting on the corner for your morning bus to the office. Two buses are scheduled to stop at the corner: first a local, then an express. You have a fixed policy of waiting for the express, because it almost always gets you to work on time, while the local very often arrives later. Now, let's consider the truth conditions of the sentence 'it is settled that if you were to take the local you wouldn't arrive on time' at various clock times that morning.

Ten minutes before you catch the local, let's suppose that the situation is in all relevant respects (traffic and road conditions, the mechanical condition of the bus, etc.) like previous ones in which the local has stopped near your office soon enough for you to arrive at work on time. At this

time, then, the sentence is false; there is a chance that if you were to take
the local you would arrive on time. But five minutes after this, let's suppose
an accident occurs on the local's route (say, a mile or so beyond your stop),
immediately creating a horrible rush-hour traffic jam. Now the sentence
becomes true; it's a sure thing that if you were to take the local, you
wouldn't arrive on time. And afterwards, of course, this continues to remain
true. So, this is a case of a possibility that is open at one time becoming
closed at a later one – only it is a *counterfactual* possibility.

It strikes us as intuitively clear that the two Edelberg inferences are
valid. In the above example, for instance, as soon as it becomes true that
it's settled that if you were to take the local you wouldn't arrive on time,
it also becomes true that if you were to take the local it would then be
settled that you would arrive late. Of course, an example doesn't suffice to
establish that an inference is valid, but it can help to make evident what is involved in an inference, then its validity, if it is valid, is a matter of intuition.

Before going any further, let us deal with a possible misunderstanding.
Though in the above example – and in others that will come later – we
are dealing with what we take to be a natural, everyday conception of
settledness, this conception is not a part of the logical theory we are proposing. Just as first-order logic is compatible with all sorts of choices of a
domain for its quantifiers – domains that contain sets, domains that don't,
domains that are finite, domains that aren't, etc. – the tense logic proposed
in this paper is compatible (or, at least, is meant to be compatible) with all
sorts of conceptions of settledness. (No doubt there are many such conceptions, from very liberal ones (for example, perhaps, one associated with
quantum theory, on which some things that have been taken to be matters of
physical law on classical theories would not be settled) to very conservative
ones (one that took the actions of everyone other than ones own to be determined, or even a strict determinism). All these are supposed to be compatible
with the logic, though one – strict determinism – is a fairly trivial special case,
in which the relation $<$ of the model structure is linear.

We are not trying to argue here for the correctness (in some extralogical
sense of 'correctness') of any of these interpretations. Indeed, it may be
that there is more than one viable conception of settledness, so that no one
interpretation is "the" correct one. The only important thing is that in
evaluating the validity of inferences – such as those under consideration –
we must be careful to apply the same conception of settledness to all the
terms of the inference.

Back now to the pathology of the formal interpretation we set up. The symptom is that it provides a way of making $L{\sim}FQ$, $FQ > LFQ$ and $L(FQ > FR)$ true, while $FQ > LFR$ is made false, as is illustrated by the following model structure.

(2.23)

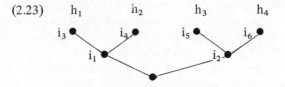

You are to assume here that the ordering relation is the strict partial order of the tree, that $i_1 \simeq i_2$ and $i_3 \simeq i_4 \simeq i_5 \simeq i_6$.

Let $V(Q, i_3) = V(Q, i_4) = F$. Then $V_{i_1}^{h_1}(FQ) = V_{i_1}^{h_2}(FQ) = F$, and so $V_{i_1}^{h_1}(L{\sim}FQ) = T$.

Let $V(Q, i_5) = V(Q, i_6) = T$ and $V(R, i_5) = T$ but $V(R, i_6) = F$. Now, $V_{i_2}^{h_3}(FQ) = V_{i_2}^{h_4}(FQ) = T$ and therefore,

(2.24) $V_{i_2}^{h_3}(LFQ) = T$.

Also, since $V_{i_2}^{h_4}(FR) = F$, we have

(2.25) $V_{i_2}^{h_3}(LFR) = F$.

Finally, let $s_1(FQ, i_1) = i_2$ and $s_2(FQ, i_1, h_1) = s_2(FQ, i_1, h_2) = h_3$. This last bit is the crucial part of our model – the part that makes the inference invalid. Notice how h_1 and h_2 are collapsed counterfactually into h_3, which is only one among two histories for i_2 on which FQ is true.

Now it is easily seen that the premisses of our inference are true on this model, but the conclusion is false. We have already seen that the first premiss $L{\sim}FQ$ is true at $\langle i_1, h_1 \rangle$.

For the second premiss we have $V_{i_1}^{h_1}(FQ > LFQ) = T$, for $V_{i_2}^{h_3}(LFQ) = T$ in view of (2.24), and $s_1(FQ, i_1) = i_2$ and $s_2(FQ, i_1, h_1) = h_3$.

To see that the third premiss is true observe that $V_{i_2}^{h_3}(FR) = T$. Since $s_2(FQ, i_1, h_1) = s_2(FQ, i_1, h_2) = h_3$, we have $V_{i_1}^{h_1}(FQ > FR) = V_{i_1}^{h_2}(FQ > FR) = T$. Hence $V_{i_1}^{h_1}(L(FQ > FR)) = T$.

Lastly, note that $V_{i_1}^{h_1}(FQ > LFR) = F$, in view of (2.25). This same model shows that the Edelberg inference is invalid.

Grant that this is a bad thing. Can we patch the theory up so that the inference becomes valid? A direct way to do it would be to rule out the kind of situation that makes counterexamples to the inference possible; we could simply require the following.

(2.26)   If $s_1(A, i) = i'$ then for all $h' \in \mathcal{H}_{i'}$ such that $V_i^{h'}(A) = T$ there is an $h \in \mathcal{H}_i$ such that $s_2(A, i, h) = h'$.

This ensures that s_2 will not create gaps in such a way as to invalidate the inference.

But besides being *ad hoc*, this condition seems to us to be ugly. What makes it so is the fact that it seems to rule out structures and assignments of truth values that don't at all seem logically impossible. Take the following case.

(2.27)

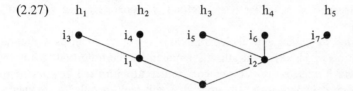

Here, let $V(P, i_3) = V(P, i_4) = F$ and $V(P, i_5) = V(P, i_6) = V(P, i_7) = T$. This makes it combinatorially impossible to match each history through i_2 in which P becomes true as an image of some history through i_1. And yet nothing seems to prohibit either the structure (2.27) or the truth assignment we have placed on it. It is true that we could rule out such cases with no effect on validity, by building copies: for instance, we could insert a copy of h_2 after i_1 to obtain enough scenarios. But this is ugly, and unless there is some independent motivation for this procedure, we find it implausible.

We note lastly that the analogue of (2.26) is not so implausible, and does not rule out the situation portrayed in (2.27), if the Stalnaker function s_2 yields, as in David Lewis' theory, a *class* of histories. The problem with this account is that it fails to validate the law of conditional excluded middle,[15] $(A > B) \vee (A > {\sim}B)$. So the difficulty that needs to be solved in this: how to validate *both* conditional excluded middle and the Edelberg inference. We present our solution to this difficulty in the next section.

3. A BETTER THEORY

This section is going to be rather technical. Readers who are not interested in the technical details may want to skim it.[16] Those who find the brief motivation we give for the technical apparatus unsatisfying may want to read Section 4 first.

Central to the theory we will present is the idea that the concept of truth for $\mathscr{R}$ should be relativised to a future choice function rather than to a history.

(3.1)   A *future choice function* is a function $\mathscr{F}$ from the set $\mathscr{K}$ of moments to $\cup \{\mathscr{H}_i / i \in \mathscr{K}\}$ such that (1) $\mathscr{F}_i \in \mathscr{H}_i$ and (2) if $i' \in \mathscr{F}_i$ and $i < i'$ then $\mathscr{F}_i = \mathscr{F}_{i'}$. We let Ω be the set of all future choice functions (for a fixed model structure).

A future choice function $\mathscr{F}$ gives at each moment i a unique history through that moment – the history that *would* be actual if i were actual. Condition (2) ensures that the histories $\mathscr{F}$ chooses at later moments are coherent with histories it chooses at earlier ones; without this condition, FA would not imply PFA.

One way to understand choice functions is to see them as a natural generalization of histories, one that is required by the transition to a tense logic in which what is true at moments copresent with i can be relevant to what is true at i. A history tells you what will happen only for moments that lie along it – for the rest it leaves the future indeterminate. A choice function, on the other hand, tells you what will happen at *all* moments. A choice function is a richer history; a history is a partial choice function.

Now in a tense logic which has only operators like P, F and L, the truth value of a formula A at a moment i depends only on histories that pass through i. You are not forced to consider moments copresent or incomparable with i. So here it is all right to think of the concept of truth as relativized to a history. But when you add conditionals to the language the truth value of a formula A at i, in general, depends also on what will happen at moments i' copresent with i. Different choices as to what *will* happen at i' affect what conditionals hold true at i. Thus in this context histories do not contain enough information; though they tell us what *will* happen, they do not tell us what *would* happen.

We implement choice functions in our semantics by defining $V_i^{\mathscr{F}}(A)$. We can keep the definitions of model structures and models as they are given in the last section, except that in the present theory the second Stalnaker function s_2 takes as arguments a formula A, a moment i and a future choice function $\mathscr{F}$ and yields as value a future choice function $s_2(A, i, \mathscr{F})$. The recursion clauses for truth functions, tenses and conditionals are adjusted to the new parameter, and except for the clause for settledness remain in essentials similar to those of classical conditional and tense logics.

In the definition below of $V_i^{\mathscr{F}}(A)$, we restrict ourselves to choice functions meeting a certain requirement: $\mathscr{F}$ must be *normal at* i, in the following sense.

(3.2) $\mathscr{F}$ is *normal at* i iff for all j $<$ i, $\mathscr{F}_j = \mathscr{F}_i$. We say that a pair $\langle i, \mathscr{F} \rangle$ is normal iff $\mathscr{F}$ is normal at i. The *normalization* of $\mathscr{F}$ to i is the choice function $\mathscr{F}'$ such that $\mathscr{F}'_j = \mathscr{F}_i$ for j $<$ i and $\mathscr{F}'_j = \mathscr{F}_j$ for j $\not< $ i.

If $\mathscr{F}$ is normal at i, $\mathscr{F}$ treats i as "actual" from the point of view of moments in the past of i. Now, the definition of satisfaction.

(3.3) $V_i^{\mathscr{F}}(Q) = V(Q, i)$.

(3.4) $V_i^{\mathscr{F}}(\sim A) = T$ iff $V_i^{\mathscr{F}}(A) = F$.

(3.5) $V_i^{\mathscr{F}}(A \supset B) = T$ iff either $V_i^{\mathscr{F}}(A) = F$ or $V_i^{\mathscr{F}}(B) = T$.

(3.6) $V_i^{\mathscr{F}}(FA) = T$ iff for some i$'\in \mathscr{F}_i$ such that i $<$ i$'$, $V_{i'}^{\mathscr{F}}(A) = T$.

(3.7) $V_i^{\mathscr{F}}(PA) = T$ iff for some i$'$ such that i$' <$ i, $V_{i'}^{\mathscr{F}}(A) = T$.

(3.8) $V_i^{\mathscr{F}}(A > B) = T$ iff either $s_1(A, i)$ is undefined or
$$V_{s_1(A, i)}^{s_2(A, i, \mathscr{F})}(B) = T.$$

The clause for L requires forethought. We should not simply say that a formula LA is true at i with respect to $\mathscr{F}$ normal at i iff A is true at i with respect to *all* choice functions $\mathscr{F}'$ normal at i. This is because LA says that A holds no matter how things *will* be. Hence, we want $\mathscr{F}'$ to differ from $\mathscr{F}$ only on moments that are after i or after some moment copresent with i. (There is also a formal reason for not accepting this account of the truth conditions of LA: on it the Edelberg inference is invalid.)

Before we state the clause for L, we need to define some ancillary concepts.

(3.9)   i is *posterior* to j iff there is a moment j$'$ copresent with j such that i $>$ j$'$ or i $=$ j$'$.

(3.10)   i is *antiposterior* to j iff either i is not posterior to j or i $\simeq$ j.[17]

(3.11) $\mathscr{F}$ *agrees with* $\mathscr{G}$ *on moments posterior to* i (symbolically, $\mathscr{F} \in$ Post $(\mathscr{G}, i)$) iff all moments j posterior to i are such that $\mathscr{F}_j = \mathscr{G}_j$.

(3.12) $\mathscr{F}$ *agrees with* $\mathscr{G}$ *on moments antiposterior to* i (symbolically,

$\mathscr{F} \in \text{APost}(\mathscr{G}, i))$ iff all moments j, k antiposterior to i are such that $j \in \mathscr{F}_k$ iff $j \in \mathscr{G}_k$.

Now clause (3.13) gives the truth conditions for LA.

(3.13) $V_i^{\mathscr{F}} (LA) = T$ iff at all choice function $\mathscr{G}$ normal at i such that $\mathscr{G} \in \text{APost}(\mathscr{G}, i)$, $V_i^{\mathscr{F}}(A) = T$.

We observe that the clause for F and P, (3.6) and (3.7), are correct only for choice function $\mathscr{F}$ normal at i; if these are generalized to *all* choice functions then the law $A \supset \text{PF}A$ becomes falsifiable. This is our motivation for restricting the recursive definition above to choice functions that are normal at a given moment. Notice that the clauses for P and F never take us to non-normal moment-choice function pairs. The same obviously holds for all other connectives, except the conditional. Here the constraints on the two Stalnaker functions ensure that $s_2(A, i, \mathscr{F})$ is always normal at $s_1(A, i)$. These constraints are as follows.

(i)   If there is an i$'$ such that $i \simeq i'$ and a $\mathscr{G} \in \Omega$ such that $\mathscr{G}$ is normal at i$'$ and $V_{i'}^{\mathscr{G}} (A) = T$ then both $s_1(A, i)$ and $s_2(A, i, \mathscr{F})$ are defined provided $\mathscr{F}$ is normal at i; and $s_2(A, i, \mathscr{F})$ is normal at $s_1(A, i)$. Otherwise both $s_1(A, i)$ and $s_2(A, i, \mathscr{F})$ are undefined.

(ii) $V_{s_1(A, i)}^{s_2(A, i, \mathscr{F})}(A) = T$.

(iii)   If $V_{s_1(A, i)}^{s_2(A, i, \mathscr{F})}(B) = V_{s_1(B, i)}^{s_2(B, i, \mathscr{F})}(A) = T$, then $s_2(A, i, \mathscr{F}) = s_2(B, i, \mathscr{F})$ and $s_1(A, i) = s_1(B, i)$.

(iv)   If $V_{s_1(A, i)}^{\mathscr{F}'}(A) = T$ where $\mathscr{F}'$ is the result of normalizing $\mathscr{F}$ to $s_1(A, i)$ then $s_2(A, i, \mathscr{F}) = \mathscr{F}'$.

(v)   If there is an $\mathscr{F} \in \Omega$ such that $\mathscr{F}$ is normal at i and $V_i^{\mathscr{F}} (A) = T$ then $s_1(A, i) = i$.

(vi)   If there are choice fuctions $\mathscr{F}$, $\mathscr{G}$ such that $\mathscr{F}$ is normal at $s_1(A, i)$ and $V_{s_1(A, i)}^{\mathscr{F}}(B) = T$, and $\mathscr{G}$ is normal at $s_1(B, i)$ and $V_{s_1(B, i)}^{\mathscr{G}}(A) = T$ then $s_1(A, i) = s_1(B, i)$.

(vii)   If $\mathscr{F} \in \text{APost}(\mathscr{G}, i)$, and $s_2(A, i, \mathscr{F})$, and $s_2(A, i, \mathscr{G})$ are defined, then $s_2(A, i, \mathscr{F}) \in \text{APost}(s_2(A, i, \mathscr{G}), i)$.

(viii)   If $\mathcal{G}$ is a choice function normal at $s_1(A, i)$ such that (a) $\mathcal{G} \in$ Post $(\mathcal{G}, i)$ where $\mathcal{G}$ is normal at i, (b) $\mathcal{G} \in$ APost($s_2(A, i, \mathcal{G}'$), i) for some choice function $\mathcal{G}' \in$ APost($\mathcal{G}, i$), and (c) $V_{s_1(A, i)}^{\mathcal{G}}(A) =$ T, then $s_2(A, i, \mathcal{F}) = \mathcal{G}$.

(ix)   If $s_1(A, i)$ is defined then $s_1(A, i) \simeq i$.

Conditions (i)–(iii) are direct analogues of ones from Stalnaker's theory. Condition (iv) requires s_2 to distort the choice function as little as possible. Conditions (v)–(vii) embody the Principle of Past Predominance. Condition (v) says that if you can preserve all of the past then you should; Condition (vi) says that you must not choose a more dissimilar past then you have to;[18] and Condition (vii) says that even at counterfactual copresent moments the past should be preserved, if possible. All these conditions have an effect on validity. Thus (iv) and (v) ensure that the inferences

(3.14)   From A and B to infer $A > B$

(3.15)   From MA and L($A \supset B$) to infer $A > B$

are valid, where M$A =_{df} {\sim}L{\sim}A$. Conditions (vi) and (vii) ensure that the inferences (3.16) and (3.17), respectively, are valid.

(3.16)   From $A > MB, B > MA, A > C$, and $A > (C > LC)$ to infer $B > C$.

(3.17)   From $A > L(A \supset B)$ to infer L($A > B$).

Condition (viii) says that the future histories at all copresent moments must be preserved by s_2 if doing so is consistent with Past Predominance. This condition helps to ensure that the Edelberg inferences are valid. (See the Appendix for proof.) Finally Condition (ix) ensures that a conditional 'If . . . then . . .' amounts to 'If . . . now then . . .'.

4. DIGGING DEEPER

We now want to discuss some problems that lead to refinements of the theory we just presented, and these in turn provide some fresh perspectives on matters of philosophical interest.

We begin with an example of Stalnaker's.[19] Suppose two coins are tossed successively, one in Chicago and the other in Bombay.

(4.1) h_1 h_2 h_3 h_4

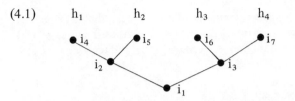

At i_1 the Chicago coin is tossed; at i_2 it has come up heads and at i_3 tails. At i_2 [i_3] the Bombay coin is tossed; at i_4 [i_6] it comes up heads and at i_5 [i_7] tails. (You are to imagine that there are no causal connections between the two tosses.) Now, intuitively, it seems settled at i_2 that if the Bombay coin will come up heads then (even) if the Chicago coin had not turned up heads the Bombay coin would (still) have come up heads. But unfortunately there is a choice function $\mathscr{F}$ at which this is false. Let $\mathscr{F}(i_1) = h_1$ and $\mathscr{F}(i_3) = h_4$. Let Q stand for the sentence 'The Chicago coin comes up heads' and R for 'The Bombay coin comes up heads'. Now the formula $FR > (P{\sim}Q > FR)$ is false at i_2 on $\mathscr{F}$, since $P{\sim}Q$ is true but FR is false on the normalization of $\mathscr{F}$ to i_3 (cf. Condition (iv) in Section 3 on Stalnaker functions). Hence $L(FR > (P{\sim}Q > FR))$ is false at i_2. This is unwelcome, because it does not seem to be an open possibility at i_2 that if the Bombay coin were going to come up heads then if the Chicago coin had turned up tails instead of heads the Bombay coin would be going to come up tails. Well, maybe in some sense it isn't a *logical* mistake to consider it an open possibility, but in so considering it we would seem to be positing some strange causal influence that makes the toss in Maharashtra depend on the toss in Illinois.

The problem arises because the second theory considers too many choice functions in the computation of LA, function like $\mathscr{F}$ which are *causally incoherent*. $\mathscr{F}$, for example, assigns different outcomes to the Bombay toss, but the only difference between i_2 and i_3 is causally irrelevant to this second toss.

What we want to do, then, is to evaluate L not with respect to all choice functions that are "logically possible," but rather with respect to a restricted class of choice functions: those that are causally coherent. Formally, it comes to this. A model structure involves, besides a set $\mathscr{K}$ and relations $\simeq$ and $<$, a set Ω^* of choice functions. We then say that $V_i(LA) = T$ if and only if $V_i^{\mathscr{F}'}(A) = T$ for all $\mathscr{F}' \in \Omega^*$ such that $\mathscr{F}' \in \mathrm{APost}\,(\mathscr{F}, i)$.

If we say only this, however, the Edelberg inference is invalidated. (We leave details to the reader; pick a class of choice functions that is lopsided.)

But the validity of the inference is restored if we impose some closure conditions on $\Omega*$.[20]

We have been led to the idea that restricted sets of choice functions may be useful in representing certain causal notions. It may be helpful in this connection to follow a different path to the same suggestion. To say that events e_1 and e_2 are causally independent is to say that whether or not e_1 occurs is independent of whether or not e_2 occurs. Thus, if e_2 *will* occur then e_2 *would* occur whether or not e_1 *does* occur. This leads to the thought that, e.g., situations in which e_1 is followed by e_2 are somehow incompatible with ones in which e_1 is *not* followed by e_2. But an analysis of this incompatibility in terms of necessity is clearly wrong, because things can *in fact* be causally independent without *necessarily* being so (in any interesting sense of 'necessarily').[21]

Such considerations (not to mention the problem of saying what an event is) make it very difficult to give an *analysis* of causal independence. But they have established that when a certain causal independence obtains, certain choice functions are excluded. If e_1 and e_2 are independent, among these will be choice functions according to which e_1 is followed by e_2 on some designated histories, but the nonoccurrence of e_1 is not followed by e_2 on other designated histories that do not differ from the first history in ways material to e_2.

If we wish to construct the kind of model theory that has proved so useful in other areas of logic, and which may help to illuminate this philosophical topic, perhaps we should look at natural classes of choice functions. These should help us to come to grips with causal independence, and perhaps even with causality.

Towards the end of Section 2, we said that our problem was to develop a logic that endorsed *both* conditional excluded middle and the Edelberg inference. This task led us into a theory that is quite complex. Now it might be helpful to stand back a bit and gain some perspective on the topic. This will be useful with another matter: how truth-value gaps enter into tensed conditionals.

There are several ways in which we might have avoided the complexities of the earlier pages. *First,* we could rule out model structures that create the problem in the first place – model structures like (2.27) which allow us to falsify the Edelberg inference. This maneuver seems to us arbitrary and implausible. (But see Van Fraassen [15] for an ingenious defense and development of this approach.)

Second, we can give up conditional excluded middle. This allows us to view the conditional as *variably strict*, not, as we have it in our theory, *variably material.*[22] On one variably strict analysis of conditionals, the second Stalnaker function s_2 yields at each A, i, h a *class* of histories $s_2(A, i, h)$. Then $A > B$ is true at $\langle i, h \rangle$ if and only if $A \supset B$ holds at $\langle s_1(A, i), h' \rangle$ for all $h' \in s_2(A, i, h)$. We can make the Edelberg inference valid if we require that for all histories $h' \in \mathcal{H}_{s_1(A, i)}$ such that $V_{s_1(A, i)}^{h'}(A) = T$ there is a history $h \in \mathcal{H}_i$ such that $h' \in s_2(A, i, h)$. On the resulting theory we can keep all the desired model structures, and also have the Edelberg inference valid, but we give up conditional exluded middle.

A *third* option is to give up the Edelberg inference, but explain its apparent validity in some more or less devious way. One such way is via supervaluations. It is clear that if we accept conditional excluded middle we have to supplement our theory with supervaluations on Stalnaker functions to account for cases such as Quine's Bizet–Verdi example. (See Stalnaker [12] for a discussion of this.) Thus instead of a single Stalnaker function s (where s is a pair $\langle s_1, s_2 \rangle$) we now have a set S of such function s compatible with the "facts" about conditionals. The truth is then what is common to all these functions: where $V_i^{\langle h, s \rangle}(A)$ is the truth value of A relative to s and h, on the theory of Section 2, we let $V_i^h(A) = T$ if $V_i^{\langle h, s \rangle}(A) = T$ for all $s \in S$, and $V_i^h(A) = F$ if $V_i^{\langle h, s \rangle}(A) = F$ for all $s \in S$; otherwise $V_i^h(A)$ is undefined. Now, we can easily make the Edelberg inferences valid, in the sense that if the premisses are true for all members of S then the conclusion is also true for all member of S. The inference will be valid in this sense if we require that whenever $\langle s_1, s_2 \rangle \in S$, $s_1(A, i) = i'$, and $V_{i'}^{h'}(A) = T$, where $i' \in h'$, then there is a $\langle s_1, s_2' \rangle \in S$, where $s_2'(A, i, h) = h'$. However, $L(A > B) \supset (A > L(A \supset B))$ will be invalid; it can fail to have a truth value. To the extent that we can discover intuitions about whether the Edelberg inference should be merely truth preserving or should provide a valid conditional, these support the latter alternative. Also, since even in those cases where you have to resort to this distinction (e.g. in explaining the validity of Convention T) it is difficult to motivate the distinction as convincingly as one would wish, it probably is a good strategy to avoid using it when it is possible to do so. This secondary consideration lends support to the theory of Section 3.

Note that truth-value gaps can arise in the third theory in two ways, for there are two parameters along which supervaluations can be introduced:

'$\mathscr{F}$' and 's'. Since choice functions are generalizations of histories, the former parameter is what yields the indeterminacy of future contingencies. Thus, a sentence like

(4.2)   This coin will come up heads on the second toss if it comes up heads on the first toss

may lack a truth value at a moment i before both tosses, because there is a choice function $\mathscr{F}_1$ (assigning i a history on which the coin comes up heads on the first toss and tails on the second) on which (4.2) is false, and there is another choice function $\mathscr{F}_2$ (assigning i a history on which the coin comes up heads on both tosses) on which (4.2) is true. And an unconditional sentence like

(4.3)   This coin will come up heads on the second toss

will lack a truth value at i for reasons that are exactly the same: $\mathscr{F}_1$, for instance, makes (4.3) false and $\mathscr{F}_2$ makes it true.

Now consider a moment j later than i, at which the coin comes up tails for the second time, and compare the "past tenses" of (4.2) and (4.3) at j.

(4.4)   It was the case that if the coin were going to come up heads on the first toss, it would come up heads on the second toss.

(4.5)   The coin was going to come up heads on the second toss.

Here we see a difference between conditionals and nonconditionals; conditionals can be *unfulfilled*, and this may cause their "past tenses" to lack a truth value. Example (4.5) is simply false at j. But, if you perform the calculations according to our theory, (4.4), which is unfulfilled at j because its antecedent did not become true, is neither true nor false. (This is because there are two choice functions normal at j, one of which makes (4.2) true at i and another of which makes (4.2) false at i.)

On our theory, then, the indeterminacy of typical unfulfilled conditionals is simply absorbed into the indeterminacy of the future, rather than requiring any indeterminacy in the Stalnaker function. This is precisely what choice functions were designed to accomplish, by allowing all sorts of counterfactual futures to be included in "the future" that is relevant to settledness at a moment.

However, there may be cases in which we would want to make the Stalnaker function indeterminate. Quine's Bizet–Verdi example is a good one: its antecedent and consequent must both be settled if true. So if this conditional is to be neither true nor false the source of this must be something other than future indeterminacy.

In the most general theory, then, we have a set Ω^* of choice functions meeting certain conditions, and a set S of Stalnaker functions. $V_i(A) = T$ if $V_i^{\langle \mathscr{F}, s \rangle}(A) = T$ for all $\mathscr{F} \in \Omega^*$ and all $s \in S$; and $V_i(A) = F$ if $V_i^{\langle \mathscr{F}, s \rangle}(A) = F$ for all $\mathscr{F} \in \Omega^*$ and all $s \in S$.

We have sketched a number of ways in which the complexities of our theory might have been avoided. What is striking about these alternatives is that they do not *solve* the problem of preserving simultaneously the Edelberg inference, conditional excluded middle and model structures of the type (2.27). This suggests strongly that the complexity is one that is needed. Our theory is complex because the problem it is designed to solve has no simple solution. Also we believe that the central concept of our theory – that of a choice function – will prove useful in an account of various causal concepts, such as that of causal independence.

APPENDIX

We show in this appendix that the Edelberg inference is valid on the second theory.

LEMMA 1. For any choice function $\mathscr{F}$ and $\mathscr{F}'$ there is a choice function $\mathscr{G}$ such that Post $(\mathscr{F}, i) \cap \text{APost}(\mathscr{F}', i) = \{\mathscr{G}\}$.

Proof. Define the function $\mathscr{G}$ as follows: (1) if moment j is posterior to i then let $\mathscr{G}_j = \mathscr{F}_j$; (2) if j is antiposterior to i and there is a k such that $i \simeq k$ and $k \in \mathscr{F}'_j$ then let $\mathscr{G}_j = \mathscr{F}_k$ (uniqueness of k is obviously guaranteed); (3) Otherwise, let $\mathscr{G}_j = \mathscr{F}'_j$. It is easily confirmed by a tedious but not ingenious calculation that Post $(\mathscr{F}, i) \cap \text{APost}(\mathscr{F}', i) = \{\mathscr{G}\}$.

LEMMA 2. Let i be a moment and $\mathscr{F}$ be a choice function such that $s_1(A, i)(= i_1)$ and $s_2(A, i, \mathscr{F})(= \mathscr{F}_1)$ are defined. Let $\mathscr{G}_1$ be a choice function normal at i_1 such that $\mathscr{G}_1 \in \text{Apost}(\mathscr{F}_1, i_1)$ and $V_{i_1}^{\mathscr{G}_1}(A) = T$. Then there is a $\mathscr{G}$ normal at i such that $\mathscr{G} \in \text{APost}(\mathscr{F}, i)$ and $s_2(A, i, \mathscr{G}) = \mathscr{G}_1$.

Proof. By Lemma 1 there is a $\mathscr{G}$ such that $\{\mathscr{G}\} = \text{Post}(\mathscr{G}_1, i) \cap \text{APost}(\mathscr{F}, i)$. Since $\mathscr{F}$ is normal at i and $\mathscr{G} \in \text{APost}(\mathscr{F}, i)$, clearly $\mathscr{G}$ is normal at i. Also $s_2(A, i, \mathscr{G}) = \mathscr{G}_1$ by Condition (vii) on Stalnaker functions (cf. Section 3) since $\mathscr{G}_1 \in$ Post $(\mathscr{G}, i)$, and $\mathscr{G}_1 \in$ APost$(\mathscr{F}_1, i)$ (Condition (ix)), and $V_{i_1}^{\mathscr{G}_1}(A) = T$.

THEOREM: The Edelberg inference is valid on the theory of Section 2.

Proof. Let $\mathscr{F}$ be a choice function normal at i such that $V_i^{\mathscr{F}}(L(A > B)) = T$. We can assume safely that $s_1(A, i)(= i_1)$ and $s_2(A, i, \mathscr{F})(= \mathscr{F}_1)$ are defined for otherwise all conditionals are vacuously true. Let $\mathscr{G}_1$ be a choice function normal at i_1 such that $\mathscr{G}_1 \in$ APost$(\mathscr{F}_1, i_1)$ and $V_{i_1}^{\mathscr{G}_1}(A) = T$. Now all the hypotheses of Lemma 2 are met. So we infer that there is a $\mathscr{G}$ normal at i such that $\mathscr{G} \in$ APost$(\mathscr{F}, i)$ and $s_2(A, i, \mathscr{G}) = \mathscr{G}_1$. Hence $V_i (A > B) = T$. Therefore, $V_{s_1(A, i)}^{s_2(A, i, \mathscr{G})}(B) = T$. That is, $V_{i_1}^{\mathscr{G}_1}(B) = T$. So at all $\mathscr{G}_1 \in$ APost$(\mathscr{F}_1, i_1)$, $V_{i_1}^{\mathscr{G}_1}(A \supset B) = T$. Thus $V_{i_1}^{\mathscr{F}_1}(L(A \supset B)) = T$, and $V_i^{\mathscr{F}}(A > L(A \supset B)) = T$.

University of Pittsburgh and
McGill University

NOTES

* This began as a paper by Thomason, written in February 1977, revised and expanded in January 1978, and presented, with comments by Gupta, at the University of Western Ontario in May 1978. The present joint version was completed in October 1978. The basic ideas took initial shape in a series of discussions between Thomason and Walter Edelberg, who deserves a great deal of credit for his insights into the topic. In particular, he was the first to see the importance of the crucial inference we call 'the Edelberg inference'. Later parts of the paper owe much to interactions with Robert Stalnaker, Bas van Fraassen, and each other.

[1] We hope that our use of terms like 'similarity' isn't misleading. We don't believe that meditating on the notion of similarity among possible worlds is likely to advance our knowledge of conditionals, or that it is very enlightening to explain the world chosen by the selection function as the most similar one in which the condition is true. We're only saying here that sometimes rather *coarse* kinds of similarity are available, and that when they are available they should be exploited.

[2] Here, we mean only to give these judgements, without attempting to justify them or to present a theory of how they should be formalized. It would be appropriate to return to these matters after the development of a model theory for tense and conditionals, but we will not attempt this in the present paper. We should mention, however, our working assumption about subjunctive mood: it has no categorematic semantic meaning. For an attempt to use pragmatics to explain mood in conditionals, see Stalnaker [11]. We suspect that this account needs to be supplemented with an explan-

ation of how mood interacts with scope. Like 'any' and 'every', indicative and sub-junctive may serve to signal "logical forms" in which operators are arranged in certain ways.

[3] To simplify things, we ignore − and will continue to ignore − necessarily false ante-cedents. Imagine that they are treated in some suitable way.

[4] Stalnaker originally called these 's-functions', by way of abbreviating 'selection function'. But we propose to call them 's-functions', by way of abbreviating 'Stalnaker function'. There is a good reason for this change. Later we will be talking a good deal about "choice-functions" and these are entirely different from s-functions.

[5] Yes, we also mean this principle to apply to cases like 'If it were 5:00...' and 'If it were Christmas ...'. That is one reason why we chose the phrase 'an alternative present' rather than 'simultaneous'.

[6] In saying 'not after' here, we wish to make clear that 'past closeness' in this context really means 'past-or-present closeness'.

[7] A similar condition is discussed in Bennett [2]. See also Lewis [4], p. 76, Lewis [5], and Slote [8]. Note that our condition of Past Predominance makes no reference to "the moment" to which the antecedent "refers". If $A > B$ is evaluated at i, then on our proposal it is closeness up to i that predominates.

[8] To say A is eternal is to say that for all h, if A is true at $\langle i, h \rangle$ for any i then A is true at $\langle i', h \rangle$ for all i' along h.

[9] Besides examples like (2.8), that are valid given Overall Similarity but invalid given Past Predominance, there are others that are valid given Past Predominance but invalid given Overall Similarity. One instance of this is (3.16), discussed in the next section.

[10] See Slote [8] for a discussion of some of these.

[11] A full discussion of the ideas presented in this paragraph requires more space than we have in the present paper. We intend to pursue these themes elsewhere.

[12] Since we do not impose a metric on the branches of our structure, there need be nothing corresponding to clock times in these structures. The relation of copresence introduced below need not be considered to stand in any simple relation to clock times, either. In effect, we are ignoring in this paper the technical and philosophical questions introduced by thinking of times as quantitative. These questions are not at all super-ficial; if conceptual problems arise in spreading time over space, what can you expect when you spread time over possible worlds? For discussions of some of the issues that can arise, see Aristotle's *Physics* 221^a 29−32 and 223^a 21−29, and Van Fraassens's account of the Leibniz−Clarke correspondence on pp. 41 of [14].

[13] Do not assume that all moments have a past moment in common. We explicitly want to allow "disconnected" moments that can be reached by a counterfactual, to provide for epistemic uses of conditionals. Example (2.17), as it would commonly be understood, may be one such use. If the Warren Commission was right, the alternative to which it takes us is epistemically but not historically possible.

[14] The Edelberg inferences turn out to be valid even without the premiss $L \sim A$. But we include the premiss because the difficulty we envisage arises for the theory only when $L \sim A$ holds.

[15] For a defence of this law see Stalnaker [12].

[16] Advice to skimmers: study carefully the definition of a future choice function, and spend some time on clause (3.13).

[17] The terminology is awkward, and the concepts will probably seem more devious

than they should be. The reason for this is that we feel it is important to state a theory without assuming that for all moments i and histories h, there is an i' such that $i \simeq i'$ and $i' \in h$. Thus there may be moments that are neither anterior nor posterior to i. Making the assumption in question would simplify things, but would lose the generality that we feel is appropriate for a tense logic. In this connection, see Note 12, above.

[18] Conditions (i)–(vi), if they were exchanged for conditions on Stalnaker functions taking *propositions* (rather than formulas) as arguments, would amount to this. There is for each normal pair $\langle i, \mathcal{F} \rangle$, a well ordering $\prec_{\langle i, \mathcal{F} \rangle}$ of normal pairs consisting of moments copresent with i and choice functions, such that (a) $\langle i, \mathcal{F} \rangle$ is the least pair under this ordering and $\langle i_1, \mathcal{F}_1 \rangle \prec_{\langle i, \mathcal{F} \rangle} \langle i_1, \mathcal{F}_2 \rangle$ for all i_1 copresent with i and all $\mathcal{F}_2$ normal at i_1, where $\mathcal{F}_1$ is the normalization of $\mathcal{F}$ to i_1; (b) if $\langle i_1, \mathcal{F}_1 \rangle \prec_{\langle i, \mathcal{F} \rangle}$ $\langle i_2, \mathcal{F}_2 \rangle$ then $\langle i_1, \mathcal{F}_1 \rangle \prec_{\langle i, \mathcal{F} \rangle} \langle i_2, \mathcal{F}_3 \rangle$ for all $\mathcal{F}_3$ normal at i_2; and (c) $a_1(A, i)$ and $s_2(A, i, \mathcal{F})$ are defined iff $s_1(A, i) = i'$ and $s_2(A, i, \mathcal{F}) = \mathcal{F}'$, where $\langle i', \mathcal{F}' \rangle$ is the least pair with respect to $\prec_{\langle i, \mathcal{F} \rangle}$ such that $V_i^{\mathcal{F}'}(A) = T$.

[19] Communicated to us in correspondence. Independently Gupta discovered an example whose import is similar to Stalnaker's and presented it in May, 1978 at London, Ontario.

[20] In particular, we must relativize everything in the theory of Section 3 to Ω^*, and require that if $\mathcal{F}_1$, $\mathcal{F}_2 \in \Omega^*$ and $i \in \mathcal{H}$ then $\mathcal{G} \in \Omega^*$, where $\mathcal{G} \in \text{Post}(\mathcal{F}_1, i) \cap$ $A'\text{Post}(\mathcal{F}_2, i)$. This last condition, as far as we can see, is a reasonable one to place on causal coherence. Justifying it, though, would lead to rather deep questions regarding the relationship of causality and time. In addition, we should also stipulate that for each $i \in \mathcal{H}$ and history h containing i there is an $\mathcal{F} \in \Omega^*$ normal at i such that $\mathcal{F}_i = h$. All this says is that no histories are ruled out by causal considerations.

[21] For instance, suppose that one evening Peter and Paul are deciding in separate places what restaurant to dine at. Peter is also deciding whether to call Paul. If he doesn't call, the decisions will be independent. If he does, they will not be. Whether Peter calls Paul is contingent. So whether the decisions about where to dine are independent is contingent.

[22] We think that the contrast between strict and material theories of the conditional is helpful. Lewis and Lewis (C. I. and D. K.) are strict theorists, Russell and Stalnaker material theorists. For Russell (cf. [7], pp. 14–15), a material conditional is one involving particular, fixed propositions, and so is shown to be true or false by looking at one case. (Russell thought of this case as the actual world, and so identified $A > B$ with $A \supset B$. He also contrasted material with what he called formal implication, but this contrast does not concern us here.) Stalnaker, like Russell and unlike Lewis, examines the truth value of $A \supset B$ in some one situation, but does not assume this to be a fixed situation, independent of A. This is what we mean by calling Stalnaker's theory variably material. C. I. Lewis takes $A > B$ to be strict, in the sense that it holds when $A \supset B$ is true in a multiplicity of situations, and he assumes this multiplicity is fixed independently of A. David Lewis relaxes this assumption, while retaining the multiplicity of situations. Note that strict theories result in a conditional that expresses some necessary connection between the antecedent and the consequent. Material theories deny that there is any such necessary connection expressed by a conditional.

BIBLIOGRAPHY

[1]   Adams, E., 'Subjunctive and Indicative Conditionals', *Foundations of Language* 6 (1970), 89–94.

[2]   Bennett, J., 'Couterfactuals and Possible Worlds', *Canadian Journal of Philosophy* 4 (1974), 381–402.

[3]   Fine, K., Review of Lewis, [4]. *Mind* **84** (1975), 451–458.

[4]   Lewis, D., *Counterfactuals,* Harvard University Press, Cambridge, Mass., 1973.

[5]   Lewis, D., 'Counterfactual Dependence and Time's Arrow,' *Nous* **13** (1979), 455–476.

[6]   Prior, A., *Past, Present and Future,* Oxford University Press, Oxford, 1967.

[7]   Russell, B., *The Principles of Mathematics,* 2nd ed., Allen and Unwin, London, 1937.

[8]   Slote, M., 'Time in Counterfactuals', *Philosophical Review* 87 (1978), 3–27.

[9]   Stalnaker, R., 'A Theory of Conditionals', in N. Rescher (ed.), *Studies in Logical Theory*, Oxford, 1968, pp. 98–112.

[10] Stalnaker, R. and R. Thomason, 'A Semantic Analysis of Conditional Logic', *Theoria* 36 (1970), 23–42.

[11] Stalnaker, R., 'Indicative Conditionals', in A. Kasher (ed.), *Language in Focus,* D. Reidel, Dordrecht, Holland, 1976, pp. 179-196.

[12] Stalnaker, R., 'A Defense of Conditional Excluded Middle', this volume, pp. 87–104.

[13] Thomason, R., 'Inderterminist Time and Truth Value Gaps', *Theoria* **36** (1970), 264–281.

[14] Van Fraassen, B., *An Introduction to the Philosophy of Space and Time,* Random House, New York, 1970.

[15] Van Fraassen, B., 'A Framework for Temporal Conditions and Chance', this volume, pp. 322–349.

BAS C. VAN FRAASSEN

A TEMPORAL FRAMEWORK FOR CONDITIONALS AND CHANCE*

In this paper I shall propose a general model for tensed language, and then explore the introduction of conditionals and of two sorts of probabilities (measures of objective chance and of subjective ignorance). On the subjects of tenses and tensed conditionals, I build on the previous work by Richmond Thomason, but I shall attempt to provide an exposition which is self-contained.

1. MODELS FOR TENSED AND TEMPORAL LANGUAGE

Some years ago I saw a French book written early in the last century which developed a system of sign communication for the deaf and dumb. I don't believe it can have been easy to learn since it contained among other things a taxonomy of about twenty distinct compound tenses (indicative and subjunctive). For that same reason, however, it must be a pioneering work in the logical study of tensed language.

The modern form of this subject received its impetus from an insight of A. N. Prior:

> Tensed propositions are propositional *functions*, with times as arguments (and propositions as values).

This insight still allows for a plethora of models. I shall propose a general model, and then use that (sort of) model for the study of temporal distinctions in conditional and probability judgments.

1.1. *Three Models*

In the simplest model, time is the real line R and the propositions are the set $K = \{T, F\}$, comprising merely the True and the False. A tensed proposition is a map of R into K. For example:

$$\text{(The earth has been created) } (t) = \begin{cases} T & \text{for} \quad t \geqslant 4004 \text{ BC} \\ F & \text{for} \quad t < 4004 \text{ BC}. \end{cases}$$

A second model enlarges K and takes the propositions to form any Boolean algebra, for convenience represented by a field of sets. Some name is needed

W. L. Harper, R. Stalnaker, and G. Pearce (eds.), Ifs, 323–340.
Copyright © 1980 by D. Reidel Publishing Company.

for the elements of these sets and a certain stroke of *PR* genius produced the name 'possible world'. Thus

(The earth has been created) $(t) = \{x \in K:$ in x, the earth's creation occurred before $t\}$.

The sentential connectives correspond to operations on tensed propositions in obvious ways:

'A is true in x at t' means '$x \in A(t)$'

$(A \& B)(t) = A(t) \cap B(t)$

$PA(t) = \{x \in K:(\exists t' < t)(x \in A(t'))\}$

for conjunction and the past tense operator; and so on. We use such special tensing connectors as:

P	"It has been (or was) the case that"
H	"It has always been the case that"
F	"It will be the case that"
G	"It will always be the case that"
L	"It has always been, is, and always will be the case that".

This last connector is sometimes referred to as a *temporal necessity*, for logically it acts like "It is necessarily the case that".

But of course, it is nothing like a tensed alethic modality such as we find in

It was at one time still possible to prevent the population explosion, but is no longer.

It will be necessary to increase the price of energy sharply by 1985, though it is not yet.

Such operators can be added to this second model; if they are not to be constant in time, we must add a time-indexed access relation R^t, reading 'xR^ty' as 'there is access from x to y, at t' or 'at t, y is possible relative to x'. Let $R^t(x) = \{y \in K: xR^ty\}$, the *access region* of x at t. We have then:

$$\square A(t) = \{x \in K: R^t(x) \subseteq A(t)\}$$

that is,

$\square A$ is true in x at t exactly if every y such that xR^ty, is also such that A is true in y at t.

which is the basic idea of the modalities studied for example by Roger Woolhouse and Hans Kamp.

I wish now to propose a third, yet more general model, whose virtues I hope to extol. In this model we have a *state-space H*. Each 'world' (or better, 'possible system') in K has, at each time t, a certain *state* (an element of H, its location or H at t). The function of h_x which gives the state $u = h_x(t)$ of world x at time t, is called the *history* or *trajectory* of x.

We can now draw certain distinctions among the propositions and the worlds that I mean to utilize. Please keep in mind that all the old notions apply:

world = element of K

proposition = subset of K (perhaps restricted to a family of subsets)

tensed proposition = map of R into the propositions.

Worlds x and y have the same history exactly if $h_x = h_y$. A proposition X is *historical* if membership in X depends only on the world's history: that is, if x is in X and y has the same history as x, then y is in X too. Other propositions I shall call *ahistorical* or *metaphysical*. A tensed proposition A may be called historical (etc.) if $A(t)$ is historical for all t.

1.2. *Ahistorical (Metaphysical) Propositions*

Metaphysical propositions become important when we discuss any sort of modality, and disputes in philosophy of science equate 'reduction' of a modality with the thesis that the modality does not lead from historical to ahistorical propositions. So let us look at some examples. The possible world picture suggests some fanciful ones:

(a) x and y are two possible worlds (neither of them actual) which have exactly the same history, but are not identical, for God created x, but y is not created.

(b) x and y are two possible worlds, which have the same history, but are not subject to all the same laws. Specifically, all laws of x are laws of y, but x is just that world with those laws in which non-law governed events *happen to* satisfy (accidentally) the laws of y.

Some will no doubt think that such realism about possible worlds, and about real necessities 'in nature', is absurd. It is good therefore that there are much more mundane examples too. If in practice I construct a model, say of a mechanical system, then I do not accommodate all its features in the state-space, but only those which vary. For example, the phase-space of classical mechanics has moments and positions as coordinates, and there are not separate coordinates, for example, for the chemical constitution of the bodies or particles studied. If two pendulum bobs are made of different alloys, but have the same mass, shape, and size, then their mechanical behaviour is the same. So two posssible systems can have the same trajectory in a given state-space although they are different in other ways. *From the trajectory you cannot tell.*

There are here two different approaches to models of language: the idea that models do or can reflect reality, and the idea that models are constructed in special ways for special, limited, purposes. On both approaches, the meta-physical/historical distinction makes sense.

1.3. *What is Settled*

The notion of what is settled was introduced into tense logic by Richmond Thomason. He paired its introduction with the view that only what is settled, at a given time, is really true then (*Wesen ist was gewesen ist?*), which I find very attractive but will omit from discussion here.

To begin we note an obvious sort of relation that two worlds can have at a time: their histories agree up to and including that time. Their pasts and also their presents, are the same but their futures may differ; each represents for the other a possible future. If something is true in world x at time t, and also true in all worlds agreeing with x through t, we say that it is *settled* (*as true*) in x at that time.

DEFINITIONS:

(a) x and y agree *through* t (briefly $x \overset{t}{\sim} y$) exactly if $h_x(t') = h_y(t')$ for all $t' \leqslant t$.

(b) $H(x, t) = \{y \in K : x \overset{t}{\sim} y\}$ is called the *t-cone* of x.

(c) $SA(t) = \{x \in K : H(x, t) \subseteq A(t)\}$.

Thus SA is true in x at t exactly if A is true at t in all worlds in the t-cone of x. Logically, S is a sort of tensed S5-necessity operator. The way to picture this is to draw a square to represent H, and in it, a tree growing upward with a long trunk. Label the point where branches begin with the symbol 't'. The lines consisting each of the trunk and one branch, represent the histories of worlds agreeing with a given one through time t.

Since the logic of what is settled was exhaustively investigated by Thomason, I shall here give only a few bits of general information. Entailment in the model is defined by

$$A_1, \ldots, A_n, \ldots \Vdash B \text{ exactly if, for every time } t,$$
$$A_1(t) \cap \ldots \cap A_n(t) \cap \ldots \subseteq B(t).$$

It will be clear that $SA \Vdash A$, for any tensed proposition A. We can call A *settled as false* at time t, in world x, if $S \neg A$ is true there then. It will be clear that SA is a historical tensed proposition, and that S is rather special in this respect: it turns even a metaphysical proposition into a historical one.

1.4. *Backward-Looking Propositions*

Now that we have the ingredients for a general model of tensed propositions, I can introduce a number of concepts that will be needed when that model is used. To begin, some historical tensed propositions are *constant*, and some are not. They are called constant if the truth-value does not change with time; as for example "It always was, is, and will be the case that something is in motion":

$$(HA \,\&\, A \,\&\, GA)\,(t) = (HA \,\&\, A \,\&\, GA)\,(t')$$

for all t and t'. Tautologies and contradictions are special sorts of constant tensed propositions, but some are contingent like this example. We can additionally distinguish *backward-looking* and *sedate* tensed propositions. A is backward-looking if membership in $A(t)$ depends solely on the world's history up to and including t. Similarly, A is sedate if x being in $A(t)$ guarantees that x is in $A(t')$ for all t' later than t (that world has, so to say, settled down into being such that A is true). Note well that a backward-looking proposition may be 'about the future', because in some respects the future may be determined by the past.

Here are some examples (I shall talk as if sentences were, rather than merely express, tensed propositions). 'It has rained' is both backward-looking and sedate. 'It rains' is backward-looking but not sedate. On the other hand, 'It will have rained' is sedate but not backward-looking. And finally, 'It will rain' is neither.

The notion of backward-looking can be applied to propositions as follows: proposition X (a set of worlds) is *backward-looking at* t exactly if every world x is such that, if x is in X, then $H(x, t)$ is part of X. The following are equivalent characterizations:

(a) X is backward-looking at t.

(b) $X \cap H(x, t)$ is either $H(x, t)$ or empty, for all x.

(c) X is the union of a set of t-cones.

(d) $X \subseteq SX$.

(e)   for all x, if $x \in X$ then $H(x, t) \subseteq X$.

Statement (e) is essentially the definition I gave above, for it says that if $x \in X$, and $x \overset{t}{\sim} y$ then $y \in X$. (If x is not in X and $x \overset{t}{\sim} y$, then y cannot be in X either, given that (e) is true, since $\overset{t}{\sim}$ is symmetric.) Now

$$SX = \{x : H(x, t) \subseteq X\}$$

so (e) is equivalent to (d). Also (e) implies that X contains the union of all the t-cones $H(x, t)$ of members x of X. It cannot contain more than that, since its members belong to those t-cones ($\overset{t}{\sim}$ is reflexive). Hence (e) implies (c); which in turn clearly implies (b). But if (b) is true, then if $x \in X$, and since $x \in H(x, t)$, it follows that all of $H(x, t)$ is part of X.

The logical principles concerning S can be elaborated by use of these notions. Because of (d) above, A is clearly backward-looking exactly if $\square(A \supset SA)$ is true (in any world, at any time, where $\square$ is pure or verbal necessity ($\square A$ is true in x at t exactly if A is true in all worlds at all times). Similarly, A is sedate exactly if $\square(A \supset GA)$ is true. The following are representative validities:

$$SA, \square(A \supset GA) \Vdash GSA$$

$$PA, \square(A \supset SA) \Vdash SPA$$

but backward-looking, at least, is a notion which we shall soon see to be important in other ways.

2. A THEORY OF TENSED CONDITIONALS

Using the general model for tensed propositions of the preceding section, we can now approach the question what tensed conditional propositions are like. I shall begin by discussing conditionals in general (without regard to tense) beginning with the theory of Robert Stalnaker, and then turn to special conditions on tensed conditionals proposed by Thomason.

2.1. *Conditionals; Weak Stalnaker Logic*

Conditional sentences are interpreted as carrying a tacit *ceteris paribus* clause. Therefore, the statement that his match will light if struck, or that it would have lit if struck, is not contradicted by the observation that it will not light (would not have lit) if it were (made) wet and struck. For the tacit 'other things being equal' meant to 'keep constant' the fact that this match is dry. Hence conditionals do not all obey the law of strengthening of the antecedent characteristic of the material and strict conditionals familiar from logic:

$$A \supset B \Vdash (A \ \& \ C) \supset B$$

$$A \rightarrow B \Vdash (A \ \& \ C) \rightarrow B$$

where I use the arrow to symbolize the general conditional which includes counterfactual ones.

Robert Stalnaker proposed a theory in which conditional propositions are modelled in a way that I can describe in the terms of the preceding section, and he provided the corresponding system of logic of conditionals; I shall call that *Stalnaker Logic*. I have previously (1976) proposed a weaker system of logic, formed by deleting one of Stalnaker's axioms; let me call that *Weak Stalnaker Logic*. I shall explain both here, and my motives for preferring the weaker system; but refer to the other publications for more detail.

Stalnaker adds to the model a *selection function s* whose job it is to give exact content to the *ceteris paribus* clause in conditionals. The conditional is then defined by

$$x \text{ is in } (Z \rightarrow Y) \text{ exactly if: } s^x Z \in Y, \text{ or } Z = \Lambda$$

where Λ is the empty set, and $s(x, Z)$ is abbreviated by superscripting x. We may read this as: $Z \rightarrow Y$ is true in world x exactly if, in the world in which Z is true but everything else is as in x, Y is also true. Conditions on this selection function which are imposed are

(a)   if $s^x Z$ is defined for any world x, then it is defined for all worlds;

(b) $s^x Z$ is a member of Z;

(c)   if x is in Z, then $(s^x Z) = x$.

These are all the conditions needed for Weak Stalnaker Logic, but Stalnaker himself imposed more:

(d)   the propositions Z such that $s^x Z$ is defined, plus the null-set, together form a field;

(e)   if $s^x Z$ is in Y and $s^x Y$ is in Z, then $(s^x Z) = (s^x Y)$.

Clause (d) is not perhaps too important; it is there to insure that if '$P \to Q$' is a sentence in the language, then there is a proposition $X \to Y$ for it to express. In simple models, this is most easily obtained by letting the domain of s^x be all the non-empty sets of worlds. In Weak Stalnaker Logic, it is always possible to let '$P \to Q$' express the same as '$P \supset Q$' if no suitable proposition $X \to Y$ is available. But I believe in any case that it is only a convention dictated by convenience to think that all sentences should, on any interpretation, express propositions; and correlatively, that it is better to study propositions directly.

The main difference lies therefore in condition (e). Its result is that Stalnaker Logic has, and Weak Stalnaker Logic does not have, the axiom

$$(P \to Q) \& (Q \to P) \& (P \to R) \cdot \supset (Q \to R).$$

It is the only axiom that relates conditionals with logically non-equivalent antecedents. Of course I agree that there should be systematic connections among such conditionals; but I do not know what they are and I believe that this axiom is much too strong. Nor have I ever seen evidence for it that I find at all convincing. So I prefer to omit it altogether.

2.2. *Tensed Conditionals*

Turning now to tensed conditionals, Thomason has laid down the following principles (see his paper) which I shall accept here without argument:

(A) $\neg S \neg A, SB \Vdash A \to B.$

(B) $S(A \to B) \Vdash A \to S(A \supset B).$

(C) $S \neg A, S(A \to B)$ does *not* imply $A \to SB.$

In this development, as in Thomason's, the troublesome part of (B) is

(B') $S \neg A, S(A \to B) \Vdash A \to S(A \supset B).$

We may also note that the following will also be validated *en passant*, as consequences

(A') $\neg S \neg A, SB \Vdash A \to SB.$

(D) $S \neg A, S(A \to B), A \to SA \Vdash A \to SB.$

(E) $A \to SA \Vdash S(A \to B) \equiv (A \to SB).$

A further principle, suggested by Anil Gupta and accepted by Thomason, I shall discuss after I have considered these.

Let us see how the model needs to be adjusted. The selection function should now be indexed not only with a world, but also a time. For in selecting another world for inspection, we must 'keep fixed' certain parts of this world, and what these parts are depends on time. For example, if I were now to add one calory of heat to this cubic centimeter of water I would raise it to $5\,°C$ – that is true if the *present* temperature of the water is $4\,°C$, and *that fact* must be the case in the selected world. So we shall have:

$$x \in (A \to B)\,(t) \text{ exactly if: } A(t) \text{ is empty, or } s_t^x(A(t)) \in B(t).$$

The question is: what constraints shall we put on the selection function, in addition to (a)–(d), to reflect this new concern with time?

The first condition I can take almost bodily from Thomason. What is settled must be respected first when we try to keep things fixed; nothing settled must be non-fixed gratuitously:

(I)   if $Y \cap H(x, t) \neq \Lambda$, then $s_t^x Y \in H(x, t)$.

In other words, we *try* first to select a word that is like x through t if we can. This condition suffices to validate (A') and hence (A).

The second further condition I need is more complicated; it is a *global* constraint on the model, with consequences for 'how many' different (sorts of) possible worlds there are. It is:

(II)   if $z = s_t^x Y$ then $H(z, t) \cap Y \subseteq \{s_t^w Y : w \overset{t}{\sim} x\}$.

This is a sort of continuity condition (as Professor Fenstad pointed out to me). If we think of worlds that agree through t as being 'close together', and think of $s_t Y$ as a map (x into $s_t^x Y$), then the condition says:

worlds which are close together come from worlds which are close together.

Another way to put the condition loosely but pictorially, is this

if z is the nearest Y-world to x at t, and z' is a Y-world, z' agreeing with z through t, then z' is the nearest Y-world at t to some x' which agrees with x through t,

though I must say that I do not take this 'nearest' metaphor seriously.

When $Y = A(t)$ and x is in $\neg S \neg A(t)$, then (II) is a consequence of (I). But when x is in $S \neg A(t)$, (II) does some work; and it validates (B). For suppose $A \to B$ is settled in x at t. Then for all $x' \overset{t}{\sim} x$, $A \to B$ is true in x' at t. But if $z = s_t^x A(t)$, and $z' \overset{t}{\sim} z$, then if $A(t)$ is true in z', we conclude that $z' = s_t^{x'} A(t)$

for such an $x' \overset{t}{\sim} x$, and hence that $B(t)$ is true in z' as well. Thus $A \supset \overset{t}{B}$ is settled in z at time t, as required.

In validating the other principles, have I accidentally contravened (C)? But I have not, since $H(z, t) - A(t)$ need not be empty; and worlds in that set need not be in $B(t)$. The reader is again advised to draw pictures; let one tree of lines represent the histories of worlds in $H(x, t)$, and another tree the histories of worlds agreeing through t with the world $z = s_t^x Y$.

2.3. Gupta's Principle

There are certainly some important differences between Thomason's models and these, but I think that differences in the sets of logical truths can be eliminated by restricting my models. For example I could stipulate that the trajectories of all members of K, in the state-space H, taken together form a tree like structure; that is, for each x and y there exists a time t 'early enough' so that $x \in H(y, t)$. A special further principle was suggested by Anil Gupta; after some discussion, it was accepted by Thomas in the form:

(F) $\Box(E \supset SE), A \rightarrow \Diamond B, B \rightarrow \Diamond A, B \rightarrow E \Vdash A \rightarrow E.$

The modality $\Box$ is the pure logical modality, but '$\Diamond$' is defined as '$\neg S \neg$'

$$\Box X = \begin{cases} K \text{ if } X = K \\ \Lambda \text{ otherwise,} \end{cases}$$

I believe that this principle becomes validated in my model if I impose, as third main condition:

(III)   if $x = s_t^w A(t)$ and some $y \overset{t}{\sim} x$ is in $B(t)$, and $z = s_t^w B(t)$ and some $v \overset{t}{\sim} z$ is in $A(t)$ then $z \overset{t}{\sim} x$.

For let us consider a world w and time t in which the premises of (F) are true, and let x and y be as in (III). Then the fact that $A \rightarrow \Diamond B$ and $B \rightarrow \Diamond A$ are true in w at t means that the antecedent of (III) is satisfied. Hence $z \overset{t}{\sim} x$. But because $w \in (B \rightarrow E)(t)$, we see that $z \in E(t)$. By the first premise of (F), z is then also in $SE(t)$. Since x is in $H(z, t)$, it follows that x is in $E(t)$ too. Hence $(A \rightarrow E)$ is true in w at t.

It may be useful to think of it this way: Gupta's principle (F) is about backward-looking propositions E, and establishes a connection between conditionals with logically non-equivalent antecedents.

2.4. *Determinism and Metaphysical Conditionals*

Be that as it may (principles A, B, and F are universal, but the evidence for them is only through particular examples) there is still a fundamental difference between Thomason's theory and the present one, which has to do with metaphysical propositions. Under quite simple conditions it turns out that in my model, some of the conditionals must be metaphysical, not historical, propositions.

We can see this by considering the question of determinism. A world x should be called *deterministic* at t exactly if any world y in $H(x, t)$ has the same history after t as well ($y \in H(x, t)$ implies $h_y = h_x$). Suppose now that

$$B(t) = \{x \in K : x \text{ is deterministic at } t\}$$
$$\neq K.$$

Then there is a world $s_t^x \neg B(t) = z$ when x is in $B(t)$. That world z is not deterministic at t, as it has a distinct $z' \overset{t}{\sim} z$. But by condition (II), $z' = s_t^x \neg B(t)$ for some $x' \overset{t}{\sim} x$. Since $z' \neq z$, also $x' \neq x$. Hence we have two deterministic worlds that have the same history throughout: $h_x = h_{x'}$. If $C(t)$ is now any proposition which contains z but not z', we find

$$(\neg B \to C) \text{ is true in } x \text{ at } t$$
$$(\neg B \to C) \text{ is false in } x' \text{ at } t$$
$$h_x = h_{x'}$$
So, $(\neg B \to C)\,(t)$ is a metaphysical proposition.

Hence, regardless of how precisely the state-space is defined, if the model has room for both deterministic and indeterministic worlds (systems) then some conditionals are metaphysical propositions.

This is something I have always suspected about conditionals, but it is surprising to find it proved.

3. ON THE COMBINATION OF CHANCE AND IGNORANCE

3.1. *The Two Probabilities*

"I once was blind but now I see" has as hidden parameter the linear continuum of time. In the same way, "Radium has a relatively short half-life" carries in present physical theory a hidden parameter of chance: the chance that a radium atom decays into radon, in any given second, equals $1/10^{12}$. Chance is a real-valued function, just as, say, time of sojourn.

In addition, there is *epistemic probability*, which is a sort of summary of what my beliefs have to say that is relevant to a truth-value. For example, the only belief I have relevant to whether a given radium atom will decay in the next second, is that chance statement above; and this may be neatly indicated by the assertion that *for me*, the probability of this decay is $1/10^{12}$.

There are two ways in which my views of chance and probability differ from those of various other discussants. I hold to a modal frequency interpretation of chance – with the emphasis on frequency – and I don't think that epistemic probability is anything like what the Bayesians call subjective probability. But formally, the disagreement can be minimized. Chance, as I understand it, does obey the laws of the probability calculus. Secondly, I think that my epistemic probability can be characterized through a *family* of probability functions, each of which is (in some sense I shall not pursue here) compatible with my belief-state. The reader will recognize this as a sort of supervaluation view; it will be convenient, and not useless, to study the ideal case in which that family has only one member, P.

In each world x, chance is a function that changes with time, and attaches to propositions; let us write

$$c_t^x(X) = r: x \text{ a world,}$$
$$t \text{ a time,}$$
$$X \text{ a proposition,}$$
$$r \text{ a real number.}$$

P too may pertain to a world (the one the subject inhabits) and will certainly change with time. But I have at present nothing useful to say about the first dependency. So I shall think of P as being linked to a given world, which is kept constant throughout the discussion.

In the fall of 1977 I discussed with Thomason the question how the probability which is the measure of my ignorance, is related to objective chance, for propositions for which both are defined. That they are related seems clear if we assume that no crystal ball is possible, or at least available. In that case, the most precise and complete information I could have about whether it will rain tomorrow, is the information that the objective chance of rain is thus or so. If I am rational, therefore, it seems that my personal opinion about rain tomorrow should be based entirely on my opinions about the chance that it will rain tomorrow.

Thomason proposed, in effect, two links. The first is that the opinion about chance that is relevant to my probability now, is about the chance there is now. This is perhaps one aspect of the above idea about there being no crystal balls, so that the chance there is now of there being rain tomorrow is the only objective guide we can have now to, for example, the chance there will be at midnight, or tomorrow morning, of rain tomorrow. The second proposed link is that the probability is the expected value of that chance. I shall formulate this proposal (emphasizing that it only applies to propositions for which both probability and chance are defined) and then give an example.

(*R-principle*): The probability at t that X is true equals the expected value of the chance at t of X: $P_t(X) = \text{Exp } c_t(X)$;

where Exp is the expected value calculated for P_t, and $c_t X$ is a random variable on the set K of worlds, taking value $c_t^x X$ as world x. The expected value is given by

$$\text{Exp} f = \sum_{x \in K} P_t(x)f(x)$$

if K is countable, and

$$\text{Exp} f = \int_K f \, dP_t$$

in the general case. In what follows I mean to defend and generalize this *R*-principle. An example first.

We have three coins, two fair and one biased 3 to 1 in favour of heads. One of these coins is picked at random and tossed. What is the probability that it comes up heads?

Everyone agrees on the answer; the reason for this is that the problem is a standard sort of exercise in elementary texts. There are three possible worlds that we might be in: the coin tossed may be fair or biased. Call these worlds x, y, z:

$$P(x) = P(y) = P(z) = 1/3,$$

$$c_t^x \text{ (Heads)} = 1/2 = c_t^y \text{ (Heads)},$$

$$c_t^z \text{ (Heads)} = 3/4,$$

$$P(\text{Heads}) = P(x) c_t^x(\text{Heads}) + P(y) c_t^y (\text{Heads}) + P(z) c_t^z(\text{Heads})$$

$$= (2/3 \times 1/2) + (1/3 \times 2/4) = 7/12.$$

'Heads' is the proposition that the coin tossed comes up heads at the given time t. More generally, if our coins were $i = x_1, \ldots, x_n$ then we would have

$$P(\text{Heads}) = \sum P(i)\,c_t^i\,(\text{Heads}): i = x_1, \ldots, x_n$$

which is just what the R-principle says. To warn you of pitfalls, let us note the important fact that *full belief in stochastic independence does not imply epistemic independence*:

$$[c_t^i(X/Y) = c_t^i(X) \text{ for all } i] \text{ does not imply } P(X/Y) = P(X).$$

For example, let Y be: the coin came up heads the first time; X be: the coin came up heads the second time. In the above example, X and Y are stochastically independent no matter which coin we toss. But if the coin comes up heads on the first toss, that makes it more likely that the coin in question is the biased one. Hence, that makes it more likely that heads also comes up on the second toss.

3.2. Representation of Chance

Chance develops in time. The chance that something has already happened is, if it makes sense at all, just zero or one depending on whether it did or did not. To be more precise, chance pertains non-trivially to what is not yet settled. It concerns the open future, the possibilities not yet closed to us.

This suggests that chance c_t^x in world x at time t is a measure, first and foremost, on the set $H(x, t)$. We may trivially extend it beyond that,

$$(1) \qquad c_t^x(X) = c_t^x(H(x, t) \cap X)$$

so that chance is, as it were, naturally conditionalized on what is settled.

Secondly, I want to say that what is settled, determines chance. If a system is such that the immediately preceding state determines present chance, we say that it has the Markov property. That is too strong a constraint, of course; I suggest only that the total sequence of states up to and including now, determines chance:

$$(2) \qquad \text{if } y \in H(x, t) \text{ then } c_t^y = c_t^x.$$

Propositions attributing chance are therefore historical, and indeed, backward-looking:

$$[c_t(X) = r] = \{x \in K : c_t^x(X) = r\}.$$

Like every other backward-looking proposition, its chance is zero or one:

(a)   If A is a backward-looking tensed proposition then
$c_t^x(A(t)) \in \{0, 1\}$.

(b)   If X is a proposition which is backward-looking at t, then
$c_t^x(X) \in \{0, 1\}$.

Of course we can write $[c_t(X) = r]$ as $[c(X) = r](t)$, showing that we deal here with a tensed proposition, but I shall keep the former for perspicacity.

Finally, there may be some advantage, and possibly compelling reasons, to restrict the domain of c_t^x to historical propositions. I don't know.

3.3. Combination with Ignorance

The main question concerning the R-principle which I have thought about, is whether it can be generalized from the absolute probability $P_t(X)$ to conditional probability $P_t(X/Y)$. It soon appeared that constraints must be put on Y. Just as for the R-principle I shall restrict the discussion to the question, what is the probability that A is true at t? If we add: given that B is true at t, we will answer that by attempting to estimate the chance at t that $A(t)$ is true, given that the world is such that $B(t)$ is true. And an answer is forthcoming provided $B(t)$ is essentially about the past and present, that is, in the terminology of Part I, *backward-looking*.

Let us begin with a discussion of the finite case, that is, the set of worlds K is finite. In that case there are only finitely many distinct histories up to time t:

(1) $K = H(x_1, t) \cup \ldots \cup H(x_n, t)$.

Let $a_i = H(x_i, t)$. The set $\{a_i\}$ is a partition of K.

Of course there is some i such that a_i represents the actual world history through time t. But for probability it does not matter which is actual, but only what we think about which is actual. Because $\{a_i\}$ is a partition of K,

(2) $P_t(X) = \sum_{i=1}^{n} P_t(a_i) P_t(X/a_i)$.

Of course, if the actual world is in a_i, then the chance of X equals $c_t^i(X)$ at t (where I abbreviate 'x_i' to 'i'). But that should then be the value of $P(X/a_i)$; hence we deduce:

(3) $P_t(X) = \sum_{i=1}^{n} P_t(a_i) c_t^i(X)$.

Since we are assuming that K is finite, a_i is finite too; say $a_i = \{y_1, \ldots, y_m\}$

where the chance of X at t in y_i is equal to that of X at t in x_i (because $y_i \overset{t}{\sim} x_i$). So (3) is equivalent to:

$$(4) \qquad P_t(X) = \sum_{y \in K} P_t(y) c_t^y(X).$$

So far we have simply followed through the reasoning for the R-principle. If we now want to conditionalize, we must move back to step (2) above. There will be no harm done if we toss out, say, worlds $x_{k+1}, \ldots, x_n$, which means discarding sets $a_{k+1}, \ldots, a_n$ altogether, and renormalize:

$$(5) \qquad P_t(X/a_1 \cup \ldots \cup a_k) = \frac{1}{P_t(a_{1k})} \sum_{i=1}^{k} P_t(a_i) P_t(X/a_i),$$

where $a_{1k} = a_1 \cup \ldots \cup a_k$. By reasoning as above, we get

$$(6) \qquad P_t(X/a_{1k}) = \frac{1}{P_t(a_{1k})} \left[\sum_{i=1}^{k} P_t(a_i) c_t^i(X) \right].$$

There is no harm done because whole sets $H(x, t)$ are discarded at once, not just parts of them. But that is true not only if we conditionalize on a_{1k}, but if we conditionalize on any backward-looking proposition. In the finite case we have therefore, by generalizing (6) and reasoning as for (4) above:

$$(7) \qquad P_t(X/E) = \frac{1}{P_t(E)} \sum \{ P_t(y) c_t^y(X) : y \in E \}$$

provided $P_t(E) \neq 0$ and E is backward-looking at t.

Generalizing also to the non-finite case, we arrive at the:

(*Basic Principle*): $P_t(X/Y) = \mathrm{Exp}_Y \, c_t(X)$

$$= \frac{1}{P_t(Y)} \int_Y c_t(X) \, dP_t$$

provided Y is backward-looking at t, and $P_t(Y) \neq 0$.

Intuitively, this says that, in appropriate cases, the conditional probability of a proposition equals the conditional expected value of its chance.

To see that the *proviso* (backward-looking) cannot be eliminated, it suffices to consider $Y = K - X$. This provides no difficulty if Y is backward-looking, but would give absurd results if we allowed the Basic Principle to apply if Y was not backward-looking. Suppose first that $Y = (K - X)$ is backward-looking, and let us consider the finite case with equation (7). Since Y is backward-looking, so is X. Hence $c_t^y(X)$ equals zero or one for each y in Y. Thus

$$\sum \{P_t(y)c_t^y(X): y \in Y\} = \sum \{P_t(y): y \in Y \cap X\}$$

$$= P_t(Y \cap X) = 0.$$

Hence in this particular case, (7) just says that $P_t(X/K - X) = 0$, as it should.

But suppose now that Y is not backward-looking. Of course, $P_t(X/Y) = P_t(X/K - X)$ is meant to be zero. But then, if (7) is correct, and $P_t(Y) \neq 0$, the sum of $P_t(y) c_t^y(X)$ must be zero. This means that $c_t^y(X)$ must be zero for all y in $Y = K - X$. In other words, X must be proposition which is such that its chance equals zero in all worlds in which it happens to be false. This is simply not something we can expect in general, and if it is not so, we would be drawing the consequence that zero equals a positive number. Hence the restriction to backward-looking Y should not be eliminated from the proviso.

The R-principle follows at once from the Basic Principle. The same is true of the principle offered by David Miller (as step one in his 'paradox') and it is true of David Lewis' generalization thereof (his 'Principal Principle') if we make Lewis' notions precise in our own way. Miller offered:

(M) $\quad P_t(X/c_t(X) = r) = r$, provided $P(c_t(x) = r) \neq 0$,

if I read him correctly (he omitted considerations of time). This principle is correct, for $[c_t(X) = r]$ is a backward-looking proposition, and such that if y is any world in $[c_t(X) = r]$ then $c_t^y(X) = r$. Hence

$$\int_Y c_t(X) \, dP_t = r \cdot P_t(Y)$$

and so the conditional expected value of X is indeed r. David Lewis' Principal Principle has in it one undefined notion, which (because time is not an explicit parameter in his models) he did not explicate. Reading that my own way, it becomes

(Lewis) $P_t(X/[c_t(X) = r] \cap E) = r$ provided E is backward-looking at t and $P_t([c_t(x) = r] \cap E) \neq 0$.

This follows from the Basic Principle by essentially the same argument as for

Miller; but in addition, in this form, is equivalent to the Basic Principle. The reason is that the set $\{[c_t(x) = r] : r \in [0, 1]\}$ is also a partition of K.

University of Toronto and
University of Southern California

NOTE

* The research for this paper was mainly done before June 1978 and supported by the Canada Council, and partly in September 1978, supported by the National Science Foundation. Parts of Section 1 and 2 were reported in the paper 'Report on Tense Logic' by van Fraassen (1980).

BIBLIOGRAPHY

Lewis, D., 'A Subjectivist's Guide to Objective Chance', circulated Princeton University, May 1978; to be published in R. Jeffrey (ed.), *Studies in Inductive Logic and Probability*, Volume 2, University of California Press.

Prior, A. N., *Time and Modality*, Oxford, 1957.

Stalnaker, R., 'A Theory of Conditionals', in N. Rescher (ed.), *Studies in Logical Theory*, American Philosophical Quarterly Monograph Series, Oxford, 1968.

Thomason, R. H., 'Indeterminist Time and Truth Value Gaps', *Theoria* **36** (1970), 264–81.

Thomason, R. G. and Gupta, A., 'A Theory of Conditionals in the Context of Branching Time', this volume, pp. 299–322.

van Fraassen, B. C., 'Foundations of Probability: A Modal Frequency Approach', pp. 344–94 in G. Toraldo di Francia (ed.), *Problems in the Foundations of Physics*, North-Holland Publishing Co., Amsterdam, 1979.

van Fraassen, B. C., 'Probabilities of Conditionals', in W. L. Harper and C. A. Hooker (eds.), *Foundations of Probability Theory, Statistical Inference, and Statistical Theories of Science*, Volume I, D. Reidel, Dordrecht, 1976, pp. 261–308.

van Fraassen, B. C., 'Rational Belief and Probability Kinematics', *Philosophy of Science*, forthcoming.

van Fraassen, B. C., 'Report on Conditionals', *Teorema* **6** (1976), 5–25.

van Fraassen, B. C., 'Report on Tense Logic', in E. Agazzi (ed.), *Modern Logic – A Survey*, D. Reidel, Dordrecht, 1980, pp. 425–38.

Woolhouse, R. S., 'Tensed Modalities', *Journal of Philosophical Logic* **2** (1972), 393–415.

INDEX

(Bibliographical information is to be found with
the notes at the end of individual papers.)

THE UNIVERSITY OF WESTERN ONTARIO
SERIES IN PHILOSOPHY OF SCIENCE

A Series of Books in Philosophy of Science, Methodology, Epistemology, Logic, History of Science, and Related Fields

8. J. M. Nicholas (ed.), *Images, Perception, and Knowledge.* Papers deriving from and related to the Philosophy of Science Workshop at Ontario, Canada, May 1974. 1977, ix + 309 pp.

9. R. E. Butts and J. Hintikka (eds.), *Logic, Foundations of Mathematics, and Computability Theory.* Part One of the Proceedings of the Fifth International Congress of Logic, Methodology and Philosophy of Science, London, Ontario, Canada, 1975. 1977, x + 406 pp.

10. R. E. Butts and J. Hintikka (eds.), *Foundational Problems in the Special Sciences.* Part Two of the Proceedings of the Fifth International Congress of Logic, Methodology and Philosophy of Science, London, Ontario, Canada, 1975. 1977, x + 427 pp.

11. R. E. Butts and J. Hintikka (eds.), *Basic Problems in Methodology and Linguistics.* Part Three of the Proceedings of the Fifth International Congress of Logic, Methodology and Philosophy of Science, London, Ontario, Canada, 1975. 1977, x + 321 pp.

12. R. E. Butts and J. Hintikka (eds.), *Historical and Philosophical Dimensions of Logic, Methodology and Philosophy of Science.* Part Four of the Proceedings of the Fifth International Congress of Logic, Methodology and Philosophy of Science, London, Ontario, Canada, 1975. 1977, x + 336 pp.

13. C. A. Hooker (ed.), *Foundations and Applications of Decision Theory,* 2 volumes. Vol. I: *Theoretical Foundations.* 1978, xxiii+442 pp. Vol. II: *Epistemic and Social Applications.* 1978, xxiii+206 pp.

14. R. E. Butts and J. C. Pitt (eds.), *New Perspectives on Galileo.* Papers deriving from and related to a workshop on Galileo held at Virginia Polytechnic Institute and State University, 1975. 1978, xvi + 262 pp.

15. W. L. Harper, R. Stalnaker, and G. Pearce (eds.), *Ifs. Conditionals, Belief, Decision, Chance, and Time.* 1980, ix + 345 pp.

16. J. C. Pitt (ed.), *Philosophy in Economics.* Papers deriving from and related to a workshop on Testability and Explanation in Economics held at Virginia Poly-Technic Institute and State University, 1979. 1981, (forthcoming).